Hydrology and Watershed Management

Hydrology and Watershed Management

by

Bhagyashree Chauhan | Geetha Jena

Wordpen Academics
Publishing Science That Matter

ISBN: 978-93-6084-764-7

Published by : **WORDPEN ACADEMICS**
4736/23, Ansari Road, Darya Ganj,
New Delhi – 110 002
Phone: +91-011-35004336
E-mail: info@wordpenacademics.com

Printed at **:** **Replika Press Pvt. Ltd.**

Exclusive Distributor: ***Bio-Green Books, New Delhi***

PREFACE

"Hydrology and Watershed Management" embarks on a comprehensive journey into the intricate dynamics of water resources, watersheds, and their management, offering readers a holistic understanding of the principles, practices, and challenges in sustainable water management. This book serves as an essential resource for water resource engineers, environmental scientists, policymakers, and practitioners seeking to address the growing complexity of water-related issues in a changing world.

Water is a precious and finite resource that sustains life, ecosystems, and human societies. As the global population grows, urbanizes, and industrializes, the demand for water continues to rise, placing unprecedented pressure on freshwater ecosystems and exacerbating water scarcity, pollution, and degradation. In this context, effective watershed management is essential for ensuring water security, biodiversity conservation, and sustainable development.

As editors of this volume, we have endeavored to compile a diverse array of perspectives, methodologies, and case studies from leading experts and practitioners in the field of hydrology and watershed management. Through a blend of theoretical insights, practical methodologies, and real-world examples, this book covers a wide range of topics, including hydrological processes, watershed modeling, water quality management, ecosystem restoration, and stakeholder engagement.

Our overarching goal in presenting this book is to inspire a paradigm shift towards more integrated, adaptive, and sustainable approaches to water resources management. By promoting holistic watershed management strategies that balance the needs of society, the economy, and the environment, we aim to empower stakeholders to address water-related challenges effectively and equitably in a rapidly changing world.

We extend our sincere gratitude to all the contributors who have generously shared their expertise, experiences, and insights in the field of hydrology and watershed management. It is our hope that this book will serve as a valuable reference and guide for policymakers, practitioners, researchers, and students, as they work together to safeguard our precious water resources for future generations.

Bhagyashree Chauhan | Geetha Jena

CONTENTS

	Preface	v
1.	Hydrology: Introduction	1
2.	Water Resources	12
3.	Water Resources in India	67
4.	Water Pollution	94
5.	Water Conservation	115
6.	Water Recycling Process	144
7.	Watershed	156
8.	Water Conservation Practices	183
9.	Rainwater Storage and Treatment	207
10.	Recycling Water and Rain Water Harvesting	236
11.	Canal Building	255
12.	Agricultural Drainage Management	285
	Bibliography	323
	Index	325

1

HYDROLOGY: INTRODUCTION

Water is one of our most precious natural resources. Without it, there would be no life on earth. Hydrology has evolved as a science in response to the need to understand the complex water system of the earth and help solve water problems. This hydrology primer gives you information about water on Earth and humans' involvement and use of water.

Hydrology is the scientific study of the movement, distribution, and management of water on Earth and other planets, including the water cycle, water resources, and environmental watershed sustainability. A practitioner of hydrology is called a hydrologist. Hydrologists are scientists studying earth or environmental science, civil or environmental engineering, and physical geography. Using various analytical methods and scientific techniques, they collect and analyze data to help solve water related problems such as environmental preservation, natural disasters, and water management.

Hydrology subdivides into surface water hydrology, groundwater hydrology (hydrogeology), and marine hydrology. Domains of hydrology include hydrometeorology, surface hydrology, hydrogeology, drainage-basin management, and water quality, where water plays the central role.

Oceanography and meteorology are not included because water is only one of many important aspects within those fields.

Hydrological research can inform environmental engineering, policy, and planning.

Water is one of our most important natural resources. Without it, there would be no life on earth. The supply of water available for our use is limited by nature.

Although there is plenty of water on earth, it is not always in the right place, at the right time and of the right quality. Adding to the problem is the increasing evidence that chemical wastes improperly discarded yesterday are showing up in our water supplies today. Hydrology has evolved as a science in response to the need to understand the complex water systems of the Earth and help solve water problems. Hydrologists play a vital role in finding solutions to water problems, and interesting and challenging careers are available to those who choose to study hydrology.

Hydrology is the study of water. In terms of environmental measurements, hydrology refers to the physical movement of a body of water, including changes in water level, flow, and other dynamic processes. Most hydrology applications involve surface water systems, such as rivers, open-channels and estuaries. However, groundwater and hydrogeological projects can be included in this section, particularly with aquifer applications.

Hydrology, particularly water flow, can affect the local environment due to changes in water quality and quantity. These changes can be man-made (e.g. a dam release) or weather-related (flooding), or due to a combination of both factors (rainfall runoff caused by poor agricultural practices). Regardless of the cause, hydrologic data assists in explaining water quality and environmental changes.

The following chapters discuss several aspects of hydrology and explain what can cause them to fluctuate, and the effects that such fluctuations will have on the local ecosystem. While these chapters are interrelated and are often tied to changes in water quality, they each have distinguishing characteristics that are worth analyzing. Each section will further detail why measuring and monitoring that particular water feature is necessary to understanding the water system as a whole.

WATER AND PEOPLE

Estimates of water use in the United States indicate that about 355 billion gallons per day (one thousand million gallons per day, abbreviated Bgal/d) were withdrawn for all uses during 2010. This total has declined about 17 percent since 1980. Fresh groundwater withdrawals (76.0 Bgal/d) during 2010 were 8 percent less than during 1980. Fresh surface-water withdrawals for 2010 were 230 Bgal/d, 18 percent less than in 1980.

Much of our water use is hidden. Think about what you had for lunch. A hamburger, for example, requires water to raise wheat for the bun, to grow hay and corn to feed the cattle and to process the bread and beef. Together with french fries and a soft drink, this all-American meal uses about 1,500 gallons of water — enough to fill a small swimming pool. How about your clothes? To grow cotton for a pair of jeans takes about 400 gallons. A shirt requires about 400 gallons. How do you get to school or to the store? To produce the amount of finished steel in a car has in the past required about 32,000 gallons of water. Similarly, the steel in a 30-pound bicycle required 480 gallons. This shows that industry must continue to strive to reduce water use through manufacturing processes that use less water, and through recycling of water.

What is Hydrology?

Hydrology is the science that encompasses the occurrence, distribution, movement and properties of the waters of the earth and their relationship with the environment within each phase of the hydrologic cycle. The water cycle, or hydrologic cycle, is a continuous process by which water is purified by evaporation and transported from the earth's surface (including the oceans) to the atmosphere and back to the land and oceans. All of the physical, chemical and biological processes involving water as it travels its various paths in the atmosphere, over and beneath the earth's surface and through growing plants, are of interest to those who study the hydrologic cycle.

There are many pathways the water may take in its continuous cycle of falling as rainfall or snowfall and returning to the atmosphere. It may be captured for millions of years in polar ice caps. It may flow to rivers and finally to the sea. It may soak into the soil to be evaporated directly from the soil surface as it dries or be transpired by growing plants. It may percolate through the soil to ground water reservoirs (aquifers) to be stored or it may flow to wells or springs or back to streams by seepage. The cycle for water may be short, or it may take millions of years.

People tap the water cycle for their own uses. Water is diverted temporarily from one part of the cycle by pumping it from the ground or drawing it from a river or lake. It is used for a variety of activities such as households, businesses and industries; for irrigation of farms and parklands; and for production of electric power. After use, water is returned to another part of the cycle: perhaps discharged downstream or allowed to soak into the ground. Used water normally is lower in quality, even after treatment, which often poses a problem for downstream users.

The hydrologist studies the fundamental transport processes to be able to describe the quantity and quality of water as it moves through the cycle (evaporation, precipitation, streamflow, infiltration, groundwater flow, and other components). The engineering hydrologist, or water resources engineer, is involved in the planning, analysis, design, construction and operation of projects for the control, utilization, and management of water resources. Water resources problems are also the concern of meteorologists, oceanographers, geologists, chemists, physicists, biologists, economists, political scientists, specialists in applied mathematics and computer science, and engineers in several fields.

What Hydrologists Do?

Hydrologists apply scientific knowledge and mathematical principles to solve water-related problems in society: problems of quantity, quality and availability. They may be concerned with finding water supplies for cities or irrigated farms, or controlling river flooding or soil erosion. Or, they may work in environmental protection: preventing or cleaning up pollution or locating sites for safe disposal of hazardous wastes.

Persons trained in hydrology may have a wide variety of job titles. Scientists and engineers in hydrology may be involved in both field investigations and office work. In the field, they may collect basic data, oversee testing of water quality, direct field crews and work with equipment. Many jobs require travel, some abroad. A hydrologist may spend considerable time doing field work in remote and rugged terrain. In the office, hydrologists do many things such as interpreting hydrologic data and performing analyses for determining possible water supplies. Much of their work relies on computers for organizing, summarizing and analyzing masses of data, and for modeling studies such as the prediction of flooding and the consequences of reservoir releases or the effect of leaking underground oil storage tanks.

The work of hydrologists is as varied as the uses of water and may range from planning multimillion dollar interstate water projects to advising homeowners about backyard drainage problems.

Surface Water

Most cities meet their needs for water by withdrawing it from the nearest river, lake or reservoir. Hydrologists help cities by collecting and analyzing the data needed to predict how much water is available from local supplies and whether it will be sufficient to meet the city's projected future needs. To do this, hydrologists study records of rainfall, snowpack depths and river flows that are collected and compiled by hydrologists in various government agencies. They inventory the extent river flow already is being used by others.

Managing reservoirs can be quite complex, because they generally serve many purposes. Reservoirs increase the reliability of local water supplies. Hydrologists use topographic maps and aerial photographs to determine where the reservoir shorelines will be and to calculate reservoir depths and storage capacity. This work ensures that, even at maximum capacity, no highways, railroads or homes would be flooded.

Deciding how much water to release and how much to store depends upon the time of year, flow predictions for the next several months, and the needs of irrigators and cities as well as downstream water-users that rely on the reservoir. If the reservoir also is used for recreation or for generation of hydroelectric power, those requirements must be considered. Decisions must be coordinated with other reservoir managers along the river. Hydrologists collect the necessary information, enter it into a computer, and run computer models to predict the results under various operating strategies. On the basis of these studies, reservoir managers can make the best decision for those involved.

The availability of surface water for swimming, drinking, industrial or other uses sometimes is restricted because of pollution. Pollution can be merely an unsightly and inconvenient nuisance, or it can be an invisible, but deadly, threat to the health of people, plants and animals.

Hydrologists assist public health officials in monitoring public water supplies to ensure that health standards are met. When pollution is discovered, environmental engineers work with hydrologists in devising the necessary sampling program. Water quality in estuaries, streams, rivers and lakes must be monitored, and the health of fish, plants and wildlife along their stretches surveyed. Related work concerns acid rain and its effects on aquatic life, and the behavior of toxic metals and organic chemicals in aquatic environments. Hydrologic and water quality mathematical models are developed and used by hydrologists for planning and management and predicting water quality effects of changed conditions. Simple analyses such as pH, turbidity, and oxygen content may be done by hydrologists in the field. Other chemical analyses require more sophisticated laboratory equipment. In the past, municipal and industrial sewage was a major source of pollution for streams and lakes. Such wastes often received only minimal treatment, or raw wastes were dumped into rivers. Today, we are more aware of the consequences of such actions, and billions of dollars must be invested in pollution-control equipment to protect the waters of the earth. Other sources of pollution are more difficult to identify and control. These include road deicing salts, storm runoff from urban areas and farmland, and erosion from construction sites.

Groundwater

Groundwater, pumped from beneath the earth's surface, is often cheaper, more convenient and less vulnerable to pollution than surface water. Therefore, it is commonly used for public water supplies. Groundwater provides the largest source of usable water storage in the United States. Underground reservoirs contain far more water than the capacity of all surface reservoirs and lakes, including the Great Lakes. In some areas, ground water may be the only option. Some municipalities survive solely on groundwater.

Hydrologists estimate the volume of water stored underground by measuring water levels in local wells and by examining geologic records from well-drilling to determine the extent, depth and thickness of water-bearing sediments and rocks. Before an investment is made in full-sized wells, hydrologists may supervise the drilling of test wells. They note the depths at which water is encountered and collect samples of soils, rock and water for laboratory analyses. They may run a variety of geophysical tests on the completed hole, keeping and accurate log of their observations and test results. Hydrologists determine the most efficient pumping rate by monitoring the extent that water levels drop in the pumped well and in its nearest neighbors. Pumping the well too fast could cause it to go dry or could interfere with neighboring wells. Along the coast, overpumping can cause saltwater intrusion. By plotting and analyzing these data, hydrologists can estimate the maximum and optimum yields of the well.

Polluted groundwater is less visible, but more insidious and difficult to clean up, than pollution in rivers and lakes. Ground water pollution most often results from improper disposal of wastes on land. Major sources include industrial and household chemicals and garbage landfills, industrial waste lagoons, tailings and process wastewater from mines, oil field brine pits, leaking underground oil storage tanks and pipelines, sewage sludge and septic systems. Hydrologists provide guidance in the location of monitoring wells around waste disposal sites and sample them at regular intervals to determine if undesirable leachate — contaminated water containing toxic or hazardous chemicals — is reaching the ground water.

In polluted areas, hydrologists may collect soil and water samples to identify the type and extent of contamination. The chemical data then are plotted on a map to show the size and direction of waste movement. In complex situations, computer modeling of water flow and waste migration provides guidance for a clean-up program. In extreme cases, remedial actions may require excavation of the polluted soil. Today, most people and industries realize that the amount of money invested in prevention is far less than that of cleanup. Hydrologists often are consulted for selection of proper sites for new waste disposal facilities. The danger of pollution is minimized by locating wells in areas of deep ground water and impermeable soils. Other practices include lining the bottom of a landfill with watertight materials, collecting any leachate with drains, and keeping the landfill surface covered as much as possible. Careful monitoring is always necessary.

History

Hydrology has been a subject of investigation and engineering for millennia. For example, about 4000 BC the Nile was dammed to improve agricultural productivity of previously barren lands. Mesopotamian towns were protected from flooding with high earthen walls. Aqueducts were built by the Greeks and Romans, while history shows that the Chinese built irrigation and flood control works. The ancient Sinhalese used hydrology to build complex irrigation works in Sri Lanka, also known for the invention of the Valve Pit which allowed construction of large reservoirs, anicuts and canals which still function.

Marcus Vitruvius, in the first century BC, described a philosophical theory of the hydrologic cycle, in which precipitation falling in the mountains infiltrated the Earth's surface and led to streams and springs in the lowlands. With the adoption of a more scientific approach, Leonardo da Vinci and Bernard Palissy independently reached an accurate representation of the hydrologic cycle. It was not until the 17th century that hydrologic variables began to be quantified.

Pioneers of the modern science of hydrology include Pierre Perrault, Edme Mariotte and Edmund Halley. By measuring rainfall, runoff, and drainage area, Perrault showed that rainfall was sufficient to account for the flow of the Seine. Mariotte combined velocity and river cross-section measurements to obtain a discharge value, again in the Seine. Halley showed that the evaporation from the Mediterranean Sea was sufficient to account for the outflow of rivers flowing into the sea.

Advances in the 18th century included the Bernoulli piezometer and Bernoulli's equation, by Daniel Bernoulli, and the Pitot tube, by Henri Pitot. The 19th century saw development in groundwater hydrology, including Darcy's law, the Dupuit-Thiem well formula, and Hagen-Poiseuille's capillary flow equation.

Rational analyses began to replace empiricism in the 20th century, while governmental agencies began their own hydrological research programs. Of particular importance were Leroy Sherman's unit hydrograph, the infiltration theory of Robert E. Horton, and C.V. Theis' aquifer test/equation describing well hydraulics.

Since the 1950s, hydrology has been approached with a more theoretical basis than in the past, facilitated by advances in the physical understanding of hydrological processes and by the advent of computers and especially geographic information systems (GIS).

Themes

The central theme of hydrology is that water circulates throughout the Earth through different pathways and at different rates. The most vivid image of this is in the evaporation of water from the ocean, which forms clouds. These clouds drift over the land and produce rain. The rainwater flows into lakes, rivers, or aquifers. The water in lakes, rivers, and aquifers then either evaporates back to the atmosphere or eventually flows back to the ocean, completing a cycle. Water changes its state of being several times throughout this cycle.

The areas of research within hydrology concern the movement of water between its various states, or within a given state, or simply quantifying the amounts in these states in a given region. Parts of hydrology concern developing methods for directly measuring these flows or amounts of water, while others concern modeling these processes either for scientific knowledge or for making a prediction in practical applications.

Groundwater

Ground water is water beneath Earth's surface, often pumped for drinking water. Groundwater hydrology (hydrogeology) considers quantifying groundwater flow and solute transport. Problems in describing the saturated zone include the characterization of aquifers in terms of flow direction, groundwater pressure and, by inference, groundwater depth (see: aquifer test). Measurements here can be made using a piezometer. Aquifers are also described in terms of hydraulic conductivity, storativity and transmissivity. There are a number of geophysical methods for characterizing aquifers. There are also problems in characterizing the vadose zone (unsaturated zone).

Infiltration

Infiltration is the process by which water enters the soil. Some of the water is absorbed, and the rest percolates down to the water table. The infiltration capacity, the maximum rate at which the soil can absorb water, depends on several factors. The layer that is already saturated provides a resistance that is proportional to its thickness, while that plus the depth of water above the soil provides the driving force (hydraulic head). Dry soil can allow rapid infiltration by capillary action; this force diminishes as the soil becomes wet. Compaction reduces the porosity and the pore sizes. Surface cover increases capacity by retarding runoff, reducing compaction and other processes. Higher temperatures reduce viscosity, increasing infiltration.

Soil moisture

Soil moisture can be measured in various ways; by capacitance probe, time domain reflectometer or Tensiometer. Other methods include solute sampling and geophysical methods.

Surface water flow

A flood hydrograph showing stage for the Shawsheen River at Wilmington.

Hydrology considers quantifying surface water flow and solute transport, although the treatment of flows in large rivers is sometimes considered as a distinct topic of hydraulics or hydrodynamics. Surface water flow can include flow both in recognizable river channels and otherwise. Methods for measuring flow once the water has reached a river include the stream gauge (see: discharge), and tracer techniques. Other topics include chemical transport as part of surface water, sediment transport and erosion.

One of the important areas of hydrology is the interchange between rivers and aquifers. Groundwater/surface water interactions in streams and aquifers can be complex and the direction of net water flux (into surface water or into the aquifer) may vary spatially along a stream channel and over time at any particular location, depending on the relationship between stream stage and groundwater levels.

Precipitation and evaporation

A standard NOAA rain gauge

In some considerations, hydrology is thought of as starting at the land-atmosphere boundary and so it is important to have adequate knowledge of both precipitation and evaporation. Precipitation can be measured in various ways: disdrometer for precipitation characteristics at a fine time scale; radar for cloud properties, rain rate estimation, hail and snow detection; rain gauge for routine accurate measurements of rain and snowfall; satellite for rainy area identification, rain rate estimation, land-cover/land-use, and soil moisture, for example.

Evaporation is an important part of the water cycle. It is partly affected by humidity, which can be measured by a sling psychrometer. It is also affected by the presence of snow, hail, and ice and can relate to dew, mist and fog. Hydrology considers evaporation of various forms: from water surfaces; as transpiration from plant surfaces in natural and agronomic ecosystems. Direct measurement of evaporation can be obtained using Simon's evaporation pan.

Remote sensing

Estimates of changes in water storage around the Tigris and Euphrates Rivers, measured by NASA's GRACE satellites. The satellites measure tiny changes in gravitational acceleration, which can then be processed to reveal movement of water due to changes in its total mass.

Remote sensing of hydrologic processes can provide information on locations where in situ sensors may be unavailable or sparse. It also enables observations over large spatial extents. Many of the variables constituting the terrestrial water balance, for example surface water storage, soil moisture, precipitation, evapotranspiration, and snow and ice, are measurable using remote sensing at various spatial-temporal resolutions and accuracies. Sources of remote sensing include land-based sensors, airborne sensors and satellite sensors which can capture microwave, thermal and near-infrared data or use lidar, for example.

Water quality

In hydrology, studies of water quality concern organic and inorganic compounds, and both dissolved and sediment material. In addition, water quality is affected by the interaction of dissolved oxygen with organic material and various chemical transformations that may take place. Measurements of water quality may involve either in-situ methods, in which analyses take place on-site, often automatically, and laboratory-based analyses and may include microbiological analysis.

Integrating measurement and modelling

Budget analyses

Parameter estimation

Scaling in time and space

Data assimilation

Quality control of data – see for example Double mass analysis

Prediction

Observations of hydrologic processes are used to make predictions of the future behavior of hydrologic systems (water flow, water quality). One of the major current concerns in hydrologic research is "Prediction in Ungauged Basins" (PUB), i.e. in basins where no or only very few data exist.

Statistical hydrology

By analyzing the statistical properties of hydrologic records, such as rainfall or river flow, hydrologists can estimate future hydrologic phenomena. When making assessments of how often relatively rare events will occur, analyses are made in terms of the return period of such events. Other quantities of interest include the average flow in a river, in a year or by season.

These estimates are important for engineers and economists so that proper risk analysis can be performed to influence investment decisions in future infrastructure and to determine the yield reliability characteristics of water supply systems. Statistical information is utilized to formulate operating rules for large dams forming part of systems which include agricultural, industrial and residential demands.

Modeling

Hydrological models are simplified, conceptual representations of a part of the hydrologic cycle. They are primarily used for hydrological prediction and for understanding hydrological processes, within the general field of scientific modeling. Two major types of hydrological models can be distinguished:

Models based on data. These models are black box systems, using mathematical and statistical concepts to link a certain input (for instance rainfall) to the model output (for instance runoff). Commonly used techniques are regression, transfer functions, and system identification. The simplest of these models may be linear models, but it is common to deploy non-linear components to represent some general aspects of a catchment's response without going deeply into the real physical processes involved. An example of such an aspect is the well-known behavior that a catchment will respond much more quickly and strongly when it is already wet than when it is dry.

Models based on process descriptions. These models try to represent the physical processes observed in the real world. Typically, such models contain representations of surface runoff, subsurface flow, evapotranspiration, and channel flow, but they can be far more complicated. Within this category, models can be divided into conceptual and deterministic. Conceptual models link simplified representations of the hydrological processes in an area, whereas deterministic models seek to resolve as much of the physics of a system as possible. These models can be subdivided into single-event models and continuous simulation models.

Recent research in hydrological modeling tries to have a more global approach to the understanding of the behavior of hydrologic systems to make better predictions and to face the major challenges in water resources management.

Transport

Water movement is a significant means by which other materials, such as soil, gravel, boulders or pollutants, are transported from place to place. Initial input to receiving waters may arise from a point source discharge or a line source or area source, such as surface runoff. Since the 1960s rather complex mathematical models have been developed, facilitated by the availability of high-speed computers. The most common pollutant classes analyzed are nutrients, pesticides, total dissolved solids and sediment.

2

WATER RESOURCES

Water resources are sources of water that are useful or potentially useful. Uses of water include agricultural, industrial, household, recreational and environmental activities. The majority of human uses require fresh water.

97 percent of the water on the Earth is salt water and only three percent is fresh water; slightly over two thirds of this is frozen in glaciers and polar ice caps. The remaining unfrozen freshwater is found mainly as groundwater, with only a small fraction present above ground or in the air.

Fresh water is a renewable resource, yet the world's supply of groundwater is steadily decreasing, with depletion occurring most prominently in Asia and North America, although it is still unclear how much natural renewal balances this usage, and whether ecosystems are threatened. The framework for allocating water resources to water users (where such a framework exists) is known as water rights.

SOURCES OF FRESH WATER

Surface water

Surface water is water in a river, lake or fresh water wetland. Surface water is naturally replenished by precipitation and naturally lost through discharge to the oceans, evaporation, evapotranspiration and groundwater recharge.

Although the only natural input to any surface water system is precipitation within its watershed, the total quantity of water in that system at any given time is also dependent on many other factors. These factors include storage capacity in lakes, wetlands and artificial reservoirs, the permeability of the soil beneath these storage bodies, the runoff characteristics of the land in the watershed, the timing of the precipitation and local evaporation rates. All of these factors also affect the proportions of water loss.

Human activities can have a large and sometimes devastating impact on these factors. Humans often increase storage capacity by constructing reservoirs and decrease it by draining wetlands. Humans often increase runoff quantities and velocities by paving areas and channelizing stream flow.

The total quantity of water available at any given time is an important consideration. Some human water users have an intermittent need for water. For example, many farms require large quantities of water in the spring, and no water at all in the winter. To supply such a farm with water, a surface water system may require a large storage capacity to collect water throughout the year and release it in a short period of time. Other users have a continuous need for water, such as a power plant that requires water for cooling. To supply such a power plant with water, a surface water system only needs enough storage capacity to fill in when average stream flow is below the power plant's need.

Nevertheless, over the long term the average rate of precipitation within a watershed is the upper bound for average consumption of natural surface water from that watershed.

Natural surface water can be augmented by importing surface water from another watershed through a canal or pipeline. It can also be artificially augmented from any of the other sources listed here, however in practice the quantities are negligible. Humans can also cause surface water to be "lost" (i.e. become unusable) through pollution.

Brazil is the country estimated to have the largest supply of fresh water in the world, followed by Russia and Canada.

Under river flow

Throughout the course of a river, the total volume of water transported downstream will often be a combination of the visible free water flow together with a substantial contribution flowing through rocks and sediments that underlie the river and its floodplain called the hyporheic zone. For many rivers in large valleys, this unseen component of flow may greatly exceed the visible flow. The hyporheic zone often forms a dynamic interface between surface water and groundwater from aquifers, exchanging flow between rivers and aquifers that may be fully charged or depleted. This is especially significant in karst areas where pot-holes and underground rivers are common.

Groundwater

Groundwater is fresh water located in the subsurface pore space of soil and rocks. It is also water that is flowing within aquifers below the water table. Sometimes it is useful to make a distinction between groundwater that is closely associated with surface water and deep groundwater in an aquifer (sometimes called "fossil water").

Groundwater can be thought of in the same terms as surface water: inputs, outputs and storage. The critical difference is that due to its slow rate of turnover, groundwater storage is generally much larger (in volume) compared to inputs than it is for surface water. This difference makes it easy for humans to use groundwater un-sustainably for a long time without severe consequences. Nevertheless, over the long term the average rate of seepage above a groundwater source is the upper bound for average consumption of water from that source.

The natural input to groundwater is seepage from surface water. The natural outputs from groundwater are springs and seepage to the oceans.

If the surface water source is also subject to substantial evaporation, a groundwater source may become saline. This situation can occur naturally under endorheic bodies of water, or artificially under irrigated farmland. In coastal areas, human use of a groundwater source may cause the direction of seepage to ocean to reverse which can also cause soil salinization. Humans can also cause groundwater to be "lost" (i.e. become unusable) through pollution. Humans can increase the input to a groundwater source by building reservoirs or detention ponds.

Frozen water

Several schemes have been proposed to make use of icebergs as a water source, however to date this has only been done for research purposes. Glacier runoff is considered to be surface water. The Himalayas, which are often called "The Roof of the World", contain some of the most extensive and rough high altitude areas on Earth as well as the greatest area of glaciers and permafrost outside of the poles. Ten of Asia's largest rivers flow from there, and more than a billion people's livelihoods depend on them. To complicate matters, temperatures there are rising more rapidly than the global average. In Nepal, the temperature has risen by 0.6 degrees Celsius over the last decade, whereas globally, the Earth has warmed approximately 0.7 degrees Celsius over the last hundred years.

Desalination

Desalination is an artificial process by which saline water (generally sea water) is converted to fresh water. The most common desalination processes are distillation and reverse osmosis. Desalination is currently expensive compared to most alternative sources of water, and only a very small fraction of total human use is satisfied by desalination. It is only economically practical for high-valued uses (such as household and industrial uses) in arid areas. The most extensive use is in the Persian Gulf.

Artificial sources of usable water

Reclaimed water

Water reclamation (also called wastewater reuse) is the process of converting municipal wastewater (sewage) or industrial wastewater into water that can be reused for a variety of purposes. Types of reuse include: urban reuse, agricultural reuse (irrigation), environmental reuse, industrial reuse, planned potable reuse, de facto wastewater reuse (unplanned potable reuse). For example, reuse may include irrigation of gardens and agricultural fields or replenishing surface water and groundwater (i.e., groundwater recharge). Reused water may also be directed toward fulfilling certain needs in residences (e.g. toilet flushing), businesses, and industry, and could even be treated to reach drinking water standards. Treated municipal wastewater reuse for irrigation is a long-established practice, especially in arid countries. Reusing wastewater as part of sustainable water management allows water to remain as an alternative water source for human activities. This can reduce scarcity and alleviate pressures on groundwater and other natural water bodies.

There are several technologies used to treat wastewater for reuse. A combination of these technologies can meet strict treatment standards and make sure that the processed water is hygienically safe, meaning free from pathogens. The following are some of the typical technologies: Ozonation, ultrafiltration, aerobic treatment (membrane bioreactor), forward osmosis, reverse osmosis, advanced oxidation. Some water demanding activities do not require high grade water. In this case, wastewater can be reused with little or no treatment.

Desalination

Desalination is a process that takes away mineral components from saline water. More generally, desalination refers to the removal of salts and minerals from a target substance, as in soil desalination, which is an issue for agriculture. Saltwater (especially sea water) is desalinated to produce water suitable for human consumption or irrigation. The by-product of the desalination process is brine. Desalination is used on many seagoing ships and submarines. Most of the modern interest in desalination is focused on cost-effective provision of fresh water for human use. Along with recycled wastewater, it is one of the few rainfall-independent water resources.

Due to its energy consumption, desalinating sea water is generally more costly than fresh water from surface water or groundwater, water recycling and water conservation. However, these alternatives are not always available and depletion of reserves is a critical problem worldwide.[13] Desalination processes are usually driven by either thermal (in the case of distillation) or mechanical (in the case of reverse osmosis) as the primary energy types.

WATER USES

Agricultural

It is estimated that 70% of worldwide water is used for irrigation, with 15-35% of irrigation withdrawals being unsustainable. It takes around 2,000 - 3,000 litres of water to produce enough food to satisfy one person's daily dietary need. This is a considerable amount, when compared to that required for drinking, which is between two and five litres. To produce food for the now over 7 billion people who inhabit the planet today requires the water that would fill a canal ten metres deep, 100 metres wide and 2100 kilometres long.

Increasing water scarcity

Around fifty years ago, the common perception was that water was an infinite resource. At that time, there were fewer than half the current number of people on the planet. People were not as wealthy as today, consumed fewer calories and ate less meat, so less water was needed to produce their food. They required a third of the volume of water we presently take from rivers. Today, the competition for water resources is much more intense. This is because there are now seven billion people on the planet, their consumption of water-thirsty meat and vegetables is rising, and there is increasing competition for water from industry, urbanisation biofuel crops, and water reliant food items. In the future, even more water will be needed to produce food because the Earth's population is forecast to rise to 9 billion by 2050. An additional 2.5 or 3 billion people, choosing to eat fewer cereals and more meat and vegetables could add an additional five million kilometres to the virtual canal mentioned above.

An assessment of water management in agriculture sector was conducted in 2007 by the International Water Management Institute in Sri Lanka to see if the world had sufficient water to provide food for its growing population. It assessed the current availability of water for agriculture on a global scale and mapped out locations suffering from water scarcity. It found that a fifth of the world's people, more than 1.2 billion, live in areas of physical water scarcity, where there is not enough water to meet all demands. One third of the worlds population does not have access to clean drinking water, which is more than 2.3 billion people. A further 1.6 billion people live in areas experiencing economic water scarcity, where the lack of investment in water or insufficient human capacity make it impossible for authorities to satisfy the demand for water. The report found that it would be possible to produce the food required in future, but that continuation of today's food production and environmental trends would lead to crises in many parts of the world. To avoid a global water crisis, farmers will have to strive to increase productivity to meet growing demands for food, while industry and cities find ways to use water more efficiently.

In some areas of the world, irrigation is necessary to grow any crop at all, in other areas it permits more profitable crops to be grown or enhances crop yield. Various irrigation methods involve different trade-offs between crop yield, water consumption and capital cost of equipment and structures. Irrigation methods such as furrow and overhead sprinkler irrigation are usually less expensive but are also typically less efficient, because much of the water evaporates, runs off or drains below the root zone. Other irrigation methods considered to be more efficient include drip or trickle irrigation, surge irrigation, and some types of sprinkler systems where the sprinklers are operated near ground level. These types of systems, while more expensive, usually offer greater potential to minimize runoff, drainage and evaporation. Any system that is improperly managed can be wasteful, all methods have the potential for high efficiencies under suitable conditions, appropriate irrigation timing and management. Some issues that are often insufficiently considered are salinization of groundwater and contaminant accumulation leading to water quality declines.

As global populations grow, and as demand for food increases in a world with a fixed water supply, there are efforts under way to learn how to produce more food with less water, through improvements in irrigation methods and technologies, agricultural water management, crop types, and water monitoring. Aquaculture is a small but growing agricultural use of water. Freshwater commercial fisheries may also be considered as agricultural uses of water, but have generally been assigned a lower priority than irrigation (see Aral Sea and Pyramid Lake).

Industrial

It is estimated that 22% of worldwide water is used in industry. Major industrial users include hydroelectric dams, thermoelectric power plants, which use water for cooling, ore and oil refineries, which use water in chemical processes, and manufacturing plants, which use water as a solvent. Water withdrawal can be very high for certain industries, but consumption is generally much lower than that of agriculture.

Water is used in renewable power generation. Hydroelectric power derives energy from the force of water flowing downhill, driving a turbine connected to a generator. This hydroelectricity is a low-cost, non-polluting, renewable energy source. Significantly, hydroelectric power can also be used for load following unlike most renewable energy sources which are intermittent. Ultimately, the energy in a hydroelectric powerplant is supplied by the sun. Heat from the sun evaporates water, which condenses as rain in higher altitudes and flows downhill. Pumped-storage hydroelectric plants also exist, which use grid electricity to pump water uphill when demand is low, and use the stored water to produce electricity when demand is high.

Hydroelectric power plants generally require the creation of a large artificial lake. Evaporation from this lake is higher than evaporation from a river due to the larger surface area exposed to the elements, resulting in much higher water consumption. The process of driving water through the turbine and tunnels or pipes also briefly removes this water from the natural environment, creating water withdrawal. The impact of this withdrawal on wildlife varies greatly depending on the design of the powerplant.

Pressurized water is used in water blasting and water jet cutters. Also, very high pressure water guns are used for precise cutting. It works very well, is relatively safe, and is not harmful to the environment. It is also used in the cooling of machinery to prevent overheating, or prevent saw blades from overheating. This is generally a very small source of water consumption relative to other uses.

Water is also used in many large scale industrial processes, such as thermoelectric power production, oil refining, fertilizer production and other chemical plant use, and natural gas extraction from shale rock. Discharge of untreated water from industrial uses is pollution. Pollution includes discharged solutes (chemical pollution) and increased water temperature (thermal pollution). Industry requires pure water for many applications and utilizes a variety of purification techniques both in water supply and discharge. Most of this pure water is generated on site, either from natural freshwater or from municipal grey water. Industrial consumption of water is generally much lower than withdrawal, due to laws requiring industrial grey water to be treated and returned to the environment. Thermoelectric powerplants using cooling towers have high consumption, nearly equal to their withdrawal, as most of the withdrawn water is evaporated as part of the cooling process. The withdrawal, however, is lower than in once-through cooling systems.

Domestic

It is estimated that 8% of worldwide water use is for domestic purposes. These include drinking water, bathing, cooking, toilet flushing, cleaning, laundry and gardening. Basic domestic water requirements have been estimated by Peter Gleick at around 50 liters per person per day, excluding water for gardens. Drinking water is water that is of sufficiently high quality so that it can be consumed or used without risk of immediate or long term harm. Such water is commonly called potable water. In most developed countries, the water supplied to domestic, commerce and industry is all of drinking water standard even though only a very small proportion is actually consumed or used in food preparation.

Recreation

Recreational water use is usually a very small but growing percentage of total water use. Recreational water use is mostly tied to reservoirs. If a reservoir is kept fuller than it would otherwise be for recreation, then the water retained could be categorized as recreational usage. Release of water from a few reservoirs is also timed to enhance whitewater boating, which also could be considered a recreational usage. Other examples are anglers, water skiers, nature enthusiasts and swimmers.

Recreational usage is usually non-consumptive. Golf courses are often targeted as using excessive amounts of water, especially in drier regions. It is, however, unclear whether recreational irrigation (which would include private gardens) has a noticeable effect on water resources. This is largely due to the unavailability of reliable data. Additionally, many golf courses utilize either primarily or exclusively treated effluent water, which has little impact on potable water availability.

Some governments, including the Californian Government, have labelled golf course usage as agricultural in order to deflect environmentalists' charges of wasting water. However, using the above figures as a basis, the actual statistical effect of this reassignment is close to zero. In Arizona, an organized lobby has been established in the form of the Golf Industry Association, a group focused on educating the public on how golf impacts the environment.

Recreational usage may reduce the availability of water for other users at specific times and places. For example, water retained in a reservoir to allow boating in the late summer is not available to farmers during the spring planting season. Water released for whitewater rafting may not be available for hydroelectric generation during the time of peak electrical demand.

Environmental

Explicit environment water use is also a very small but growing percentage of total water use. Environmental water may include water stored in impoundments and released for environmental purposes (held environmental water), but more often is water retained in waterways through regulatory limits of abstraction. Environmental water usage includes watering of natural or artificial wetlands, artificial lakes intended to create wildlife habitat, fish ladders, and water releases from reservoirs timed to help fish spawn, or to restore more natural flow regimes

Like recreational usage, environmental usage is non-consumptive but may reduce the availability of water for other users at specific times and places. For example, water release from a reservoir to help fish spawn may not be available to farms upstream, and water retained in a river to maintain waterway health would not be available to water abstractors downstream.

Water stress

The concept of water stress is relatively simple: According to the World Business Council for Sustainable Development, it applies to situations where there is not enough water for all uses, whether agricultural, industrial or domestic. Defining thresholds for stress in terms of available water per capita is more complex, however, entailing assumptions about water use and its efficiency. Nevertheless, it has been proposed that when annual per capita renewable freshwater availability is less than 1,700 cubic meters, countries begin to experience periodic or regular water stress. Below 1,000 cubic meters, water scarcity begins to hamper economic development and human health and well-being.

Population growth

In 2000, the world population was 6.2 billion. The UN estimates that by 2050 there will be an additional 3.5 billion people with most of the growth in developing countries that already suffer water stress. Thus, water demand will increase unless there are corresponding increases in water conservation and recycling of this vital resource. In building on the data presented here by the UN, the World Bank goes on to explain that access to water for producing food will be one of the main challenges in the decades to come. Access to water will need to be balanced with the importance of managing water itself in a sustainable way while taking into account the impact of climate change, and other environmental and social variables.

Expansion of business activity

Business activity ranging from industrialization to services such as tourism and entertainment continues to expand rapidly. This expansion requires increased water services including both supply and sanitation, which can lead to more pressure on water resources and natural ecosystem.

Rapid urbanization

The trend towards urbanization is accelerating. Small private wells and septic tanks that work well in low-density communities are not feasible within high-density urban areas. Urbanization requires significant investment in water infrastructure in order to deliver water to individuals and to process the concentrations of wastewater - both from individuals and from business. These polluted and contaminated waters must be treated or they pose unacceptable public health risks.

In 60% of European cities with more than 100,000 people, groundwater is being used at a faster rate than it can be replenished. Even if some water remains available, it costs increasingly more to capture it.

Climate change

Climate change could have significant impacts on water resources around the world because of the close connections between the climate and hydrological cycle. Rising temperatures will increase evaporation and lead to increases in precipitation, though there will be regional variations in rainfall. Both droughts and floods may become more frequent in different regions at different times, and dramatic changes in snowfall and snow melt are expected in mountainous areas. Higher temperatures will also affect water quality in ways that are not well understood. Possible impacts include increased eutrophication. Climate change could also mean an increase in demand for farm irrigation, garden sprinklers, and perhaps even swimming pools. There is now ample evidence that increased hydrologic variability and change in climate has and will continue have a profound impact on the water sector through the hydrologic cycle, water availability, water demand, and water allocation at the global, regional, basin, and local levels.

Depletion of aquifers

Due to the expanding human population, competition for water is growing such that many of the world's major aquifers are becoming depleted. This is due both for direct human consumption as well as agricultural irrigation by groundwater. Millions of pumps of all sizes are currently extracting groundwater throughout the world. Irrigation in dry areas such as northern China, Nepal and India is supplied by groundwater, and is being extracted at an unsustainable rate. Cities that have experienced aquifer drops between 10 to 50 meters include Mexico City, Bangkok, Beijing, Madras and Shanghai.

Pollution and watcr protcction

Water pollution is one of the main concerns of the world today. The governments of numerous countries have striven to find solutions to reduce this problem. Many pollutants threaten water supplies, but the most widespread, especially in developing countries, is the discharge of raw sewage into natural waters; this method of sewage disposal is the most common method in underdeveloped countries, but also is prevalent in quasi-developed countries such as China, India, Nepal and Iran. Sewage, sludge, garbage, and even toxic pollutants are all dumped into the water. Even if sewage is treated, problems still arise. Treated sewage forms sludge, which may be placed in landfills, spread out on land, incinerated or dumped at sea. In addition to sewage, nonpoint source pollution such as agricultural runoff is a significant source of pollution in some parts of the world, along with urban stormwater runoff and chemical wastes dumped by industries and governments.

Water and conflicts

Competition for water has widely increased, and it has become more difficult to conciliate the necessities for water supply for human consumption, food production, ecosystems and other uses. Water administration is frequently involved in contradictory and complex problems. Approximately 10% of the worldwide annual runoff is used for human necessities. Several areas of the world are flooded, while others have such low precipitations that human life is almost impossible. As population and development increase, raising water demand, the possibility of problems inside a certain country or region increases, as it happens with others outside the region.

Over the past 25 years, politicians, academics and journalists have frequently predicted that disputes over water would be a source of future wars. Commonly cited quotes include: that of former Egyptian Foreign Minister and former Secretary-General of the United Nations Boutrous Ghali, who forecast, "The next war in the Middle East will be fought over water, not politics"; his successor at the UN, Kofi Annan, who in 2001 said, "Fierce competition for fresh water may well become a source of conflict and wars in the future," and the former Vice President of the World Bank, Ismail Serageldin, who said the wars of the next century will be over water unless significant changes in governance occurred. The water wars hypothesis had its roots in earlier research carried out on a small number of transboundary rivers such as the Indus, Jordan and Nile. These particular rivers became the focus because they had experienced water-related disputes. Specific events cited as evidence include Israel's bombing of Syria's attempts to divert the Jordan's headwaters, and military threats by Egypt against any country building dams in the upstream waters of the Nile. However, while some links made between conflict and water were valid, they did not necessarily represent the norm.

The only known example of an actual inter-state conflict over water took place between 2500 and 2350 BC between the Sumerian states of Lagash and Umma. Water stress has most often led to conflicts at local and regional levels. Tensions arise most often within national borders, in the downstream areas of distressed river basins. Areas such as the lower regions of China's Yellow River or the Chao Phraya River in Thailand, for example, have already been experiencing water stress for several years. Water stress can also exacerbate conflicts and political tensions which are not directly caused by water. Gradual reductions over time in the quality and/or quantity of fresh water can add to the instability of a region by depleting the health of a population, obstructing economic development, and exacerbating larger conflicts.

Shared water resources can promote collaboration

Water resources that span international boundaries are more likely to be a source of collaboration and cooperation than war. Scientists working at the International Water Management Institute have been investigating the evidence behind water war predictions. Their findings show that, while it is true there has been conflict related to water in a handful of international basins, in the rest of the world's approximately 300 shared basins the record has been largely positive. This is exemplified by the hundreds of treaties in place guiding equitable water use between nations sharing water resources. The institutions created by these agreements can, in fact, be one of the most important factors in ensuring cooperation rather than conflict.

The International Union for the Conservation of Nature (IUCN) published the book Share: Managing water across boundaries. One chapter covers the functions of trans-boundary institutions and how they can be designed to promote cooperation, overcome initial disputes and find ways of coping with the uncertainty created by climate change. It also covers how the effectiveness of such institutions can be monitored.

Water Shortages

In 2025, water shortages will be more prevalent among poorer countries where resources are limited and population growth is rapid, such as the Middle East, Africa, and parts of Asia. By 2025, large urban and peri-urban areas will require new infrastructure to provide safe water and adequate sanitation. This suggests growing conflicts with agricultural water users, who currently consume the majority of the water used by humans.

Generally speaking the more developed countries of North America, Europe and Russia will not see a serious threat to water supply by the year 2025, not only because of their relative wealth, but more importantly their populations will be better aligned with available water resources. North Africa, the Middle East, South Africa and northern China will face very severe water shortages due to physical scarcity and a condition of overpopulation relative to their carrying capacity with respect to water supply. Most of South America, Sub-Saharan Africa, Southern China and India will face water supply shortages by 2025; for these latter regions the causes of scarcity will be economic constraints to developing safe drinking water, as well as excessive population growth.

Economic Considerations

Water supply and sanitation require a huge amount of capital investment in infrastructure such as pipe networks, pumping stations and water treatment works. It is estimated that Organisation for Economic Co-operation and Development (OECD) nations need to invest at least USD 200 billion per year to replace aging water infrastructure to guarantee supply, reduce leakage rates and protect water quality.

International attention has focused upon the needs of the developing countries. To meet the Millennium Development Goals targets of halving the proportion of the population lacking access to safe drinking water and basic sanitation by 2015, current annual investment on the order of USD 10 to USD 15 billion would need to be roughly doubled. This does not include investments required for the maintenance of existing infrastructure.

Once infrastructure is in place, operating water supply and sanitation systems entails significant ongoing costs to cover personnel, energy, chemicals, maintenance and other expenses. The sources of money to meet these capital and operational costs are essentially either user fees, public funds or some combination of the two. An increasing dimension to consider is the flexibility of the water supply system.

Ocean

The ocean (also the sea or the world ocean) is the body of salt water which covers approximately 71% of the surface of the Earth. It is also "any of the large bodies of water into which the great ocean is divided". A common definition lists five oceans, in descending order by area, the Pacific, Atlantic, Indian, Southern (Antarctic), and Arctic Oceans.

Seawater covers approximately 361,000,000 km2 (139,000,000 sq mi) and is customarily divided into several principal oceans and smaller seas, with the ocean as a whole covering approximately 71% of Earth's surface and 90% of the Earth's biosphere. The world ocean contains 97% of Earth's water, and oceanographers have stated that less than 20% of the oceans have been mapped. The total volume is approximately 1.35 billion cubic kilometers (320 million cu mi) with an average depth of nearly 3,700 meters (12,100 ft).

As the world's ocean is the principal component of Earth's hydrosphere, it is integral to life, forms part of the carbon cycle, and influences climate and weather patterns. The ocean is the habitat of 230,000 known species, but because much of it is unexplored, the number of species in the ocean is much larger, possibly over two million. The origin of Earth's oceans is unknown; a sizable quantity of water would have been in the material that formed the Earth. Water molecules would have escaped Earth's gravity more easily when it was less massive during its formation due to atmospheric escape. Oceans are thought to have formed in the Hadean eon and may have been the cause for the emergence of life. There are numerous environmental issues for oceans which include for example marine pollution, overfishing, ocean acidification and other effects of climate change on oceans.

Pacific Ocean

The Pacific Ocean is the largest and deepest of Earth's oceanic divisions. It extends from the Arctic Ocean in the north to the Southern Ocean (or, depending on definition, to Antarctica) in the south and is bounded by the continents of Asia and Australia in the west and the Americas in the east.

At 165,250,000 square kilometers (63,800,000 square miles) in the area (as defined with a southern Antarctic border), this largest division of the World Ocean—and, in turn, the hydrosphere—covers about 46% of Earth's water surface and about 32% of its total surface area, making it larger than all of Earth's land area combined (148,000,000 square kilometers). The centers of both the Water Hemisphere and the Western Hemisphere are in the Pacific Ocean. Ocean circulation (caused by the Coriolis effect) subdivides it[citation needed] into two largely independent volumes of water, which meet at the equator: the North(ern) Pacific Ocean and South(ern) Pacific Ocean. The Galápagos and Gilbert Islands, while straddling the equator, are deemed wholly within the South Pacific.

Its mean depth is 4,000 meters (13,000 feet). Challenger Deep in the Mariana Trench, located in the western north Pacific, is the deepest point in the world, reaching a depth of 10,928 meters (35,853 feet). The Pacific also contains the deepest point in the Southern Hemisphere, the Horizon Deep in the Tonga Trench, at 10,823 meters (35,509 feet). The third deepest point on Earth, the Sirena Deep, is also located in the Mariana Trench.

The western Pacific has many major marginal seas, including the South China Sea, the East China Sea, the Sea of Japan, the Sea of Okhotsk, the Philippine Sea, the Coral Sea, and the Tasman Sea.

- Australia
- Brunei
- Cambodia
- Canada
- Chile
- China
- Colombia
- Cook Islands
- Costa Rica
- Ecuador
- El Salvador
- Federated States of Micronesia
- Fiji

- Guatemala
- Honduras
- Indonesia
- Japan
- Kiribati
- Malaysia
- Marshall Islands
- Mexico
- Nauru
- New Zealand
- Nicaragua
- Niue
- North Korea
- Palau
- Panama
- Papua New Guinea
- Peru
- Philippines
- Russia
- Samoa
- Singapore
- Solomon Islands
- South Korea
- Taiwan
- Thailand
- Timor-Leste
- Tonga
- Tuvalu
- United States
- Vanuatu
- Vietnam
- Territories
- American Samoa (US)
- Baker Island (US)

- Clipperton Island (France)
- Coral Sea Islands (Australia)
- French Polynesia (France)
- Guam (US)
- Hong Kong (China)
- Howland Island (US)
- Jarvis Island (US)
- Johnston Island (US)
- Kingman Reef (US)
- Macau (China)
- Midway Atoll (US)
- New Caledonia (France)
- Norfolk Island (Australia)
- Northern Mariana Islands (US)
- Palmyra Atoll (US)
- Pitcairn Islands (UK)
- Tokelau (New Zealand)
- Wallis and Futuna (France)
- Wake Island (US)

TARAWA ATOLL IN THE REPUBLIC OF KIRIBATI

The Pacific Ocean has most of the islands in the world. There are about 25,000 islands in the Pacific Ocean. The islands entirely within the Pacific Ocean can be divided into three main groups known as Micronesia, Melanesia and Polynesia. Micronesia, which lies north of the equator and west of the International Date Line, includes the Mariana Islands in the northwest, the Caroline Islands in the center, the Marshall Islands to the east and the islands of Kiribati in the southeast. Melanesia, to the southwest, includes New Guinea, the world's second largest island after Greenland and by far the largest of the Pacific islands. The other main Melanesian groups from north to south are the Bismarck Archipelago, the Solomon Islands, Santa Cruz, Vanuatu, Fiji and New Caledonia.

The largest area, Polynesia, stretching from Hawaii in the north to New Zealand in the south, also encompasses Tuvalu, Tokelau, Samoa, Tonga and the Kermadec Islands to the west, the Cook Islands, Society Islands and Austral Islands in the center, and the Marquesas Islands, Tuamotu, Mangareva Islands, and Easter Island to the east. Islands in the Pacific Ocean are of four basic types: continental islands, high islands, coral reefs and uplifted coral platforms. Continental islands lie outside the andesite line and include New Guinea, the islands of New Zealand, and the Philippines. Some of these islands are structurally associated with nearby continents. High islands are of volcanic origin, and many contain active volcanoes. Among these are Bougainville, Hawaii, and the Solomon Islands.

The coral reefs of the South Pacific are low-lying structures that have built up on basaltic lava flows under the ocean's surface. One of the most dramatic is the Great Barrier Reef off northeastern Australia with chains of reef patches. A second island type formed of coral is the uplifted coral platform, which is usually slightly larger than the low coral islands. Examples include Banaba (formerly Ocean Island) and Makatea in the Tuamotu group of French Polynesia.

- Ladrilleros Beach in Colombia on the coast of Chocó natural region
- Tahuna maru islet, French Polynesia
- Los Molinos on the coast of Southern Chile
- Water characteristics

The volume of the Pacific Ocean, representing about 50.1 percent of the world's oceanic water, has been estimated at some 714 million cubic kilometers (171 million cubic miles). Surface water temperatures in the Pacific can vary from -1.4 °C (29.5 °F), the freezing point of sea water, in the poleward areas to about 30 °C (86 °F) near the equator. Salinity also varies latitudinally, reaching a maximum of 37 parts per thousand in the southeastern area. The water near the equator, which can have a salinity as low as 34 parts per thousand, is less salty than that found in the mid-latitudes because of abundant equatorial precipitation throughout the year. The lowest counts of less than 32 parts per thousand are found in the far north as less evaporation of seawater takes place in these frigid areas. The motion of Pacific waters is generally clockwise in the Northern Hemisphere (the North Pacific gyre) and counter-clockwise in the Southern Hemisphere. The North Equatorial Current, driven westward along latitude 15°N by the trade winds, turns north near the Philippines to become the warm Japan or Kuroshio Current. Turning eastward at about 45°N, the Kuroshio forks and some water moves northward as the Aleutian Current, while the rest turns southward to rejoin the North Equatorial Current. The Aleutian Current branches as it approaches North America and forms the base of a counter-clockwise circulation in the Bering Sea. Its southern arm becomes the chilled slow, south-flowing California Current. The South Equatorial Current, flowing west along the equator, swings southward east of New Guinea, turns east at about 50°S, and joins the main westerly circulation of the South Pacific, which includes the Earth-circling Antarctic Circumpolar Current. As it approaches the Chilean coast, the South Equatorial Current divides; one branch flows around Cape Horn and the other turns north to form the Peru or Humboldt Current.

Climate

The climate patterns of the Northern and Southern Hemispheres generally mirror each other. The trade winds in the southern and eastern Pacific are remarkably steady while conditions in the North Pacific are far more varied with, for example, cold winter temperatures on the east coast of Russia contrasting with the milder weather off British Columbia during the winter months due to the preferred flow of ocean currents.

In the tropical and subtropical Pacific, the El Niño Southern Oscillation (ENSO) affects weather conditions. To determine the phase of ENSO, the most recent three-month sea surface temperature average for the area approximately 3,000 km (1,900 mi) to the southeast of Hawaii is computed, and if the region is more than 0.5 °C (0.9 °F) above or below normal for that period, then an El Niño or La Niña is considered in progress.

Typhoon Tip at global peak intensity on 12 October 1979

In the tropical western Pacific, the monsoon and the related wet season during the summer months contrast with dry winds in the winter which blow over the ocean from the Asian landmass. Worldwide, tropical cyclone activity peaks in late summer, when the difference between temperatures aloft and sea surface temperatures is the greatest; however, each particular basin has its own seasonal patterns. On a worldwide scale, May is the least active month, while September is the most active month. November is the only month in which all the tropical cyclone basins are active. The Pacific hosts the two most active tropical cyclone basins, which are the northwestern Pacific and the eastern Pacific. Pacific hurricanes form south of Mexico, sometimes striking the western Mexican coast and occasionally the southwestern United States between June and October, while typhoons forming in the northwestern Pacific moving into southeast and east Asia from May to December. Tropical cyclones also form in the South Pacific basin, where they occasionally impact island nations. In the arctic, icing from October to May can present a hazard for shipping while persistent fog occurs from June to December. A climatological low in the Gulf of Alaska keeps the southern coast wet and mild during the winter months. The Westerlies and associated jet stream within the Mid-Latitudes can be particularly strong, especially in the Southern Hemisphere, due to the temperature difference between the tropics and Antarctica, which records the coldest temperature readings on the planet. In the Southern hemisphere, because of the stormy and cloudy conditions associated with extratropical cyclones riding the jet stream, it is usual to refer to the Westerlies as the Roaring Forties, Furious Fifties and Shrieking Sixties according to the varying degrees of latitude.

Geology

Ring of Fire. The Pacific is ringed by many volcanoes and oceanic trenches.

The ocean was first mapped by Abraham Ortelius; he called it Maris Pacifici following Ferdinand Magellan's description of it as "a pacific sea" during his circumnavigation from 1519 to 1522. To Magellan, it seemed much more calm (pacific) than the Atlantic. The andesite line is the most significant regional distinction in the Pacific. A petrologic boundary, it separates the deeper, mafic igneous rock of the Central Pacific Basin from the partially submerged continental areas of felsic igneous rock on its margins. The andesite line follows the western edge of the islands off California and passes south of the Aleutian arc, along the eastern edge of the Kamchatka Peninsula, the Kuril Islands, Japan, the Mariana Islands, the Solomon Islands, and New Zealand's North Island.

Ulawun stratovolcano situated on the island of New Britain, Papua New Guinea

The dissimilarity continues northeastward along the western edge of the Andes Cordillera along South America to Mexico, returning then to the islands off California. Indonesia, the Philippines, Japan, New Guinea, and New Zealand lie outside the andesite line.

Mount Saint Helens in 2020

Within the closed loop of the andesite line are most of the deep troughs, submerged volcanic mountains, and oceanic volcanic islands that characterize the Pacific basin. Here basaltic lavas gently flow out of rifts to build huge dome-shaped volcanic mountains whose eroded summits form island arcs, chains, and clusters. Outside the andesite line, volcanism is of the explosive type, and the Pacific Ring of Fire is the world's foremost belt of explosive volcanism. The Ring of Fire is named after the several hundred active volcanoes that sit above the various subduction zones.

The Pacific Ocean is the only ocean which is mostly bounded by subduction zones. Only the Antarctic and Australian coasts have no nearby subduction zones.

Geological history

The Pacific Ocean was born 750 million years ago at the breakup of Rodinia, although it is generally called the Panthalassa until the breakup of Pangea, about 200 million years ago. The oldest Pacific Ocean floor is only around 180 Ma old, with older crust subducted by now.

Seamount chains

The Pacific Ocean contains several long seamount chains, formed by hotspot volcanism. These include the Hawaiian–Emperor seamount chain and the Louisville Ridge.

Economy

The exploitation of the Pacific's mineral wealth is hampered by the ocean's great depths. In shallow waters of the continental shelves off the coasts of Australia and New Zealand, petroleum and natural gas are extracted, and pearls are harvested along the coasts of Australia, Japan, Papua New Guinea, Nicaragua, Panama, and the Philippines, although in sharply declining volume in some cases.

Fishing

Fish are an important economic asset in the Pacific. The shallower shoreline waters of the continents and the more temperate islands yield herring, salmon, sardines, snapper, swordfish, and tuna, as well as shellfish. Overfishing has become a serious problem in some areas. For example, catches in the rich fishing grounds of the Okhotsk Sea off the Russian coast have been reduced by at least half since the 1990s as a result of overfishing.

Nuclear waste

Prime Minister Suga declined to drink the bottle of Fukushima's treated radioactive water that he was holding, which would otherwise be discharged to the Pacific. 2020.

From 1946 to 1958, Marshall Islands served as the Pacific Proving Grounds for the United States and was the site of 67 nuclear tests on various atolls. Several nuclear weapons were lost in the Pacific Ocean, including one-megaton bomb lost during the 1965 Philippine Sea A-4 incident.

In 2021, the dumping of radioactive water of the Fukushima nuclear plant into the Pacific Ocean over a course of 30 years was finally approved by the Japanese Cabinet. The Cabinet asserted the radioactive water would have been diluted to drinkable standard. Apart from dumping, leakage of tritium into the Pacific was estimated to be between 20 and 40 trillion Bqs from 2011 to 2013, according to the Fukushima plant.

Environmental issues

The quantity of small plastic fragments floating in the north-east Pacific Ocean increased a hundredfold between 1972 and 2012. The ever-growing Great Pacific garbage patch between California and Japan is three times the size of France. An estimated 80,000 metric tons of plastic inhabit the patch, totaling 1.8 trillion pieces.

Marine debris on a Hawaiian coast

Marine pollution is a generic term for the harmful entry into the ocean of chemicals or particles. The main culprits are those using the rivers for disposing of their waste. The rivers then empty into the ocean, often also bringing chemicals used as fertilizers in agriculture. The excess of oxygen-depleting chemicals in the water leads to hypoxia and the creation of a dead zone.

Marine debris, also known as marine litter, is human-created waste that has ended up floating in a lake, sea, ocean, or waterway. Oceanic debris tends to accumulate at the center of gyres and coastlines, frequently washing aground where it is known as beach litter.

In addition, the Pacific Ocean has served as the crash site of satellites, including Mars 96, Fobos-Grunt, and Upper Atmosphere Research Satellite.

Atlantic Ocean

The Atlantic Ocean is the second-largest of the world's oceans, with an area of about 106,460,000 km2 (41,100,000 sq mi). It covers approximately 20 percent of Earth's surface and about 29 percent of its water surface area. It is known to separate the "Old World" from the "New World" in the European perception of the World.

The Atlantic Ocean occupies an elongated, S-shaped basin extending longitudinally between Europe and Africa to the east, and the Americas to the west. As one component of the interconnected World Ocean, it is connected in the north to the Arctic Ocean, to the Pacific Ocean in the southwest, the Indian Ocean in the southeast, and the Southern Ocean in the south (other definitions describe the Atlantic as extending southward to Antarctica). The Atlantic Ocean is divided into two parts, by the Equatorial Counter Current, with the North(ern) Atlantic Ocean and the South(ern) Atlantic Ocean at about 8°N.

Scientific explorations of the Atlantic include the Challenger expedition, the German Meteor expedition, Columbia University's Lamont-Doherty Earth Observatory and the United States Navy Hydrographic Office.

The oldest known mentions of an "Atlantic" sea come from Stesichorus around mid-sixth century BC (Sch. A. R. 1. 211): Atlantikôi pelágei and in The Histories of Herodotus around 450 BC (Hdt. 1.202.4): Atlantis thalassa where the name refers to "the sea beyond the pillars of Heracles" which is said to be part of the sea that surrounds all land. In these uses, the name refers to Atlas, the Titan in Greek mythology, who supported the heavens and who later appeared as a frontispiece in Medieval maps and also lent his name to modern atlases. On the other hand, to early Greek sailors and in Ancient Greek mythological literature such as the Iliad and the Odyssey, this all-encompassing ocean was instead known as Oceanus, the gigantic river that encircled the world; in contrast to the enclosed seas well known to the Greeks: the Mediterranean and the Black Sea. In contrast, the term "Atlantic" originally referred specifically to the Atlas Mountains in Morocco and the sea off the Strait of Gibraltar and the North African coast. The Greek word thalassa has been reused by scientists for the huge Panthalassa ocean that surrounded the supercontinent Pangaea hundreds of millions of years ago.

An alternative origin is derived from Native America language from the word Atlan of Nahuatl.

The term "Aethiopian Ocean", derived from Ancient Ethiopia, was applied to the Southern Atlantic as late as the mid-19th century. During the Age of Discovery, the Atlantic was also known to English cartographers as the Great Western Ocean.

The Pond is a term often used by British and American speakers in reference to the Northern Atlantic Ocean, as a form of meiosis, or sarcastic understatement. The term dates to as early as 1640, first appearing in print in pamphlet released during the reign of Charles I, and reproduced in 1869 in Nehemiah Wallington's Historical Notices of Events Occurring Chiefly in The Reign of Charles I, where "great Pond" is used in reference to the Atlantic Ocean by Francis Windebank, Charles I's Secretary of State.

The International Hydrographic Organization (IHO) defined the limits of the oceans and seas in 1953, but some of these definitions have been revised since then and some are not used by various authorities, institutions, and countries, see for example the CIA World Factbook. Correspondingly, the extent and number of oceans and seas vary.

The Atlantic Ocean is bounded on the west by North and South America. It connects to the Arctic Ocean through the Denmark Strait, Greenland Sea, Norwegian Sea and Barents Sea. To the east, the boundaries of the ocean proper are Europe: the Strait of Gibraltar (where it connects with the Mediterranean Sea—one of its marginal seas—and, in turn, the Black Sea, both of which also touch upon Asia) and Africa.

In the southeast, the Atlantic merges into the Indian Ocean. The 20° East meridian, running south from Cape Agulhas to Antarctica defines its border. In the 1953 definition it extends south to Antarctica, while in later maps it is bounded at the 60° parallel by the Southern Ocean.

The Atlantic has irregular coasts indented by numerous bays, gulfs and seas. These include the Baltic Sea, Black Sea, Caribbean Sea, Davis Strait, Denmark Strait, part of the Drake Passage, Gulf of Mexico, Labrador Sea, Mediterranean Sea, North Sea, Norwegian Sea, almost all of the Scotia Sea, and other tributary water bodies. Including these marginal seas the coast line of the Atlantic measures 111,866 km (69,510 mi) compared to 135,663 km (84,297 mi) for the Pacific.

Including its marginal seas, the Atlantic covers an area of 106,460,000 km2 (41,100,000 sq mi) or 23.5% of the global ocean and has a volume of 310,410,900 km3 (74,471,500 cu mi) or 23.3% of the total volume of the earth's oceans. Excluding its marginal seas, the Atlantic covers 81,760,000 km2 (31,570,000 sq mi) and has a volume of 305,811,900 km3 (73,368,200 cu mi). The North Atlantic covers 41,490,000 km2 (16,020,000 sq mi) (11.5%) and the South Atlantic 40,270,000 km2 (15,550,000 sq mi) (11.1%). The average depth is 3,646 m (11,962 ft) and the maximum depth, the Milwaukee Deep in the Puerto Rico Trench, is 8,376 m (27,480 ft).

- Biggest seas in Atlantic Ocean
- Sargasso Sea - 3.5 million km2
- Caribbean Sea - 2.754 million km2
- Mediterranean Sea - 2.510 million km2
- Gulf of Guinea - 2.35 million km2
- Gulf of Mexico - 1.550 million km2
- Norwegian Sea - 1.383 million km2
- Hudson Bay - 1.23 million km2
- Greenland Sea - 1.205 million km2
- Argentine Sea - 1 million km2
- Labrador Sea - 841,000 km2
- Irminger Sea - 780,000 km2
- Baffin Bay - 689,000 km2
- North Sea - 575,000 km2
- Black Sea - 436,000 km2
- Baltic Sea - 377,000 km2
- Libyan Sea - 350,000 km2
- Levantine Sea - 320,000 km2
- Celtic Sea - 300,000 km2
- Tyrrhenian Sea - 275,000 km2
- Gulf of Saint Lawrence - 226,000 km2
- Bay of Biscay - 223,000 km2
- Aegean Sea - 214,000 km2
- Ionian Sea - 169,000 km2
- Balearic Sea - 150,000 km2
- Adriatic Sea - 138,000 km2
- Gulf of Bothnia - 116,300 km2
- Sea of Crete - 95,000 km2
- Gulf of Maine - 93,000 km2

- Ligurian Sea - 80,000 km2
- English Channel - 75,000 km2
- James Bay - 68,300 km2
- Bothnian Sea - 66,000 km2
- Gulf of Sidra - 57,000 km2
- Sea of the Hebrides - 47,000 km2
- Irish Sea - 46,000 km2
- Sea of Azov - 39,000 km2
- Bothnian Bay - 36,800 km2
- Gulf of Venezuela - 17,840 km2
- Bay of Campeche - 16,000 km2
- Gulf of Lion - 15,000 km2
- Sea of Marmara - 11,350 km2
- Wadden Sea - 10,000 km2
- Archipelago Sea - 8,300 km2

BATHYMETRY

False color map of ocean depth in the Atlantic basin

The bathymetry of the Atlantic is dominated by a submarine mountain range called the Mid-Atlantic Ridge (MAR). It runs from 87°N or 300 km (190 mi) south of the North Pole to the subantarctic Bouvet Island at 54°S.

Mid-Atlantic Ridge

The MAR divides the Atlantic longitudinally into two halves, in each of which a series of basins are delimited by secondary, transverse ridges. The MAR reaches above 2,000 m (6,600 ft) along most of its length, but is interrupted by larger transform faults at two places: the Romanche Trench near the Equator and the Gibbs Fracture Zone at 53°N. The MAR is a barrier for bottom water, but at these two transform faults deep water currents can pass from one side to the other.

The MAR rises 2–3 km (1.2–1.9 mi) above the surrounding ocean floor and its rift valley is the divergent boundary between the North American and Eurasian plates in the North Atlantic and the South American and African plates in the South Atlantic. The MAR produces basaltic volcanoes in Eyjafjallajökull, Iceland, and pillow lava on the ocean floor. The depth of water at the apex of the ridge is less than 2,700 m (1,500 fathoms; 8,900 ft) in most places, while the bottom of the ridge is three times as deep.

The MAR is intersected by two perpendicular ridges: the Azores–Gibraltar Transform Fault, the boundary between the Nubian and Eurasian plates, intersects the MAR at the Azores Triple Junction, on either side of the Azores microplate, near the 40°N. A much vaguer, nameless boundary, between the North American and South American plates, intersects the MAR near or just north of the Fifteen-Twenty Fracture Zone, approximately at 16°N.

In the 1870s, the Challenger expedition discovered parts of what is now known as the Mid-Atlantic Ridge, or:

An elevated ridge rising to an average height of about 1,900 fathoms [3,500 m; 11,400 ft] below the surface traverses the basins of the North and South Atlantic in a meridianal direction from Cape Farewell, probably its far south at least as Gough Island, following roughly the outlines of the coasts of the Old and the New Worlds.

The remainder of the ridge was discovered in the 1920s by the German Meteor expedition using echo-sounding equipment.[32] The exploration of the MAR in the 1950s led to the general acceptance of seafloor spreading and plate tectonics.

Most of the MAR runs under water but where it reaches the surfaces it has produced volcanic islands. While nine of these have collectively been nominated a World Heritage Site for their geological value, four of them are considered of "Outstanding Universal Value" based on their cultural and natural criteria: Þingvellir, Iceland; Landscape of the Pico Island Vineyard Culture, Portugal; Gough and Inaccessible Islands, United Kingdom; and Brazilian Atlantic Islands: Fernando de Noronha and Atol das Rocas Reserves, Brazil.

Ocean floor

For a list of fracture zones in the Atlantic, see List of fracture zones § Atlantic Ocean.

Continental shelves in the Atlantic are wide off Newfoundland, southernmost South America, and north-eastern Europe. In the western Atlantic carbonate platforms dominate large areas, for example, the Blake Plateau and Bermuda Rise. The Atlantic is surrounded by passive margins except at a few locations where active margins form deep trenches: the Puerto Rico Trench (8,376 m or 27,480 ft maximum depth) in the western Atlantic and South Sandwich Trench (8,264 m or 27,113 ft) in the South Atlantic. There are numerous submarine canyons off north-eastern North America, western Europe, and north-western Africa. Some of these canyons extend along the continental rises and farther into the abyssal plains as deep-sea channels.

In 1922 a historic moment in cartography and oceanography occurred. The USS Stewart used a Navy Sonic Depth Finder to draw a continuous map across the bed of the Atlantic. This involved little guesswork because the idea of sonar is straightforward with pulses being sent from the vessel, which bounce off the ocean floor, then return to the vessel.[33] The deep ocean floor is thought to be fairly flat with occasional deeps, abyssal plains, trenches, seamounts, basins, plateaus, canyons, and some guyots. Various shelves along the margins of the continents constitute about 11% of the bottom topography with few deep channels cut across the continental rise.

The mean depth between 60°N and 60°S is 3,730 m (12,240 ft), or close to the average for the global ocean, with a modal depth between 4,000 and 5,000 m (13,000 and 16,000 ft).

In the South Atlantic the Walvis Ridge and Rio Grande Rise form barriers to ocean currents. The Laurentian Abyss is found off the eastern coast of Canada.

Map displaying a looping line with arrows indicating that water flows eastward in the far Southern Ocean, angling northeast of Australia, turning sough-after passing Alaska, then crossing the mid-Pacific to flow north of Australia, continuing west below Africa, then turning northwest until reaching eastern Canada, then angling east to southern Europe, then finally turning south just below Greenland and flowing down the Americas' eastern coast, and resuming its flow eastward to complete the circle

Path of the thermohaline circulation. Purple paths represent deep-water currents, while blue paths represent surface currents.

Surface water temperatures, which vary with latitude, current systems, and season and reflect the latitudinal distribution of solar energy, range from below -2 °C (28 °F) to over 30 °C (86 °F). Maximum temperatures occur north of the equator, and minimum values are found in the polar regions. In the middle latitudes, the area of maximum temperature variations, values may vary by 7–8 °C (13–14 °F).

From October to June the surface is usually covered with sea ice in the Labrador Sea, Denmark Strait, and Baltic Sea.

The Coriolis effect circulates North Atlantic water in a clockwise direction, whereas South Atlantic water circulates counter-clockwise. The south tides in the Atlantic Ocean are semi-diurnal; that is, two high tides occur every 24 lunar hours. In latitudes above 40° North some east–west oscillation, known as the North Atlantic oscillation, occurs.

Salinity

On average, the Atlantic is the saltiest major ocean; surface water salinity in the open ocean ranges from 33 to 37 parts per thousand (3.3–3.7%) by mass and varies with latitude and season. Evaporation, precipitation, river inflow and sea ice melting influence surface salinity values. Although the lowest salinity values are just north of the equator (because of heavy tropical rainfall), in general, the lowest values are in the high latitudes and along coasts where large rivers enter. Maximum salinity values occur at about 25° north and south, in subtropical regions with low rainfall and high evaporation.

The high surface salinity in the Atlantic, on which the Atlantic thermohaline circulation is dependent, is maintained by two processes: the Agulhas Leakage/Rings, which brings salty Indian Ocean waters into the South Atlantic, and the "Atmospheric Bridge", which evaporates subtropical Atlantic waters and exports it to the Pacific.

There are five intermediate waters: four low-salinity waters formed at subpolar latitudes and one high-salinity formed through evaporation. Arctic Intermediate Water, flows from north to become the source for North Atlantic Deep Water south of the Greenland-Scotland sill. These two intermediate waters have different salinity in the western and eastern basins. The wide range of salinities in the North Atlantic is caused by the asymmetry of the northern subtropical gyre and the large number of contributions from a wide range of sources: Labrador Sea, Norwegian-Greenland Sea, Mediterranean, and South Atlantic Intermediate Water.

The North Atlantic Deep Water (NADW) is a complex of four water masses, two that form by deep convection in the open ocean — Classical and Upper Labrador Sea Water — and two that form from the inflow of dense water across the Greenland-Iceland-Scotland sill — Denmark Strait and Iceland-Scotland Overflow Water. Along its path across Earth the composition of the NADW is affected by other water masses, especially Antarctic Bottom Water and Mediterranean Overflow Water. The NADW is fed by a flow of warm shallow water into the northern North Atlantic which is responsible for the anomalous warm climate in Europe. Changes in the formation of NADW have been linked to global climate changes in the past. Since man-made substances were introduced into the environment, the path of the NADW can be traced throughout its course by measuring tritium and radiocarbon from nuclear weapon tests in the 1960s and CFCs.

Gyres

Map showing 5 circles. The first is between western Australia and eastern Africa. The second is between eastern Australia and western South America. The third is between Japan and western North America. Of the two in the Atlantic, one is in hemisphere.

- North Atlantic
- GyreNorth Atlantic
- GyreNorth Atlantic
- GyreIndian
- Ocean

- GyreNorth
- Pacific
- gyreSouth
- Pacific
- gyreSouth Atlantic

The clockwise warm-water North Atlantic Gyre occupies the northern Atlantic, and the counter-clockwise warm-water South Atlantic Gyre appears in the southern Atlantic.

In the North Atlantic, surface circulation is dominated by three inter-connected currents: the Gulf Stream which flows north-east from the North American coast at Cape Hatteras; the North Atlantic Current, a branch of the Gulf Stream which flows northward from the Grand Banks; and the Subpolar Front, an extension of the North Atlantic Current, a wide, vaguely defined region separating the subtropical gyre from the subpolar gyre. This system of currents transport warm water into the North Atlantic, without which temperatures in the North Atlantic and Europe would plunge dramatically.

In the subpolar gyre of the North Atlantic warm subtropical waters are transformed into colder subpolar and polar waters. In the Labrador Sea this water flows back to the subtropical gyre.

North of the North Atlantic Gyre, the cyclonic North Atlantic Subpolar Gyre plays a key role in climate variability. It is governed by ocean currents from marginal seas and regional topography, rather than being steered by wind, both in the deep ocean and at sea level. The subpolar gyre forms an important part of the global thermohaline circulation. Its eastern portion includes eddying branches of the North Atlantic Current which transport warm, saline waters from the subtropics to the north-eastern Atlantic. There this water is cooled during winter and forms return currents that merge along the eastern continental slope of Greenland where they form an intense (40–50 Sv) current which flows around the continental margins of the Labrador Sea. A third of this water becomes part of the deep portion of the North Atlantic Deep Water (NADW). The NADW, in its turn, feeds the meridional overturning circulation (MOC), the northward heat transport of which is threatened by anthropogenic climate change. Large variations in the subpolar gyre on a decade-century scale, associated with the North Atlantic oscillation, are especially pronounced in Labrador Sea Water, the upper layers of the MOC.

The South Atlantic is dominated by the anti-cyclonic southern subtropical gyre. The South Atlantic Central Water originates in this gyre, while Antarctic Intermediate Water originates in the upper layers of the circumpolar region, near the Drake Passage and the Falkland Islands. Both these currents receive some contribution from the Indian Ocean. On the African east coast, the small cyclonic Angola Gyre lies embedded in the large subtropical gyre. The southern subtropical gyre is partly masked by a wind-induced Ekman layer. The residence time of the gyre is 4.4–8.5 years. North Atlantic Deep Water flows southward below the thermocline of the subtropical gyre.

Sargasso Sea

The Sargasso Sea in the western North Atlantic can be defined as the area where two species of Sargassum (S. fluitans and natans) float, an area 4,000 km (2,500 mi) wide and encircled by the Gulf Stream, North Atlantic Drift, and North Equatorial Current. This population of seaweed probably originated from Tertiary ancestors on the European shores of the former Tethys Ocean and has, if so, maintained itself by vegetative growth, floating in the ocean for millions of years.

Sargassum fish (Histrio histrio)

Other species endemic to the Sargasso Sea include the sargassum fish, a predator with algae-like appendages which hovers motionless among the Sargassum. Fossils of similar fishes have been found in fossil bays of the former Tethys Ocean, in what is now the Carpathian region, that were similar to the Sargasso Sea. It is possible that the population in the Sargasso Sea migrated to the Atlantic as the Tethys closed at the end of the Miocene around 17 Ma. The origin of the Sargasso fauna and flora remained enigmatic for centuries. The fossils found in the Carpathians in the mid-20th century often called the "quasi-Sargasso assemblage", finally showed that this assemblage originated in the Carpathian Basin from where it migrated over Sicily to the Central Atlantic where it evolved into modern species of the Sargasso Sea.

The location of the spawning ground for European eels remained unknown for decades. In the early 19th century it was discovered that the southern Sargasso Sea is the spawning ground for both the European and American eel and that the former migrate more than 5,000 km (3,100 mi) and the latter 2,000 km (1,200 mi). Ocean currents such as the Gulf Stream transport eel larvae from the Sargasso Sea to foraging areas in North America, Europe, and Northern Africa. Recent but disputed research suggests that eels possibly use Earth's magnetic field to navigate through the ocean both as larvae and as adults.

Climate

Map of Caribbean showing seven approximately parallel westward-pointing arrows that extend from east of the Virgin Islands to Cuba. The southern arrows bend northward just east of the Dominican Republic before straightening out again.

Waves in the trade winds in the Atlantic Ocean—areas of converging winds that move along the same track as the prevailing wind—create instabilities in the atmosphere that may lead to the formation of hurricanes.

Tropical wet and dry climate in San Andrés island caribbean, Colombia.

Climate is influenced by the temperatures of the surface waters and water currents as well as winds. Because of the ocean's great capacity to store and release heat, maritime climates are more moderate and have less extreme seasonal variations than inland climates. Precipitation can be approximated from coastal weather data and air temperature from water temperatures.

The oceans are the major source of the atmospheric moisture that is obtained through evaporation. Climatic zones vary with latitude; the warmest zones stretch across the Atlantic north of the equator. The coldest zones are in high latitudes, with the coldest regions corresponding to the areas covered by sea ice. Ocean currents influence the climate by transporting warm and cold waters to other regions. The winds that are cooled or warmed when blowing over these currents influence adjacent land areas.

The Gulf Stream and its northern extension towards Europe, the North Atlantic Drift is thought to have at least some influence on climate. For example, the Gulf Stream helps moderate winter temperatures along the coastline of southeastern North America, keeping it warmer in winter along the coast than inland areas. The Gulf Stream also keeps extreme temperatures from occurring on the Florida Peninsula. In the higher latitudes, the North Atlantic Drift, warms the atmosphere over the oceans, keeping the British Isles and north-western Europe mild and cloudy, and not severely cold in winter like other locations at the same high latitude. The cold water currents contribute to heavy fog off the coast of eastern Canada (the Grand Banks of Newfoundland area) and Africa's north-western coast. In general, winds transport moisture and air over land areas.

Natural hazards

Every winter, the Icelandic Low produces frequent storms. Icebergs are common from early February to the end of July across the shipping lanes near the Grand Banks of Newfoundland. The ice season is longer in the polar regions, but there is little shipping in those areas.

Hurricanes are a hazard in the western parts of the North Atlantic during the summer and autumn. Due to a consistently strong wind shear and a weak Intertropical Convergence Zone, South Atlantic tropical cyclones are rare.

Geology and plate tectonics

The Atlantic Ocean is underlain mostly by dense mafic oceanic crust made up of basalt and gabbro and overlain by fine clay, silt and siliceous ooze on the abyssal plain. The continental margins and continental shelf mark lower density, but greater thickness felsic continental rock that is often much older than that of the seafloor. The oldest oceanic crust in the Atlantic is up to 145 million years and situated off the west coast of Africa and east coast of North America, or on either side of the South Atlantic.

In many places, the continental shelf and continental slope are covered in thick sedimentary layers. For instance, on the North American side of the ocean, large carbonate deposits formed in warm shallow waters such as Florida and the Bahamas, while coarse river outwash sands and silt are common in shallow shelf areas like the Georges Bank. Coarse sand, boulders, and rocks were transported into some areas, such as off the coast of Nova Scotia or the Gulf of Maine during the Pleistocene ice ages.

CENTRAL ATLANTIC

Opening of the Central Atlantic 200-170 Ma

The break-up of Pangaea began in the Central Atlantic, between North America and Northwest Africa, where rift basins opened during the Late Triassic and Early Jurassic. This period also saw the first stages of the uplift of the Atlas Mountains. The exact timing is controversial with estimates ranging from 200 to 170 Ma.

The opening of the Atlantic Ocean coincided with the initial break-up of the supercontinent Pangaea, both of which were initiated by the eruption of the Central Atlantic Magmatic Province (CAMP), one of the most extensive and voluminous large igneous provinces in Earth's history associated with the Triassic–Jurassic extinction event, one of Earth's major extinction events. Theoliitic dikes, sills, and lava flows from the CAMP eruption at 200 Ma have been found in West Africa, eastern North America, and northern South America. The extent of the volcanism has been estimated to 4.5×106 km2 (1.7×106 sq mi) of which 2.5×106 km2 (9.7×105 sq mi) covered what is now northern and central Brazil.

The formation of the Central American Isthmus closed the Central American Seaway at the end of the Pliocene 2.8 Ma ago. The formation of the isthmus resulted in the migration and extinction of many land-living animals, known as the Great American Interchange, but the closure of the seaway resulted in a "Great American Schism" as it affected ocean currents, salinity, and temperatures in both the Atlantic and Pacific. Marine organisms on both sides of the isthmus became isolated and either diverged or went extinct.

North Atlantic

Geologically, the Northern Atlantic is the area delimited to the south by two conjugate margins, Newfoundland and Iberia, and to the north by the Arctic

Eurasian Basin. The opening of the Northern Atlantic closely followed the margins of its predecessor, the Iapetus Ocean, and spread from the Central Atlantic in six stages: Iberia–Newfoundland, Porcupine–North America, Eurasia–Greenland, Eurasia–North America. Active and inactive spreading systems in this area are marked by the interaction with the Iceland hotspot.

Seafloor spreading led to the extension of the crust and formations of troughs and sedimentary basins. The Rockall Trough opened between 105 and 84 million years ago although along the rift failed along with one leading into the Bay of Biscay.

Spreading began opening the Labrador Sea around 61 million years ago, continuing until 36 million years ago. Geologists distinguish two magmatic phases. One from 62 to 58 million years ago predates the separation of Greenland from northern Europe while the second from 56 to 52 million years ago happened as the separation occurred.

Iceland began to form 62 million years ago due to a particularly concentrated mantle plume. Large quantities of basalt erupted at this time period are found on Baffin Island, Greenland, the Faroe Islands, and Scotland, with ash falls in Western Europe acting as a stratigraphic marker. The opening of the North Atlantic caused significant uplift of continental crust along the coast. For instance, in spite of 7 km thick basalt, Gunnbjorn Field in East Greenland is the highest point on the island, elevated enough that it exposes older Mesozoic sedimentary rocks at its base, similar to old lava fields above sedimentary rocks in the uplifted Hebrides of western Scotland.

South Atlantic

The opening of the South Atlantic

West Gondwana (South America and Africa) broke up in the Early Cretaceous to form the South Atlantic. The apparent fit between the coastlines of the two continents was noted on the first maps that included the South Atlantic and it was also the subject of the first computer-assisted plate tectonic reconstructions in 1965. This magnificent fit, however, has since then proven problematic and later reconstructions have introduced various deformation zones along the shorelines to accommodate the northward-propagating break-up. Intra-continental rifts and deformations have also been introduced to subdivide both continental plates into sub-plates.

Geologically the South Atlantic can be divided into four segments: Equatorial segment, from 10°N to the Romanche Fracture Zone (RFZ);; Central segment, from RFZ to Florianopolis Fracture Zone (FFZ, north of Walvis Ridge and Rio Grande Rise); Southern segment, from FFZ to the Agulhas-Falkland Fracture Zone (AFFZ); and Falkland segment, south of AFFZ.

In the southern segment the Early Cretaceous (133–130 Ma) intensive magmatism of the Paraná–Etendeka Large Igneous Province produced by the Tristan hotspot resulted in an estimated volume of 1.5×106 to 2.0×106 km3 (3.6×105 to 4.8×105 cu mi). It covered an area of 1.2×106 to 1.6×106 km2 (4.6×105 to 6.2×105 sq mi) in Brazil, Paraguay, and Uruguay and 0.8×105 km2 (3.1×104 sq mi) in Africa. Dyke swarms in Brazil, Angola, eastern Paraguay, and Namibia, however, suggest the LIP originally covered a much larger area and also indicate failed rifts in all these areas. Associated offshore basaltic flows reach as far south as the Falkland Islands and South Africa. Traces of magmatism in both offshore and onshore basins in the central and southern segments have been dated to 147–49 Ma with two peaks between 143 and 121 Ma and 90–60 Ma.

In the Falkland segment rifting began with dextral movements between the Patagonia and Colorado sub-plates between the Early Jurassic (190 Ma) and the Early Cretaceous (126.7 Ma). Around 150 Ma sea-floor spreading propagated northward into the southern segment. No later than 130 Ma rifting had reached the Walvis Ridge–Rio Grande Rise.

In the central segment rifting started to break Africa in two by opening the Benue Trough around 118 Ma. Rifting in the central segment, however, coincided with the Cretaceous Normal Superchron (also known as the Cretaceous quiet period), a 40 Ma period without magnetic reversals, which makes it difficult to date sea-floor spreading in this segment.

The equatorial segment is the last phase of the break-up, but, because it is located on the Equator, magnetic anomalies cannot be used for dating. Various estimates date the propagation of sea-floor spreading in this segment to the period 120–96 Ma. This final stage, nevertheless, coincided with or resulted in the end of continental extension in Africa.

About 50 Ma the opening of the Drake Passage resulted from a change in the motions and separation rate of the South American and Antarctic plates. First small ocean basins opened and a shallow gateway appeared during the Middle Eocene. 34–30 Ma a deeper seaway developed, followed by an Eocene–Oligocene climatic deterioration and the growth of the Antarctic ice sheet.

Closure of the Atlantic

n embryonic subduction margin is potentially developing west of Gibraltar. The Gibraltar Arc in the western Mediterranean is migrating westward into the Central Atlantic where it joins the converging African and Eurasian plates. Together these three tectonic forces are slowly developing into a new subduction system in the eastern Atlantic Basin. Meanwhile, the Scotia Arc and Caribbean Plate in the western Atlantic Basin are eastward-propagating subduction systems that might, together with the Gibraltar system, represent the beginning of the closure of the Atlantic Ocean and the final stage of the Atlantic Wilson cycle.

Indian Ocean

The Indian Ocean is the third-largest of the world's oceanic divisions, covering 70,560,000 km2 (27,240,000 sq mi) or 19.8% of the water on Earth's surface. It is bounded by Asia to the north, Africa to the west and Australia to the east. To the south it is bounded by the Southern Ocean or Antarctica, depending on the definition in use. Along its core, the Indian Ocean has some large marginal or regional seas such as the Arabian Sea, the Laccadive Sea, the Somali Sea, Bay of Bengal, and the Andaman Sea.

The Indian Ocean has been known by its present name since at least 1515 when the Latin form Oceanus Orientalis Indicus ("Indian Eastern Ocean") is attested, named for India, which projects into it. It was earlier known as the Eastern Ocean, a term that was still in use during the mid-18th century (see map), as opposed to the Western Ocean (Atlantic) before the Pacific was surmised.

Conversely, Chinese explorers in the Indian Ocean during the 15th century called it the Western Oceans. The ocean has also been known as the Hind Mahasagar, Hindu Ocean and Indic Ocean in various languages. In Ancient Greek geography, the Indian Ocean region known to the Greeks was called the Erythraean Sea.

Extent and data

The borders of the Indian Ocean, as delineated by the International Hydrographic Organization in 1953 included the Southern Ocean but not the marginal seas along the northern rim, but in 2000 the IHO delimited the Southern Ocean separately, which removed waters south of 60°S from the Indian Ocean but included the northern marginal seas. Meridionally, the Indian Ocean is delimited from the Atlantic Ocean by the 20° east meridian, running south from Cape Agulhas, and from the Pacific Ocean by the meridian of 146°49'E, running south from the southernmost point of Tasmania. The northernmost extent of the Indian Ocean (including marginal seas) is approximately 30° north in the Persian Gulf.

The Indian Ocean covers 70,560,000 km2 (27,240,000 sq mi), including the Red Sea and the Persian Gulf but excluding the Southern Ocean, or 19.5% of the world's oceans; its volume is 264,000,000 km3 (63,000,000 cu mi) or 19.8% of the world's oceans' volume; it has an average depth of 3,741 m (12,274 ft) and a maximum depth of 7,906 m (25,938 ft).

All of the Indian Ocean is in the Eastern Hemisphere and the centre of the Eastern Hemisphere, the 90th meridian east, passes through the Ninety East Ridge.

Coasts and shelves

In contrast to the Atlantic and Pacific, the Indian Ocean is enclosed by major landmasses and an archipelago on three sides and does not stretch from pole to pole, and can be likened to an embayed ocean. It is centered on the Indian Peninsula. Although this subcontinent has played a significant role in its history, the Indian Ocean has foremostly been a cosmopolitan stage, interlinking diverse regions by innovations, trade, and religion since early in human history.

The active margins of the Indian Ocean have an average depth (land to shelf break) of 19 ± 0.61 km (11.81 ± 0.38 mi) with a maximum depth of 175 km (109 mi). The passive margins have an average depth of 47.6 ± 0.8 km (29.58 ± 0.50 mi). [13] The average width of the slopes of the continental shelves are 50.4–52.4 km (31.3–32.6 mi) for active and passive margins respectively, with a maximum depth of 205.3–255.2 km (127.6–158.6 mi).

Australia, Indonesia, and India are the three countries with the longest shorelines and exclusive economic zones. The continental shelf makes up 15% of the Indian Ocean. More than two billion people live in countries bordering the Indian Ocean, compared to 1.7 billion for the Atlantic and 2.7 billion for the Pacific (some countries border more than one ocean).

Rivers

The Indian Ocean drainage basin covers 21,100,000 km2 (8,100,000 sq mi), virtually identical to that of the Pacific Ocean and half that of the Atlantic basin, or 30% of its ocean surface (compared to 15% for the Pacific). The Indian Ocean drainage basin is divided into roughly 800 individual basins, half that of the Pacific, of which 50% are located in Asia, 30% in Africa, and 20% in Australasia. The rivers of the Indian Ocean are shorter on average (740 km (460 mi)) than those of the other major oceans. The largest rivers are (order 5) the Zambezi, Ganges-Brahmaputra, Indus, Jubba, and Murray rivers and (order 4) the Shatt al-Arab, Wadi Ad Dawasir (a dried-out river system on the Arabian Peninsula) and Limpopo rivers.

Marginal seas

Marginal seas, gulfs, bays and straits of the Indian Ocean include:

Along the east coast of Africa, the Mozambique Channel separates Madagascar from mainland Africa, while the Sea of Zanj is located north of Madagascar.

On the northern coast of the Arabian Sea, Gulf of Aden is connected to the Red Sea by the strait of Bab-el-Mandeb. In the Gulf of Aden, the Gulf of Tadjoura is located in Djibouti and the Guardafui Channel separates Socotra island from the Horn of Africa. The northern end of the Red Sea terminates in the Gulf of Aqaba and Gulf of Suez. The Indian Ocean is artificially connected to the Mediterranean

Sea without ship lock through the Suez Canal, which is accessible via the Red Sea. The Arabian Sea is connected to the Persian Gulf by the Gulf of Oman and the Strait of Hormuz. In the Persian Gulf, the Gulf of Bahrain separates Qatar from the Arabic Peninsula.

Along the west coast of India, the Gulf of Kutch and Gulf of Khambat are located in Gujarat in the northern end while the Laccadive Sea separates the Maldives from the southern tip of India. The Bay of Bengal is off the east coast of India. The Gulf of Mannar and the Palk Strait separates Sri Lanka from India, while the Adam's Bridge separates the two. The Andaman Sea is located between the Bay of Bengal and the Andaman Islands.

In Indonesia, the so-called Indonesian Seaway is composed of the Malacca, Sunda and Torres Straits. The Gulf of Carpentaria of located on the Australian north coast while the Great Australian Bight constitutes a large part of its southern coast.

- Arabian Sea - 3.862 million km2
- Bay of Bengal - 2.172 million km2
- Andaman Sea - 797,700 km2
- Laccadive Sea - 786,000 km2
- Mozambique Channel - 700,000 km2
- Timor Sea - 610,000 km2
- Red Sea - 438,000 km2
- Gulf of Aden - 410,000 km2
- Persian Gulf - 251,000 km2
- Flores Sea - 240,000 km2
- Molucca Sea - 200,000 km2
- Oman Sea - 181,000 km2
- Great Australian Bight - 45,926 km2
- Gulf of Aqaba - 239 km2
- Gulf of Khambhat
- Gulf of Kutch
- Gulf of Suez
- Climate

During summer, warm continental masses draw moist air from the Indian Ocean hence producing heavy rainfall. The process is reversed during winter, resulting in dry conditions.

Several features make the Indian Ocean unique. It constitutes the core of the large-scale Tropical Warm Pool which, when interacting with the atmosphere, affects the climate both regionally and globally. Asia blocks heat export and prevents the ventilation of the Indian Ocean thermocline. That continent also drives the Indian Ocean monsoon, the strongest on Earth, which causes large-scale seasonal variations in ocean currents, including the reversal of the Somali Current and Indian Monsoon Current. Because of the Indian Ocean Walker circulation there are no continuous equatorial easterlies. Upwelling occurs near the Horn of Africa and the Arabian Peninsula in the Northern Hemisphere and north of the trade winds in the Southern Hemisphere. The Indonesian Throughflow is a unique Equatorial connection to the Pacific.

The climate north of the equator is affected by a monsoon climate. Strong north-east winds blow from October until April; from May until October south and west winds prevail. In the Arabian Sea, the violent Monsoon brings rain to the Indian subcontinent. In the southern hemisphere, the winds are generally milder, but summer storms near Mauritius can be severe. When the monsoon winds change, cyclones sometimes strike the shores of the Arabian Sea and the Bay of Bengal. Some 80% of the total annual rainfall in India occurs during summer and the region is so dependent on this rainfall that many civilisations perished when the Monsoon failed in the past. The huge variability in the Indian Summer Monsoon has also occurred pre-historically, with a strong, wet phase 33,500–32,500 BP; a weak, dry phase 26,000–23,500 BC; and a very weak phase 17,000–15,000 BP, corresponding to a series of dramatic global events: Bølling-Allerød, Heinrich, and Younger Dryas.

Air pollution in South Asia spread over the Bay of Bengal and beyond.

The Indian Ocean is the warmest ocean in the world. Long-term ocean temperature records show a rapid, continuous warming in the Indian Ocean, at about 1.2 °C (34.2 °F) (compared to 0.7 °C (33.3 °F) for the warm pool region) during 1901–2012. Research indicates that human induced greenhouse warming, and changes in the frequency and magnitude of El Niño (or the Indian Ocean Dipole), events are a trigger to this strong warming in the Indian Ocean.

South of the Equator (20-5°S), the Indian Ocean is gaining heat from June to October, during the austral winter, while it is losing heat from November to March, during the austral summer.

In 1999, the Indian Ocean Experiment showed that fossil fuel and biomass burning in South and Southeast Asia caused air pollution (also known as the Asian brown cloud) that reach as far as the Intertropical Convergence Zone at 60°S. This pollution has implications on both a local and global scale.

OCEANOGRAPHY

40% of the sediment of the Indian Ocean is found in the Indus and Ganges fans. The oceanic basins adjacent to the continental slopes mostly contain terrigenous sediments. The ocean south of the polar front (roughly 50° south latitude) is high in biologic productivity and dominated by non-stratified sediment composed mostly of siliceous oozes. Near the three major mid-ocean ridges the ocean floor is relatively young and therefore bare of sediment, except for the Southwest Indian Ridge due to its ultra-slow spreading rate.

The ocean's currents are mainly controlled by the monsoon. Two large gyres, one in the northern hemisphere flowing clockwise and one south of the equator moving anticlockwise (including the Agulhas Current and Agulhas Return Current), constitute the dominant flow pattern. During the winter monsoon (November–February), however, circulation is reversed north of 30°S and winds are weakened during winter and the transitional periods between the monsoons.

The Indian Ocean contains the largest submarine fans of the world, the Bengal Fan and Indus Fan, and the largest areas of slope terraces and rift valleys.

The inflow of deep water into the Indian Ocean is 11 Sv, most of which comes from the Circumpolar Deep Water (CDW). The CDW enters the Indian Ocean through the Crozet and Madagascar basins and crosses the Southwest Indian Ridge at 30°S. In the Mascarene Basin the CDW becomes a deep western boundary current before it is met by a re-circulated branch of itself, the North Indian Deep Water. This mixed water partly flows north into the Somali Basin whilst most of it flows clockwise in the Mascarene Basin where an oscillating flow is produced by Rossby waves.

Water circulation in the Indian Ocean is dominated by the Subtropical Anticyclonic Gyre, the eastern extension of which is blocked by the Southeast Indian Ridge and the 90°E Ridge. Madagascar and the Southwest Indian Ridge separate three cells south of Madagascar and off South Africa. North Atlantic Deep Water reaches into the Indian Ocean south of Africa at a depth of 2,000–3,000 m (6,600–9,800 ft) and flows north along the eastern continental slope of Africa. Deeper than NADW, Antarctic Bottom Water flows from Enderby Basin to Agulhas Basin across deep channels (<4,000 m (13,000 ft)) in the Southwest Indian Ridge, from where it continues into the Mozambique Channel and Prince Edward Fracture Zone.

North of 20° south latitude the minimum surface temperature is 22 °C (72 °F), exceeding 28 °C (82 °F) to the east. Southward of 40° south latitude, temperatures drop quickly.

The Bay of Bengal contributes more than half (2,950 km3 (710 cu mi)) of the runoff water to the Indian Ocean. Mainly in summer, this runoff flows into the Arabian Sea but also south across the Equator where it mixes with fresher seawater from the Indonesian Throughflow. This mixed freshwater joins the South Equatorial

Current in the southern tropical Indian Ocean. Sea surface salinity is highest (more than 36 PSU) in the Arabian Sea because evaporation exceeds precipitation there. In the Southeast Arabian Sea salinity drops to less than 34 PSU. It is the lowest (c. 33 PSU) in the Bay of Bengal because of river runoff and precipitation. The Indonesian Throughflow and precipitation results in lower salinity (34 PSU) along the Sumatran west coast. Monsoonal variation results in eastward transportation of saltier water from the Arabian Sea to the Bay of Bengal from June to September and in westerly transport by the East India Coastal Current to the Arabian Sea from January to April.

An Indian Ocean garbage patch was discovered in 2010 covering at least 5 million square kilometres (1.9 million square miles). Riding the southern Indian Ocean Gyre, this vortex of plastic garbage constantly circulates the ocean from Australia to Africa, down the Mozambique Channel, and back to Australia in a period of six years, except for debris that gets indefinitely stuck in the centre of the gyre. The garbage patch in the Indian Ocean will, according to a 2012 study, decrease in size after several decades to vanish completely over centuries. Over several millennia, however, the global system of garbage patches will accumulate in the North Pacific.

There are two amphidromes of opposite rotation in the Indian Ocean, probably caused by Rossby wave propagation.

Icebergs drift as far north as 55° south latitude, similar to the Pacific but less than in the Atlantic where icebergs reach up to 45°S. The volume of iceberg loss in the Indian Ocean between 2004 and 2012 was 24 Gt. Since the 1960s, anthropogenic warming of the global ocean combined with contributions of freshwater from retreating land ice causes a global rise in sea level. Sea level increases in the Indian Ocean too, except in the south tropical Indian Ocean where it decreases, a pattern most likely caused by rising levels of greenhouse gases.

Marine life

A dolphin off Western Australia and a swarm of surgeonfish near Maldives Islands represents the well-known, exotic fauna of the warmer parts of the Indian Ocean. King Penguins on a beach in the Crozet Archipelago near Antarctica attract fewer tourists.

Among the tropical oceans, the western Indian Ocean hosts one of the largest concentrations of phytoplankton blooms in summer, due to the strong monsoon winds. The monsoonal wind forcing leads to a strong coastal and open ocean upwelling, which introduces nutrients into the upper zones where sufficient light is available for photosynthesis and phytoplankton production. These phytoplankton blooms support the marine ecosystem, as the base of the marine food web, and eventually the larger fish species. The Indian Ocean accounts for the second-largest share of the most economically valuable tuna catch. Its fish are of great and growing importance to the bordering countries for domestic consumption and export. Fishing fleets from Russia, Japan, South Korea, and Taiwan also exploit the Indian Ocean, mainly for shrimp and tuna.

Research indicates that increasing ocean temperatures are taking a toll on the marine ecosystem. A study on the phytoplankton changes in the Indian Ocean indicates a decline of up to 20% in the marine plankton in the Indian Ocean, during the past six decades. The tuna catch rates have also declined 50–90% during the past half-century, mostly due to increased industrial fisheries, with the ocean warming adding further stress to the fish species.

Right: The coelacanth (here a model from Oxford), thought extinct for million years, was rediscovered in the 20th century. The Indian Ocean species is blue whereas the Indonesian species is brown.

Coral reefs, sea grass beds, and mangrove forests are the most productive ecosystems of the Indian Ocean — coastal areas produce 20 tones per square kilometre of fish. These areas, however, are also being urbanised with populations often exceeding several thousand people per square kilometre and fishing techniques become more effective and often destructive beyond sustainable levels while the increase in sea surface temperature spreads coral bleaching.

Mangroves covers 80,984 km2 (31,268 sq mi) in the Indian Ocean region, or almost half of the world's mangrove habitat, of which 42,500 km2 (16,400 sq mi) is located in Indonesia, or 50% of mangroves in the Indian Ocean. Mangroves originated in the Indian Ocean region and have adapted to a wide range of its habitats but it is also where it suffers its biggest loss of habitat.

In 2016 six new animal species were identified at hydrothermal vents in the Southwest Indian Ridge: a "Hoff" crab, a "giant peltospirid" snail, a whelk-like snail, a limpet, a scaleworm and a polychaete worm.

The West Indian Ocean coelacanth was discovered in the Indian Ocean off South Africa in the 1930s and in the late 1990s another species, the Indonesian coelacanth, was discovered off Sulawesi Island, Indonesia. Most extant coelacanths have been found in the Comoros. Although both species represent an order of lobe-finned fishes known from the Early Devonian (410 mya) and though extinct 66 mya, they are morphologically distinct from their Devonian ancestors. Over millions of years, coelacanths evolved to inhabit different environments — lungs adapted for shallow, brackish waters evolved into gills adapted for deep marine waters.

Biodiversity

The origin of this diversity is debated; the break-up of Gondwana can explain vicariance older than 100 mya, but the diversity on the younger, smaller islands must have required a Cenozoic dispersal from the rims of the Indian Ocean to the islands. A "reverse colonisation", from islands to continents, apparently occurred more recently; the chameleons, for example, first diversified on Madagascar and then colonised Africa. Several species on the islands of the Indian Ocean are textbook cases of evolutionary processes; the dung beetles, day geckos, and lemurs are all examples of adaptive radiation.[citation needed] Many bones (250 bones per square metre) of recently extinct vertebrates have been found in the Mare aux Songes swamp in Mauritius, including bones of the Dodo bird (Raphus cucullatus) and Cylindraspis giant tortoise. An analysis of these remains suggests a process of aridification began in the southwest Indian Ocean began around 4,000 years ago.

Mammalian megafauna once widespread in the MPA was driven to near extinction in the early 20th century. Some species have been successfully recovered since then — the population of white rhinoceros (Ceratotherium simum simum) increased from less than 20 individuals in 1895 to more than 17,000 as of 2013. Other species are still dependent of fenced areas and management programs, including black rhinoceros (Diceros bicornis minor), African wild dog (Lycaon pictus), cheetah (Acynonix jubatus), elephant (Loxodonta africana), and lion (Panthera leo).

This biodiversity hotspot (and namesake ecoregion and "Endemic Bird Area") is a patchwork of small forested areas, often with a unique assemblage of species within each, located within 200 km (120 mi) from the coast and covering a total area of c. 6,200 km2 (2,400 sq mi). It also encompasses coastal islands, including Zanzibar and Pemba, and Mafia.

Coral reefs of the Maldives

This area, one of the only two hotspots that are entirely arid, includes the Ethiopian Highlands, the East African Rift valley, the Socotra islands, as well as some small islands in the Red Sea and areas on the southern Arabic Peninsula. Endemic and threatened mammals include the dibatag (Ammodorcas clarkei) and Speke's gazelle (Gazella spekei); the Somali wild ass (Equus africanus somaliensis) and hamadryas baboon (Papio hamadryas). It also contains many reptiles. In Somalia, the centre of the 1,500,000 km2 (580,000 sq mi) hotspot, the landscape is dominated by Acacia-Commiphora deciduous bushland, but also includes the Yeheb nut (Cordeauxia edulus) and species discovered more recently such as the Somali cyclamen (Cyclamen somalense), the only cyclamen outside the Mediterranean. Warsangli linnet (Carduelis johannis) is an endemic bird found only in northern Somalia. An unstable political regime has resulted in overgrazing which has produced one of the most degraded hotspots where only c. 5 % of the original habitat remains.

Encompassing the west coast of India and Sri Lanka, until c. 10,000 years ago a landbridge connected Sri Lanka to the Indian Subcontinent, hence this region shares a common community of species.

Aldabra giant tortoise from the islands of the Aldabra Atoll in the Seychelles

Indo-Burma encompasses a series of mountain ranges, five of Asia's largest river systems, and a wide range of habitats. The region has a long and complex geological history, and long periods rising sea levels and glaciations have isolated ecosystems and thus promoted a high degree of endemism and speciation. The region includes two centres of endemism: the Annamite Mountains and the northern highlands on the China-Vietnam border. Several distinct floristic regions, the Indian, Malesian, Sino-Himalayan, and Indochinese regions, meet in a unique way in Indo-Burma and the hotspot contains an estimated 15,000–25,000 species of vascular plants, many of them endemic.

Sundaland; 25,000 (15,000 endemic) species of plants; 771 (146) birds; 449 (244) reptiles; 950 (350) freshwater fishes; 258 (210) amphibians; and 397 (219) mammals.

Sundaland encompasses 17,000 islands of which Borneo and Sumatra are the largest. Endangered mammals include the Bornean and Sumatran orangutans, the proboscis monkey, and the Javan and Sumatran rhinoceroses.

Wallacea; 10,000 (1,500 endemic) species of plants; 650 (265) birds; 222 (99) reptiles; 250 (50) freshwater fishes; 49 (33) amphibians; and 244 (144) mammals.

Southwest Australia; 5,571 (2,948 endemic) species of plants; 285 (10) birds; 177 (27) reptiles; 20 (10) freshwater fishes; 32 (22) amphibians; and 55 (13) mammals.

Stretching from Shark Bay to Israelite Bay and isolated by the arid Nullarbor Plain, the southwestern corner of Australia is a floristic region with a stable climate in which one of the world's largest floral biodiversity and an 80% endemism has evolved. From June to September it is an explosion of colours and the Wildflower Festival in Perth in September attracts more than half a million visitors.

Geology

Left: The oldest ocean floor of the Indian Ocean formed c. 150 Ma when the Indian Subcontinent and Madagascar broke-up from Africa. Right: The India–Asia collision c. 40 Ma completed the closure of the Tethys Ocean (grey areas north of India). Geologically, the Indian Ocean is the ocean floor that opened up south of India.

As the youngest of the major oceans, the Indian Ocean has active spreading ridges that are part of the worldwide system of mid-ocean ridges. In the Indian Ocean these spreading ridges meet at the Rodrigues Triple Point with the Central Indian Ridge, including the Carlsberg Ridge, separating the African Plate from the Indian Plate; the Southwest Indian Ridge separating the African Plate from the Antarctic Plate; and the Southeast Indian Ridge separating the Australian Plate from the Antarctic Plate. The Central Indian Ridge is intercepted by the Owen Fracture Zone. Since the late 1990s, however, it has become clear that this traditional definition of the Indo-Australian Plate cannot be correct; it consists of three plates — the Indian Plate, the Capricorn Plate, and Australian Plate — separated by diffuse boundary zones. Since 20 Ma the African Plate is being divided by the East African Rift System into the Nubian and Somalia plates.

There are only two trenches in the Indian Ocean: the 6,000 km (3,700 mi)-long Java Trench between Java and the Sunda Trench and the 900 km (560 mi)-long Makran Trench south of Iran and Pakistan.

A series of ridges and seamount chains produced by hotspots pass over the Indian Ocean. The Réunion hotspot (active 70–40 million years ago) connects Réunion and the Mascarene Plateau to the Chagos-Laccadive Ridge and the Deccan Traps in north-western India; the Kerguelen hotspot (100–35 million years ago) connects the Kerguelen Islands and Kerguelen Plateau to the Ninety East Ridge and the Rajmahal Traps in north-eastern India; the Marion hotspot (100–70 million years ago) possibly connects Prince Edward Islands to the Eighty Five East Ridge. These hotspot tracks have been broken by the still active spreading ridges mentioned above. There are fewer seamounts in the Indian Ocean than in the Atlantic and Pacific. These are typically deeper than 3,000 m (9,800 ft) and located north of 55°S and west of 80°E. Most originated at spreading ridges but some are now located in basins far away from these ridges. The ridges of the Indian Ocean form ranges of seamounts, sometimes very long, including the Carlsberg Ridge, Madagascar Ridge, Central Indian Ridge, Southwest Indian Ridge, Chagos-Laccadive Ridge, 85°E Ridge, 90°E Ridge, Southeast Indian Ridge, Broken Ridge, and East Indiaman Ridge. The Agulhas Plateau and Mascarene Plateau are the two major shallow areas.

The opening of the Indian Ocean began c. 156 Ma when Africa separated from East Gondwana. The Indian Subcontinent began to separate from Australia-Antarctica 135–125 Ma and as the Tethys Ocean north of India began to close 118–84 Ma the Indian Ocean opened behind it.

Southern Ocean

The Southern Ocean, also known as the Antarctic Ocean,[note 4] comprises the southernmost waters of the World Ocean, generally taken to be south of 60° S latitude and encircling Antarctica. As such, it is regarded as the second-smallest of the five principal oceanic divisions: smaller than the Pacific, Atlantic, and Indian oceans but larger than the Arctic Ocean. Over the past 30 years, the Southern Ocean has been subject to rapid climate change, which has led to changes in the marine ecosystem.

By way of his voyages in the 1770s, James Cook proved that waters encompassed the southern latitudes of the globe. Since then, geographers have disagreed on the Southern Ocean's northern boundary or even existence, considering the waters as various parts of the Pacific, Atlantic, and Indian Oceans, instead. However, according to Commodore John Leech of the International Hydrographic Organization (IHO), recent oceanographic research has discovered the importance of Southern Circulation, and the term Southern Ocean has been used to define the body of water which lies south of the northern limit of that circulation. This remains the current official policy of the IHO, since a 2000 revision of its definitions including the Southern Ocean as the waters south of the 60th parallel has not yet been adopted. Others regard the seasonally-fluctuating Antarctic Convergence as the natural boundary. This oceanic zone is where cold, northward flowing waters from the Antarctic mix with warmer Subantarctic waters.

The maximum depth of the Southern Ocean, using the definition that it lies south of 60th parallel, was surveyed by the Five Deeps Expedition in early February 2019. The expedition's multibeam sonar team identified the deepest point at 60° 28' 46"S, 025° 32' 32"W, with a depth of 7,434 metres (24,390 ft). The expedition leader and chief submersible pilot Victor Vescovo, has proposed naming this deepest point in the Southern Ocean the "Factorian Deep", based on the name of the manned submersible DSV Limiting Factor, in which he successfully visited the bottom for the first time on February 3, 2019.

Borders and names for oceans and seas were internationally agreed when the International Hydrographic Bureau, the precursor to the IHO, convened the First International Conference on 24 July 1919. The IHO then published these in its Limits of Oceans and Seas, the first edition being 1928. Since the first edition, the limits of the Southern Ocean have moved progressively southwards; since 1953, it has been omitted from the official publication and left to local hydrographic offices to determine their own limits.

The IHO included the ocean and its definition as the waters south of the 60th parallel south in its 2000 revisions, but this has not been formally adopted, due to continuing impasses about some of the content, such as the naming dispute over the Sea of Japan. The 2000 IHO definition, however, was circulated in a draft edition in 2002, and is used by some within the IHO and by some other organizations such as the CIA World Factbook and Merriam-Webster.

The Australian Government regards the Southern Ocean as lying immediately south of Australia (see § Australian standpoint).

The National Geographic Society recognized the ocean officially in June 2021. [13] Prior to this, it depicted it in a typeface different from the other world oceans; instead, it shows the Pacific, Atlantic, and Indian Oceans extending to Antarctica on both its print and online maps. Map publishers using the term Southern Ocean on their maps include Hema Maps and GeoNova.

Geography

The Southern Ocean, geologically the youngest of the oceans, was formed when Antarctica and South America moved apart, opening the Drake Passage, roughly 30 million years ago. The separation of the continents allowed the formation of the Antarctic Circumpolar Current.

With a northern limit at 60°S, the Southern Ocean differs from the other oceans in that its largest boundary, the northern boundary, does not abut a landmass (as it did with the first edition of Limits of Oceans and Seas). Instead, the northern limit is with the Atlantic, Indian and Pacific Oceans.

One reason for considering it as a separate ocean stems from the fact that much of the water of the Southern Ocean differs from the water in the other oceans. Water gets transported around the Southern Ocean fairly rapidly because of the Antarctic Circumpolar Current which circulates around Antarctica. Water in the Southern Ocean south of, for example, New Zealand, resembles the water in the Southern Ocean south of South America more closely than it resembles the water in the Pacific Ocean.

The Southern Ocean has typical depths of between 4,000 and 5,000 m (13,000 and 16,000 ft) over most of its extent with only limited areas of shallow water. The Southern Ocean's greatest depth of 7,236 m (23,740 ft) occurs at the southern end of the South Sandwich Trench, at 60°00'S, 024°W. The Antarctic continental shelf appears generally narrow and unusually deep, its edge lying at depths up to 800 m (2,600 ft), compared to a global mean of 133 m (436 ft).

Equinox to equinox in line with the sun's seasonal influence, the Antarctic ice pack fluctuates from an average minimum of 2.6 million square kilometres (1.0×106 sq mi) in March to about 18.8 million square kilometres (7.3×106 sq mi) in September, more than a sevenfold increase in area.

Sub-divisions of the Southern Ocean

Sub-divisions of oceans are geographical features such as "seas", "straits", "bays", "channels", and "gulfs". There are many sub-divisions of the Southern Ocean defined in the never-approved 2002 draft fourth edition of the IHO publication Limits of Oceans and Seas. In clockwise order these include (with sector):

- Weddell Sea (57°18'W – 12°16'E)
- King Haakon VII Sea[note 5] (20°W – 45°E)
- Lazarev Sea (0° – 14°E)
- Riiser-Larsen Sea (14° – 30°E)
- Cosmonauts Sea (30° – 50°E)
- Cooperation Sea (59°34' – 85°E)
- Davis Sea (82° – 96°E)
- Mawson Sea (95°45' – 113°E)
- D'Urville Sea (140°E)
- Sea (150° – 170°E)
- Sea (166°E – 155°W)
- Sea (102°20' – 126°W)
- Sea (57°18' – 102°20'W)
- Part of the Drake Passage[note 6] (54° – 68°W)
- Bransfield Strait (54° – 62°W)
- Part of the Scotia Sea[note 7] (26°30' – 65°W)

A number of these such as the 2002 Russian-proposed "Cosmonauts Sea", "Cooperation Sea", and "Somov (mid-1950s Russian polar explorer) Sea" are not included in the 1953 IHO document which remains currently in force, because they received their names largely originated from 1962 onward. Leading geographic authorities and atlases do not use these latter three names, including the 2014 10th edition World Atlas from the United States' National Geographic Society and the 2014 12th edition of the British Times Atlas of the World, but Soviet and Russian-issued maps do.

Biggest seas in Southern Ocean

Top large seas:

- Weddell Sea – 2,800,000 km2 (1,100,000 sq mi)
- Somov Sea – 1,150,000 km2 (440,000 sq mi)
- Riiser-Larsen Sea – 1,138,000 km2 (439,000 sq mi)
- Lazarev Sea – 929,000 km2 (359,000 sq mi)
- Scotia Sea – 900,000 km2 (350,000 sq mi)
- Cosmonauts Sea – 699,000 km2 (270,000 sq mi)
- Ross Sea – 637,000 km2 (246,000 sq mi)
- Bellingshausen Sea – 487,000 km2 (188,000 sq mi)
- Mawson Sea – 333,000 km2 (129,000 sq mi)

- Cooperation Sea – 258,000 km2 (100,000 sq mi)
- Amundsen Sea – 98,000 km2 (38,000 sq mi)
- Davis Sea – 21,000 km2 (8,100 sq mi)

D'URVILLE SEA

King Haakon VII Sea

Natural Resources

The Southern Ocean probably contains large, and possibly giant, oil and gas fields on the continental margin. Placer deposits, accumulation of valuable minerals such as gold, formed by gravity separation during sedimentary processes are also expected to exist in the Southern Ocean.

Manganese nodules are expected to exist in the Southern Ocean. Manganese nodules are rock concretions on the sea bottom formed of concentric layers of iron and manganese hydroxides around a core. The core may be microscopically small and is sometimes completely transformed into manganese minerals by crystallization. Interest in the potential exploitation of polymetallic nodules generated a great deal of activity among prospective mining consortia in the 1960s and 1970s. The icebergs that form each year around in the Southern Ocean hold enough fresh water to meet the needs of every person on Earth for several months. For several decades there have been proposals, none yet to be feasible or successful, to tow Southern Ocean icebergs to more arid northern regions (such as Australia) where they can be harvested.

Natural hazards

Icebergs can occur at any time of year throughout the ocean. Some may have drafts up to several hundred meters; smaller icebergs, iceberg fragments and sea-ice (generally 0.5 to 1 m thick) also pose problems for ships. The deep continental shelf has a floor of glacial deposits varying widely over short distances.

Sailors know latitudes from 40 to 70 degrees south as the "Roaring Forties", "Furious Fifties" and "Shrieking Sixties" due to high winds and large waves that form as winds blow around the entire globe unimpeded by any land-mass. Icebergs, especially in May to October, make the area even more dangerous. The remoteness of the region makes sources of search and rescue scarce.

Antarctic Circumpolar Current and Antarctic Convergence

The Antarctic Circumpolar Current moves perpetually eastward – chasing and joining itself, and at 21,000 km (13,000 mi) in length – it comprises the world's longest ocean current, transporting 130 million cubic metres per second (4.6×109 cu ft/s) of water – 100 times the flow of all the world's rivers.

Several processes operate along the coast of Antarctica to produce, in the Southern Ocean, types of water masses not produced elsewhere in the oceans of the Southern Hemisphere. One of these is the Antarctic Bottom Water, a very cold, highly saline, dense water that forms under sea ice.

Associated with the Circumpolar Current is the Antarctic Convergence encircling Antarctica, where cold northward-flowing Antarctic waters meet the relatively warmer waters of the subantarctic, Antarctic waters predominantly sink beneath subantarctic waters, while associated zones of mixing and upwelling create a zone very high in nutrients. These nurture high levels of phytoplankton with associated copepods and Antarctic krill, and resultant foodchains supporting fish, whales, seals, penguins, albatrosses and a wealth of other species.

The Antarctic Convergence is considered to be the best natural definition of the northern extent of the Southern Ocean.

UPWELLING IN THE SOUTHERN OCEAN

Upwelling

Large-scale upwelling is found in the Southern Ocean. Strong westerly (eastward) winds blow around Antarctica, driving a significant flow of water northwards. This is actually a type of coastal upwelling. Since there are no continents in a band of open latitudes between South America and the tip of the Antarctic Peninsula, some of this water is drawn up from great depths. In many numerical models and observational syntheses, the Southern Ocean upwelling represents the primary means by which deep dense water is brought to the surface. Shallower, wind-driven upwelling is also found off the west coasts of North and South America, northwest and southwest Africa, and southwest and southeast Australia, all associated with oceanic subtropical high pressure circulations.

Some models of the ocean circulation suggest that broad-scale upwelling occurs in the tropics, as pressure driven flows converge water toward the low latitudes where it is diffusively warmed from above. The required diffusion coefficients, however, appear to be larger than are observed in the real ocean. Nonetheless, some diffusive upwelling does probably occur.

Location of the Southern Ocean gyres.

Ross and Weddell Gyres

The Ross Gyre and Weddell Gyre are two gyres that exist within the Southern Ocean. The gyres are located in the Ross Sea and Weddell Sea respectively, and both rotate clockwise. The gyres are formed by interactions between the Antarctic Circumpolar Current and the Antarctic Continental Shelf.

Sea ice has been noted to persist in the central area of the Ross Gyre. There is some evidence that global warming has resulted in some decrease of the salinity of the waters of the Ross Gyre since the 1950s.

Due to the Coriolis effect acting to the left in the Southern Hemisphere and the resulting Ekman transport away from the centres of the Weddell Gyre, these regions are very productive due to upwelling of cold, nutrient rich water.

Climate

Sea temperatures vary from about -2 to 10 °C (28 to 50 °F). Cyclonic storms travel eastward around the continent and frequently become intense because of the temperature contrast between ice and open ocean. The ocean-area from about latitude 40 south to the Antarctic Circle has the strongest average winds found anywhere on Earth. In winter the ocean freezes outward to 65 degrees south latitude in the Pacific sector and 55 degrees south latitude in the Atlantic sector, lowering surface temperatures well below 0 degrees Celsius. At some coastal points, however, persistent intense drainage winds from the interior keep the shoreline ice-free throughout the winter.

Clouds over the Southern Ocean, with continent labels

Climate change

The Southern Ocean is one of the regions in which rapid climate change is the most visiblly taking place. In this region, small perturbations in temperature lead to major environmental perturbation. The effects of climate change in the Southern Ocean are expected to manifest themselves in a regional and diverse manner. This will include changes in the climate and weather patterns across different time-scales with alterations to the long interdecadal background signals such as the El Niño–Southern Oscillation (ENSO). Increasing ocean temperatures and changes in the extent and seasonality of sea ice affect the biological productivity and community of this ecosystem. The magitude and exact manifestation of these changes could lead to different populations of the same species responding and adapting differently to climate change depending on the region of the Southern Ocean they inhabit.

Biodiversity

Orca (*Orcinus orca*) hunting a Weddell seal in the Southern Ocean

Animals

A variety of marine animals exist and rely, directly or indirectly, on the phytoplankton in the Southern Ocean. Antarctic sea life includes penguins, blue whales, orcas, colossal squids and fur seals. The emperor penguin is the only penguin that breeds during the winter in Antarctica, while the Adélie penguin breeds farther south than any other penguin. The rockhopper penguin has distinctive feathers around the eyes, giving the appearance of elaborate eyelashes. King penguins, chinstrap penguins, and gentoo penguins also breed in the Antarctic.

The Antarctic fur seal was very heavily hunted in the 18th and 19th centuries for its pelt by sealers from the United States and the United Kingdom. The Weddell seal, a "true seal", is named after Sir James Weddell, commander of British sealing expeditions in the Weddell Sea. Antarctic krill, which congregates in large schools, is the keystone species of the ecosystem of the Southern Ocean, and is an important food organism for whales, seals, leopard seals, fur seals, squid, icefish, penguins, albatrosses and many other birds.

The benthic communities of the seafloor are diverse and dense, with up to 155,000 animals found in 1 square metre (10.8 sq ft). As the seafloor environment is very similar all around the Antarctic, hundreds of species can be found all the way around the mainland, which is a uniquely wide distribution for such a large community. Deep-sea gigantism is common among these animals.

A census of sea life carried out during the International Polar Year and which involved some 500 researchers was released in 2010. The research is part of the global Census of Marine Life (CoML) and has disclosed some remarkable findings. More than 235 marine organisms live in both polar regions, having bridged the gap of 12,000 km (7,500 mi). Large animals such as some cetaceans and birds make the round trip annually. More surprising are small forms of life such as mudworms, sea cucumbers and free-swimming snails found in both polar oceans. Various factors may aid in their distribution – fairly uniform temperatures of the deep ocean at the poles and the equator which differ by no more than 5 °C (9.0 °F), and the major current systems or marine conveyor belt which transport egg and larva stages. However, among smaller marine animals generally assumed to be the same in the Antarctica and the Arctic, more detailed studies of each population have often—but not always—revealed differences, showing that they are closely related cryptic species rather than a single bipolar species.

A WANDERING ALBATROSS (DIOMEDEA EXULANS) ON SOUTH GEORGIA

Birds

The rocky shores of mainland Antarctica and its offshore islands provide nesting space for over 100 million birds every spring. These nesters include species of albatrosses, petrels, skuas, gulls and terns. The insectivorous South Georgia pipit is endemic to South Georgia and some smaller surrounding islands. Freshwater ducks inhabit South Georgia and the Kerguelen Islands.

The flightless penguins are all located in the Southern Hemisphere, with the greatest concentration located on and around Antarctica. Four of the 18 penguin species live and breed on the mainland and its close offshore islands. Another four species live on the subantarctic islands. Emperor penguins have four overlapping layers of feathers, keeping them warm. They are the only Antarctic animal to breed during the winter.

Fish

There are relatively few fish species in few families in the Southern Ocean. The most species-rich family are the snailfish (Liparidae), followed by the cod icefish (Nototheniidae) and eelpout (Zoarcidae). Together the snailfish, eelpouts and notothenioids (which includes cod icefish and several other families) account for almost 9/10 of the more than 320 described fish species of the Southern Ocean (tens of undescribed species also occur in the region, especially among the snailfish). Southern Ocean snailfish are generally found in deep waters, while the icefish also occur in shallower waters.

Icefish

Fish of the Notothenioidei suborder, such as this young icefish, are mostly restricted to the Antarctic and Subantarctic

Cod icefish (Nototheniidae), as well as several other families, are part of the Notothenioidei suborder, collectively sometimes referred to as icefish. The suborder contains many species with antifreeze proteins in their blood and tissue, allowing them to live in water that is around or slightly below 0 °C (32 °F). Antifreeze proteins are also known from Southern Ocean snailfish.

The crocodile icefish (family Channichthyidae), also known as white-blooded fish, are only found in the Southern Ocean. They lack hemoglobin in their blood, resulting in their blood being colourless. One Channichthyidae species, the mackerel icefish (Champsocephalus gunnari), was once the most common fish in coastal waters less than 400 metres (1,312 ft) deep, but was overfished in the 1970s and 1980s. Schools of icefish spend the day at the seafloor and the night higher in the water column eating plankton and smaller fish.

There are two species from the genus Dissostichus, the Antarctic toothfish (Dissostichus mawsoni) and the Patagonian toothfish (Dissostichus eleginoides). These two species live on the seafloor 100–3,000 metres (328–9,843 ft) deep, and can grow to around 2 metres (7 ft) long weighing up to 100 kilograms (220 lb), living up to 45 years. The Antarctic toothfish lives close to the Antarctic mainland, whereas the Patagonian toothfish lives in the relatively warmer subantarctic waters. Toothfish are commercially fished, and overfishing has reduced toothfish populations.

Another abundant fish group is the genus Notothenia, which like the Antarctic toothfish have antifreeze in their bodies.

An unusual species of icefish is the Antarctic silverfish (Pleuragramma antarcticum), which is the only truly pelagic fish in the waters near Antarctica.

Weddell seals (Leptonychotes weddellii) are the most southerly of Antarctic mammals.

Mammals

Seven pinniped species inhabit Antarctica. The largest, the elephant seal (Mirounga leonina), can reach up to 4,000 kilograms (8,818 lb), while females of the smallest, the Antarctic fur seal (Arctocephalus gazella), reach only 150 kilograms (331 lb). These two species live north of the sea ice, and breed in harems on beaches. The other four species can live on the sea ice. Crabeater seals (Lobodon carcinophagus) and Weddell seals (Leptonychotes weddellii) form breeding colonies, whereas leopard seals (Hydrurga leptonyx) and Ross seals (Ommatophoca rossii) live solitary lives. Although these species hunt underwater, they breed on land or ice and spend a great deal of time there, as they have no terrestrial predators.

The four species that inhabit sea ice are thought to make up 50% of the total biomass of the world's seals. Crabeater seals have a population of around 15 million, making them one of the most numerous large animals on the planet. The New Zealand sea lion (Phocarctos hookeri), one of the rarest and most localised pinnipeds, breeds almost exclusively on the subantarctic Auckland Islands, although historically it had a wider range. Out of all permanent mammalian residents, the Weddell seals live the furthest south.

There are 10 cetacean species found in the Southern Ocean; six baleen whales, and four toothed whales. The largest of these, the blue whale (Balaenoptera musculus), grows to 24 metres (79 ft) long weighing 84 tonnes. Many of these species are migratory, and travel to tropical waters during the Antarctic winter.

Antarctic krill (Euphausia superba) are a keystone species of the food web.

INVERTEBRATES

Arthropods

Five species of krill, small free-swimming crustaceans, have been found in the Southern Ocean. The Antarctic krill (Euphausia superba) is one of the most abundant animal species on earth, with a biomass of around 500 million tonnes. Each individual is 6 centimetres (2.4 in) long and weighs over 1 gram (0.035 oz). The swarms that form can stretch for kilometres, with up to 30,000 individuals per 1 cubic metre (35 cu ft), turning the water red. Swarms usually remain in deep water during the day, ascending during the night to feed on plankton. Many larger animals depend on krill for their own survival. During the winter when food is scarce, adult Antarctic krill can revert to a smaller juvenile stage, using their own body as nutrition.

Many benthic crustaceans have a non-seasonal breeding cycle, and some raise their young in a brood pouch. Glyptonotus antarcticus is an unusually large benthic isopod, reaching 20 centimetres (8 in) in length weighing 70 grams (2.47 oz). Amphipods are abundant in soft sediments, eating a range of items, from algae to other animals. The amphipods are highly diverse with more than 600 recognized species found south of the Antarctic Convergence and there are indications that many undescribed species remain. Among these are several "giants", such as the iconic epimeriids that are up to 8 cm (3.1 in) long.

Slow moving sea spiders are common, sometimes growing as large as a human hand. They feed on the corals, sponges, and bryozoans that litter the seabed.

A female warty squid (Moroteuthis ingens)

Others

Many aquatic molluscs are present in Antarctica. Bivalves such as Adamussium colbecki move around on the seafloor, while others such as Laternula elliptica live in burrows filtering the water above. There are around 70 cephalopod species in the Southern Ocean, the largest of which is the colossal squid (Mesonychoteuthis hamiltoni), which at up to 14 metres (46 ft) is among the largest invertebrate in the world. Squid makes up most of the diet of some animals, such as grey-headed albatrosses and sperm whales, and the warty squid (Moroteuthis ingens) is one of the subantarctic's most preyed upon species by vertebrates.

The sea urchin genus Abatus burrow through the sediment eating the nutrients they find in it. Two species of salps are common in Antarctic waters, Salpa thompsoni and Ihlea racovitzai. Salpa thompsoni is found in ice-free areas, whereas Ihlea racovitzai is found in the high latitude areas near ice. Due to their low nutritional value, they are normally only eaten by fish, with larger animals such as birds and marine mammals only eating them when other food is scarce.

Antarctic sponges are long-lived, and sensitive to environmental changes due to the specificity of the symbiotic microbial communities within them. As a result, they function as indicators of environmental health.

ENVIRONMENT

Current Issues

Increased solar ultraviolet radiation resulting from the Antarctic ozone hole has reduced marine primary productivity (phytoplankton) by as much as 15% and has started damaging the DNA of some fish.[109] Illegal, unreported and unregulated fishing, especially the landing of an estimated five to six times more Patagonian toothfish than the regulated fishery, likely affects the sustainability of the stock. Long-line fishing for toothfish causes a high incidence of seabird mortality.

International Agreements

All international agreements regarding the world's oceans apply to the Southern Ocean. In addition, it is subject to these agreements specific to the region:

The Southern Ocean Whale Sanctuary of the International Whaling Commission (IWC) prohibits commercial whaling south of 40 degrees south (south of 60 degrees south between 50 degrees and 130 degrees west). Japan regularly does not recognize this provision, because the sanctuary violates IWC charter. Since the scope of the sanctuary is limited to commercial whaling, in regard to its whaling permit and whaling for scientific research, a Japanese fleet carried out an annual whale-hunt in the region. On 31 March 2014, the International Court of Justice ruled that Japan's whaling program, which Japan has long claimed is for scientific purposes, was a cloak for commercial whaling, and no further permits would be granted.

Convention for the Conservation of Antarctic Seals is part of the Antarctic Treaty System. It was signed at the conclusion of a multilateral conference in London on 11 February 1972.

Convention for the Conservation of Antarctic Marine Living Resources (CCAMLR) is part of the Antarctic Treaty System. The Convention was entered into force on 7 April 1982 and has its goal is to preserve marine life and environmental integrity in and near Antarctica. It was established in large part to concerns that an increase in krill catches in the Southern Ocean could have a serious impact on populations of other marine life which are dependent upon krill for food.

Many nations prohibit the exploration for and the exploitation of mineral resources south of the fluctuating Antarctic Convergence, which lies in the middle of the Antarctic Circumpolar Current and serves as the dividing line between the very cold polar surface waters to the south and the warmer waters to the north. The Antarctic Treaty covers the portion of the globe south of sixty degrees south; it prohibits new claims to Antarctica.

The Convention for the Conservation of Antarctic Marine Living Resources applies to the area south of 60° South latitude as well as the areas further north up to the limit of the Antarctic Convergence.

Economy

Between 1 July 1998 and 30 June 1999, fisheries landed 119,898 tonnes, of which 85% consisted of krill and 14% of Patagonian toothfish. International agreements came into force in late 1999 to reduce illegal, unreported, and unregulated fishing, which in the 1998–99 season landed five to six times more Patagonian toothfish than the regulated fishery.

Ports and harbors

Severe cracks in an ice pier in use for four seasons at McMurdo Station slowed cargo operations in 1983 and proved a safety hazard. Major operational ports include: Rothera Station, Palmer Station, Villa Las Estrellas, Esperanza Base, Mawson Station, McMurdo Station, and offshore anchorages in Antarctica.

Few ports or harbors exist on the southern (Antarctic) coast of the Southern Ocean, since ice conditions limit use of most shores to short periods in midsummer; even then some require icebreaker escort for access. Most Antarctic ports are operated by government research stations and, except in an emergency, remain closed to commercial or private vessels; vessels in any port south of 60 degrees south are subject to inspection by Antarctic Treaty observers.

The Southern Ocean's southernmost port operates at McMurdo Station at 77°50'S 166°40'E. Winter Quarters Bay forms a small harbor, on the southern tip of Ross Island where a floating ice pier makes port operations possible in summer. Operation Deep Freeze personnel constructed the first ice pier at McMurdo in 1973. Based on the original 1928 IHO delineation of the Southern Ocean (and the 1937 delineation if the Great Australian Bight is considered integral), Australian ports and harbors between Cape Leeuwin and Cape Otway on the Australian mainland and along the west coast of Tasmania would also be identified as ports and harbors existing in the Southern Ocean. These would include the larger ports and harbors of Albany, Thevenard, Port Lincoln, Whyalla, Port Augusta, Port Adelaide, Portland, Warrnambool, and Macquarie Harbour.

Even though organizers of several yacht races define their routes as involving the Southern Ocean, the actual routes don't enter the actual geographical boundaries of the Southern Ocean. The routes involve instead South Atlantic, South Pacific and Indian Ocean.

3

WATER RESOURCES IN INDIA

Water resources in India includes information on precipitation, surface and groundwater storage and hydropower potential. India experiences an average precipitation of 1,170 millimetres (46 in) per year, or about 4,000 cubic kilometres (960 cu mi) of rains annually or about 1,720 cubic metres (61,000 cu ft) of fresh water per person every year. India accounts for 18% of the world population and about 4% of the world's water resources. One of the solutions to solve the country's water woes is to create Indian Rivers Inter-link.c Some 80 percent of its area experiences rains of 750 millimetres (30 in) or more a year. However, this rain is not uniform in time or geography. Most of the rains occur during its monsoon seasons (June to September), with the north east and north receiving far more rains than India's west and south. Other than rains, the melting of snow over the Himalayas after winter season feeds the northern rivers to varying degrees.

The southern rivers, however experience more flow variability over the year. For the Himalayan basin, this leads to flooding in some months and water scarcity in others. Despite extensive river system, safe clean drinking water as well as irrigation water supplies for sustainable agriculture are in shortage across India, in part because it has, as yet, harnessed a small fraction of its available and recoverable surface water resource. India harnessed 761 cubic kilometres (183 cu mi) (20 percent) of its water resources in 2010, part of which came from unsustainable use of groundwater. Of the water it withdrew from its rivers and groundwater wells, India dedicated about 688 cubic kilometres (165 cu mi) to irrigation, 56 cubic kilometres (13 cu mi) to municipal and drinking water applications and 17 cubic kilometres (4.1 cu mi) to industry.

Vast area of India is under tropical climate which is conducive throughout the year for agriculture due to favourable warm and sunny conditions provided perennial water supply is available to cater to the high rate of evapotranspiration from the cultivated land. Though the overall water resources are adequate to meet all the requirements of the country, the water supply gaps due to temporal and spatial distribution of water resources are to be bridged by interlinking the rivers of India. The total water resources going waste to the sea are nearly 1200 billion cubic meters after sparing moderate environmental / salt export water requirements of all rivers. Food security in India is possible by achieving water security first which in turn is possible with energy security to supply the electricity for the required water pumping as part of its rivers interlinking.

Instead of opting for centralised mega water transfer projects which would take long time to give results, it would be cheaper alternative to deploy extensively shade nets over the cultivated lands for using the locally available water sources efficiently to crops throughout the year. Plants need less than 2% of total water for metabolism requirements and rest 98% is for cooling purpose through transpiration. Shade nets or polytunnels installed over the agriculture lands suitable for all weather conditions would reduce the potential evaporation drastically by reflecting the excessive and harmful sun light without falling on the cropped area.

Drought, floods and shortage of drinking water

The precipitation pattern in India varies dramatically across distance and over calendar months. Much of the precipitation in India, about 85%, is received during summer months through monsoons in the Himalayan catchments of the Ganges-Brahmaputra-Meghna basin. The north eastern region of the country receives heavy precipitation, in comparison with the north western, western and southern parts. The uncertainty in onset of annual monsoon, sometimes marked by prolonged dry spells and fluctuations in seasonal and annual rainfall is a serious problem for the country. Large area of the country is not put to use for agriculture due to local water scarcity or poor water quality. The nation sees cycles of drought years and flood years, with large parts of west and south experiencing more deficits and large variations, resulting in immense hardship particularly the poorest farmers and rural populations. Dependence on erratic rains and lack of irrigation water supply regionally leads to crop failures and farmer suicides. Despite abundant rains during June–September, some regions in other seasons see shortages of drinking water. Some years, the problem temporarily becomes too much rainfall, and weeks of havoc from floods.

Surface and ground water storage

India currently stores only 6% of its annual rainfall or 253 billion cubic metres (8.9×1012 cu ft), while developed nations strategically store 250% of the annual rainfall in arid river basins. India also relies excessively on groundwater resources, which accounts for over 50 percent of irrigated area with 20 million tube wells installed. India has built nearly 5,000 major or medium dams, barrages, etc. to store the river waters and enhance ground water recharging. The important dams (59 nos) have an aggregate gross storage capacity of 170 billion cubic metres (6.0×1012 cu ft). About 15 percent of India's food is being produced using rapidly depleting / mining groundwater resources. The end of the era of massive expansion in groundwater use is going to demand greater reliance on surface water supply systems.

India is not running out of water whereas water is running out of India without extracting its full potential benefits. Land based water reservoirs construction is very costly after meeting the land & property compensation and rehabilitation expenditures. To create adequate water storage, fresh water coastal reservoirs located on the sea area near the river deltas, is the suitable option socioeconomically without land and forest submergence problems.

Hydro power potential

Indian rivers have fairly good hydro power potential when they descend from their source mountains (Himalayas, Western Ghats, Aravali Range, Vindhya Mountains, Eastern Ghats, etc) before the water consumption or flowing to the sea. The hydro power potential keeps on varying depending on the technological developments including alternate power sources, priorities and limitations.

Rivers

Map of India based on survey of rivers of India.

River Ganga

The Ganges is one of the major rivers of the Indian subcontinent which flows from east through the Gangetic Plain of northern India into the Bangladesh. The river rises around 2,510 km in the western Himalayas in the Uttarakhand state of India and drains into the Sunderbans delta in the Bay of Bengal. Do you know that it is considered the longest holy river by Hindus and worshipped as the goddess Ganga in Hinduism?

The Ganges is highly polluted. Pollution threatens not only humans, but also more than 140 fish species, 90 amphibian species and the endangered Ganges river dolphin. The Ganges is a major source of global ocean plastic pollution. The levels of fecal coliform bacteria from human waste in the waters of the river near Varanasi are more than 100 times the Indian government's official limit. The Ganga Action Plan, an environmental initiative to clean up the river, has been a major failure thus far,[a][b] due to rampant corruption, lack of will on behalf of the government and its bureaucracy, lack of technical expertise,[c] poor environmental planning,[d] and lack of support from religious authorities. Historically also it is important as many former provincial or imperial capitals like Patliputra, Allahabad, Kannauj, Murshidabad, Calcutta etc. has been located on its banks. The basins of Ganga drain around 1,000,000 square kilometres and support one of the world's highest densities of humans. Let us study more about Ganga River and from which place it is originated..

The Ganges River Catchment Basin covers an area of 390,000 sq miles (1,000,000 sq km) and supplies to one of the maximum populated areas in the world. The average depth of the Ganges River is 16 m or 52 feet and the highest depth is 30 m or 100 feet. The river has been proclaimed as the National river of India. The first Prime Minister of India, Pandit Jawaharlal Nehru, cited a number of emblematic interpretations regarding the Ganges on the Indian subcontinent in his famous book, the Discovery of India (published in 1946). The Ganges passes through the states of Uttarakhand, Uttar Pradesh, Bihar, Jharkhand, and West Bengal.

Yamuna River

Yamuna River, also called Jumna, major river of northern India, primarily in Uttarakhand and Uttar Pradesh states. It is one of the country's most-sacred rivers.

The Yamuna rises on the slopes of the Bandarpunch massif in the Great Himalayas near Yamnotri (Jamnotri) in western Uttarakhand. It flows in a southerly direction swiftly through the Himalayan foothills and, exiting Uttarakhand, onto the Indo-Gangetic Plain, along the border between Uttar Pradesh and Haryana state to the west. The Eastern and Western Yamuna canals are fed from the river at that point.

The source of Yamuna lies in the Yamunotri Glacier at an elevation of 6, 387 metres (20, 955 ft), on the south western slopes of Banderpooch peaks, which lie in theMussoorie range of Lower Himalayas, in the Uttarkashi district, Uttarakhand, north ofHaridwar. Yamunotri temple, a shrine dedicated to the goddess, Yamuna is one of the holiest shrines in Hinduism, and part of the Chota Char Dham Yatra circuit. Also standing close to the temple, on its 13-kilometre (8 mi) trek route, that follows the right bank of the river, lies the Markendeya Tirtha, where the sage Markandeya wrote the Markandeya Purana.

From here it flows southwards, for about 200 kilometres (120 mi) through the Lower Himalayas and the Shivalik Hills Range and morainicdeposited are found in its steep Upper Yamuna village, highlighted with geomorphic features such as interlocking spurs, steep rock benches, gorges and stream terraces. Large terraces formed over a long period of time can be seen in the lower course of the river, like ones near Naugoan. An important part of its early catchment areatotalling 2, 320 square kilometres (900 sq mi) lies inHimachal Pradesh, and an important tributary draining the Upper Catchment Area is the Tons, Yamuna's largest and longest tributary, which rises from the Hari-ki-dun valley and holds water more than the main stream, which it merges after Kalsi near Dehradun. The entire drainage system of the river stretches all the way between Giri-Sutlej catchment in Himachal and Yamuna-Bhilangna catchment in Garhwal, indeed the southern ridge of Shimla is also drained into this system. Kalanag (6, 387 metres (20, 955 ft)) is the highest point of the entire Yamuna basin.

Other tributaries in the region are the Giri, Rishi Ganga, Kunta, Hanuman Ganga and Bata tributaries, which drain the Upper Catchment Area of the vast Yamuna basin. Thereafter the river descends on to the plains of Doon Valley, at Dak Pathar near Dehradun. Here through the Dakpathar Barrage, the water is diverted into a canal for power generation, little further down where Yamuna is met by the Assan River, lies the Asan Barrage, which hosts a Bird Sanctuary as well. After passing the Sikh pilgrimage town of Paonta Sahib, it reaches Tajewala in Yamuna Nagar district, of Haryana, where a dam built in 1873, is the originating place of two important canals, the Western Yamuna Canal and Eastern Yamuna Canal, which irrigate the states of Haryana and Uttar Pradesh. The Western Yamuna Canal (WYC) crosses Yamuna Nagar, Karnal and Panipat before reaching the Haiderpur treatment plant, which supplies part of municipal water supply to Delhi, further it also receives waste water from Yamuna Nagar and Panipat cities. Yamuna is replenished again after this by seasonal streams and groundwater accrual, in fact during the dry season, it remains dry in many stretches from Tajewala till Delhi, where it enters near Palla village after traversing 224 kilometres (139 mi).

The Yamuna also creates natural state borders between the Himachal Pradesh and Uttarakhand states, and further down between the state of Haryana, Delhi and Uttar Pradesh. Along with the Ganges to which run almost parallel after it touches the Indo-Gangetic plain, the largestalluvial fertile plain in the world, it creates the Ganges-Yamuna Doab region spread across 69, 000 square kilometres (27, 000 sq mi), one-third of the entire plain, and today known for its agricultural outputs, prominent among them is the cultivation of Basmati Rice. The plain itself supports one-third of India's population through its farming.

State	Catchment	% of catchment
	area (km2)	area
Uttar Pradesh	74, 208	21. 5 %
and Uttarakhand		
Himachal	5, 799	1. 6
Pradesh		

Haryana	21, 265	6. 5
Rajasthan	102, 883	29. 8
Madhya	14, 023	40. 6
Pradesh		
Delhi	1, 485	0. 4

Subsequently, it flows through the states of Delhi, Haryana and Uttar Pradesh] before merging with theGanges at a sacred spot known as Triveni Sangam in Allahabad after traversing a distance of 1, 376 kilometers (855 mi). Here pilgrims travel by boats to platforms erected mid stream to offer prayers. During the Kumbh Mela, held every 12 years, the ghats around the Sangam are venue of large congregation of people, who take dip in the sacred waters of the confluence. The cities of Baghpat, Delhi, Noida, Mathura, Agra, Firozabad, Etawah, Kalpi, Hamirpur, Allahabad lie on its banks. At Etawah, it meets it another important tributary, Chambal, followed by a host of tributaries further down, including, Sindh, the Betwa, and Ken.

Important tributaries

- Tons River, Yamuna's largest and longest tributary, rises in the 20, 720 ft (6, 315 m) high Bandarpoonch mountain, and has a large basin inHimachal Pradesh. It meets Yamuna below Kalsi near Dehradun, Uttarakhand.
- Hindon River, originates in the Saharanpur District, from Upper Shivalik in Lower Himalayan Range, is entirely rainfed and has acatchment area of 7, 083 square kilometres (2, 735 sq mi), traverses 400 kilometres (250 mi) through Muzaffarnagar District, Meerut District, Baghpat District, Ghaziabad, Noida, Greater Noida, before joining Yamuna just outside Delhi.
- Ken River, flows through Bundelkhand region of Madhya Pradesh and Uttar Pradesh, it originates near village Ahirgawan in Jabalpur district and travels a distance of 427 kilometres (265 mi), before merging with the Yamuna at Chilla village, near Fatehpur in Uttar Pradesh, and has an overall drainage basin of 28, 058 square kilometres (10, 833 sq mi).
- Chambal River, known as Charmanvati in ancient times, flows through Rajasthan and Madhya Pradesh, with a drainage basin of 143, 219 square kilometres (55, 297 sq mi) and traverses a total distance of 960 kilometres (600 mi), from its source in Vindhya Range, near Mhowand supports hydro-power generation at Gandhi Sagar dam, Rana Pratap Sagar dam and Jawahar Sagar dam, before merging into the Yamuna south east of Sohan Goan, in Etawah district, shortly thereafter followed by another tributary, the Sindh River.

- Sasur Khaderi River, known as Sasur Khaderi is a tributary in Fatehpur district.

Son River

Son River of central India is the largest of the Ganges' southern tributaries.Son River, Son also spelled Sone, principal southern tributary of the Ganges (Ganga) River, rising in Madhya Pradesh state, central India. It flows north past Manpur and then turns northeast. The river cuts through the Kaimur Range and joins the Ganges above Patna, after a 487-mile (784-km) course. The Son valley is geologically almost a continuation of that of the Narmada River to the southwest. It is largely forested and sparsely populated. The valley is bordered by the Kaimur Range to the north and the Chota Nagpur plateau to the south. The river's flow is seasonal, and the Son is unimportant for navigation. Dams have been constructed on some of its tributaries. At Dehri, in Bihar state, are the headworks of the Son canal system.

The Son originates near Amarkantak in Madhya Pradesh, just east of the headwater of theNarmada River, and flows north-northwest through Madhya Pradesh state before turning sharply eastward where it encounters the southwest-northeast-running Kaimur Range. The Son parallels the Kaimur hills, flowing east-northeast through Uttar Pradesh, Jharkhand and Bihar states to join the Ganges just above Patna. Geologically, the lower valley of the Son is an extension of the Narmada Valley, and the Kaimur Range an extension of the Vindhya Range. Dehri on sone is the major town situated on Son River.

The Son river at 784 kilometres (487 mi) long, is one of the largest rivers of India. Its chief tributaries are the Rihand and the North Koel. The Son has a steep gradient (35–55 cm per km) with quick run-off and ephemeral regimes, becoming a roaring river with the rain-waters in thecatchment area but turning quickly into a fordable stream. The Son, being wide and shallow,

leaves disconnected pools of water in the remaining part of the year. The channel of the Son is very wide (about 5 km at Dehri on sone) but the floodplain is narrow, only 3 to 5 kilometres (2 to 3 mi) wide. In the past, the Son has been notorious for changing course, as it is traceable from several old beds near its east bank. In modern times this tendency has been checked with the anicut at Dehri, and now more so with the Indrapuri Barrage.

In Bihar, this river forms border line between Bhojpuri andMagahi speaking regions.

Sir John Houlton, the British administrator, describes the Son as follows, "After passing the steep escarpments of the Kaimur range, it flows straight across the plain to the Ganges. For much of this distance it is over two miles – and at one point, opposite Tilothu – three miles wide. In the dry weather there is vast expanse of sand, with a stream not more than a hundred yards wide, and the hot west winds pile up the sand on the east bank, making natural embankments. After heavy rain in the hills even this wide bed cannot carry the waters of the Son and disastrous floods in Shahabad, Gaya, and Patna are not uncommon."

Gandaki River

The story of Gandaki* contains the universal theme of the Ultimate justifying sinners by their faith in the divine and changing them into new, reborn, creations. Another central universal element is the union of opposite elements, the male and female. It is also a story about those who in spite of their worldly position want to seek divine truth. And it's the story of the Divine preventing the believer at the last minute to perform a horrible act of faith. Themes found in other world religions too.

The Kali Gandaki river source is at the border with Tibet at an elevation of 6, 268 metres (20, 564 ft) at the Nhubine Himal Glacier in the Mustang region of Nepal.

The headwaters stream on some maps is named the Chhuama Khola and then, nearing Lo Manthang, the Nhichung Khola or Choro Khola. The Kali Gandaki then flows southwest (with the name of Mustang Khola on old, outdated maps) through a sheer-sided, deep canyon before widening at the steel footbridge at Chele, where part of its flow funnels through a rock tunnel, and from this point the now wide river is called the Kali Gandaki on all maps. InKagbeni a major tributary named Johng Khola, Kak Khola or Krishnaa descends fromMuktinath.

The river then flows southward through a steep gorge known as the Kali Gandaki Gorge, or Andha Galchi, between the mountains Dhaulagiri, elevation 8, 167 metres (26, 795 ft) to the west and Annapurna I, elevation 8, 091 metres (26, 545 ft) to the east. If one measures the depth of a canyon by the difference between the river height and the heights of the highest peaks on either side, this gorge is the world's deepest. The portion of the river directly between Dhaulagiri and Annapurna I, 7 kilometres (4 mi) downstream from Tukuche), is at an elevation of 2, 520 metres (8, 270 ft), which is 5, 571 metres (18, 278 ft) lower than Annapurna I. The river is older than the Himalayas. As tectonic activity forces the mountains higher, the river has cut through the uplift.

South of the gorge, the river is joined by Rahughat Khola at Galeshwor, Myagdi Khola atBeni, Modi Khola near Kushma and Badigaad at Rudrabeni above Ridi Bazaar. The river then turns east to run along the northern edge of the Mahabharat Range. The largesthydroelectricity project in Nepal is located along this stretch of the river. Turning south again and breaking through the Mahabharats, Kali Gandaki is then joined by a major tributary, the Trishuli, at

Devighat, then by the East Rapti River draining the Inner Terai valley known as Chitwan. The Gandaki then crosses the outermost foothills of the Himalayas—Sivalik Hills—into the Terai plains of Nepal. From Devighat, the river flows southwest of Gaindakot town. The river later curves back towards the southeast as it enters India where it is called the Gandak.

Below Gaindakot the river is known as the Narayani or Sapt Gandaki (Seven Gandakis), for seven tributaries rising in the Himalaya or further north along the main Ganges-Brahmaputra divide. These are the Kali Gandaki, the Trishuli River, and the five main tributaries of the Trishuli known as the Daraudi, Seti, Madi, Marsyandi and Budhi.

Koshi River

The Koshi or Kosi River drains the northern slopes of the Himalayas in the Tibet Autonomous Region and the southern slopes in Nepal. From a major confluence of tributaries north of the Chatra Gorge onwards, the Koshi River is also known as Saptakoshi for its seven upper tributaries. These include the Tamur Koshi originating from the Kanchenjunga area in the east, Arun River from Tibet and Sun Koshi from the Gosainthan area farther west. The Sun Koshi's tributaries from east to west areDudh Koshi, Bhote Koshi, Tamba Koshi and Indravati Koshi. The Saptakoshi crosses into northern Bihar where it branches into distributaries before joining the Gangesnear Kursela in Katihar district.

The Saptakoshi is 720 km (450 mi) long and drains an area of about 61, 000 km2(24, 000 sq mi) in Tibet, Nepal and Bihar. In the past, several authors proposed that the river has shifted its course for more than 133 km (83 mi) from east to west during the last 200 years. But a review of 28 historical maps dating 1760 to 1960 revealed a slight eastward shift for a long duration, and that the shifting was random and oscillating in nature.

The river basin is surrounded by ridges which separate it from the Yarlung Tsangpo River in the north, the Gandaki in the west and the Mahananda in the east. The river is joined by major tributaries in the Mahabharat Range approximately 48 km (30 mi) north of the Indo-Nepal border. Below the Siwaliks, the river has built up a megafansome 15, 000 km2 (5, 800 sq mi) in extent, breaking into more than twelve distinct channels, all with shifting courses due to flooding. Kamalâ, Bâgmati (Kareh) andBudhi Gandak are major tributaries of Koshi in India, besides minor tributaries such as Bhutahi Balân.

Its unstable nature has been attributed to the heavy silt it carries during the monsoonseason and flooding in India has extreme effects. Fishing is an important enterprise on the river but fishing resources are being depleted and youth are leaving for other areas of work.

Brahmaputra

Brahmaputra River, Bengali Jamuna, Tibetan Tsangpo, Chinese (Pinyin) Yarlung Zangbo Jiang or (Wade-Giles romanization) Ya-lu-tsang-pu Chiang, major river of Central and South Asia. It flows some 1,800 miles (2,900 km) from its source in the Himalayas to its confluence with the Ganges (Ganga) River, after which the mingled waters of the two rivers empty into the Bay of Bengal.The Brahmaputra, also called Tsangpo-Brahmaputra, is a trans-boundary river and one of the major rivers of Asia. Majuli is the riverine island formed by River Brahmaputra in Assam in India.

Along its course the Brahmaputra passes through the Tibet Autonomous Region of China, the Indian states of Arunachal Pradesh and Assam, and Bangladesh. For most of its length, the river serves as an important inland waterway. It is not, however, navigable between the mountains of Tibet and the plains of India. In its lower course the river is both a creator and a destroyer—depositing huge quantities of fertile alluvial soil but also causing disastrous and frequent floods.

With its origin in the Angsi Glacier, located on the northern side of the Himalayas inBurang County of Tibet as the Yarlung Tsangpo River, it flows across southern Tibetto break through the Himalayas in great gorges (including the Yarlung Tsangpo Grand Canyon) and into Arunachal Pradesh (India), where it is known as Dihang or Siang. It flows southwest through the Assam Valley as Brahmaputra and south throughBangladesh as the Jamuna (not to be mistaken with Yamuna of India). In the vastGanges Delta, it merges with the Padma, the popular name of the river Ganges in Bangladesh, and finally the Meghna and from here it is known as Meghna before emptying into the Bay of Bengal.

About 1, 800 mi (2, 900 km) long, the Brahmaputra is an important river for irrigationand transportation. The average depth of the river is 124 ft (38 m) and maximum depth is 380 ft (120 m). The river is prone to catastrophic flooding in spring when the Himalayan snows melt. The average discharge of the river is about 19, 300 m3/s (680, 000 cu ft/s), and floods can reach over 100, 000 m3/s (3, 500, 000 cu ft/s). It is a classic example of a braided river and is highly susceptible to channel migration andavulsion. It is also one of the few rivers in the world that exhibit a tidal bore. It is navigable for most of its length.

The river drains the Himalaya east of the Indo-Nepal border, south-central portion of the Tibetan plateau above the Ganges basin, south-eastern portion of Tibet, thePatkai-Bum hills, the northern slopes of the Meghalaya hills, the Assam plains, and the northern portion of Bangladesh. The basin, especially south of Tibet, is characterized by high levels of rainfall. Kangchenjunga (8, 586 m) is the only peak above 8, 000 m, hence is the highest point within the Brahmaputra basin.

The Brahmaputra's upper course was long unknown, and its identity with the Yarlung Tsangpo was only established by exploration in 1884–86. This river is often called Tsangpo-Brahmaputra river.

The lower reaches are sacred to Hindus. While most rivers on the Indian subcontinent have female names, this river has a rare male name, as it means "son of Brahma" inSanskrit (putra means "son").

Tibet

The Brahmaputra River (also called as "Burlung-Buthur" by the Bodo people of Assam), called Yarlung Tsangpo in Tibetan language, originates on the Angsi Glacier located on the northern side of the Himalayas in Burang County of Tibet and not Chema-Yungdung glacier, which was previously identified by geographer Swami Pranavananda in the 1930s. The river is 3, 848 km (2, 391 mi) long, and its drainage area is 712, 035 km2 (274, 918 sq mi) according to the new findings, while previous documents showed its length varied from 2, 900 to 3, 350 km and its drainage area between 520, 000 and 1. 73 million km2. This finding has been given by Mr. Liu Shaochuang, a researcher with the Institute of Remote Sensing Applications under the analysis using expeditions and satellite imagery from the Chinese Academy of Sciences (CAS).

From its source, the river runs for nearly 700 mi (1, 100 km) in a generally easterly direction between the main range of the Himalayas to the south and the Kailas Range to the north. Throughout its upper course, the river is generally known as the Tsangpo ("Purifier"); it is also known by its Chinese name (Yarlung Zangbo) and by other local Tibetan names.

In Tibet, the Tsangpo receives a number of tributaries. The most important left-bank tributaries are the Raka Zangbo (Raka Tsangpo), which joins the river west of Xigazê (Shigatse), and the Lhasa (Kyi), which flows past the Tibetan capital of Lhasa and joins the Tsangpo at the Nyang Qu (Gyamda) River joins the river from the north at Zela (Tsela Dzong). On the right bank, a second river called the Nyang Qu (Nyang Chu) meets the Tsangpo at Xigazê.

After passing Pi (Pe) in Tibet, the river turns suddenly to the north and northeast and cuts a course through a succession of great narrow gorges between the mountainous massifs of Gyala Peri and Namjagbarwa (Namcha Barwa) in a series of rapids and cascades. Thereafter, the river turns south and southwest and flows through a deep gorge (the "Grand Canyon" of the Tsangpo) across the eastern extremity of the Himalayas with canyon walls that extend upward for 16, 500 ft (5, 000 m) and more on each side. During that stretch, the river enters northern Arunachal Pradesh state in northeastern India, where it is known as the Dihang (or Siang) River, and turns more southerly.

Assam and adjoining region

The Brahmaputra enters India in the state of Arunachal Pradesh, where it is called Siang. It makes a very rapid descent from its original height in Tibet, and finally appears in the plains, where it is called Dihang. It flows for about 35 km (22 mi) and is joined by the Dibang River and the Lohit River at the head of the Assam Valley. Below the Lohit, the river is called Brahmaputra, enters the state of Assam, and becomes very wide—as wide as 10 km (6 mi) in parts of Assam. It is joined in Sonitpur by the Kameng River (or Jia Bhoreli).

The Dihang, winding out of the mountains, turns toward the southeast and descends into a low-lying basin as it enters northeastern Assam state. Just west of the town of Sadiya, the river again turns to the southwest and is joined by two mountain streams, the Lohit and the Dibang. Below that confluence, about 900 mi (1, 450 km) from the Bay of Bengal, the river becomes known conventionally as the Brahmaputra ("Son of Brahma"). In Assam, the river is mighty, even in the dry season, and during the rains, its banks are more than 5 mi (8 km) apart. As the river follows its braided 450-mi (700-km) course through the valley, it receives several rapidly rushing Himalayan streams, including the Subansiri, Kameng, Bhareli, Dhansiri, Manas, Champamati, Saralbhanga, and Sankosh Rivers. The main tributaries from the hills and from the plateau to the south are the Burhi Dihing, the Disang, the Dikhu, and the Kopili.

Between Dibrugarh and Lakhimpur Districts, the river divides into two channels—the northern Kherkutia channel and the southern Brahmaputra channel. The two channels join again about 100 km (62 mi) downstream, forming the Majuli island, which was, until some time back, the largest river island in the world. At Guwahati, near the ancient pilgrimage centre of Hajo, the Brahmaputra cuts through the rocks of the Shillong Plateau, and is at its narrowest at 1 km (1, 100 yd) bank-to-bank. Due to the river's narrow width, the Battle of Saraighat was fought here in March 1671. The first combined rail/road bridge across the Brahmaputra was opened to traffic in April 1962 at Saraighat.

The environment of the Brahmaputra floodplains in Assam have been described as the Brahmaputra Valley semi-evergreen forests ecoregion.

Bangladesh

In Bangladesh, the Brahmaputra is joined by the Teesta River (or Tista), one of its largest tributaries. Below the Tista, the Brahmaputra splits into two distributary branches. The western branch, which contains the majority of the river's flow, continues due south as the Jamuna (Jomuna) to merge with the lower Ganges, called the Padma River (Pôdda). The eastern branch, formerly the larger, but now much smaller, is called the lower or old Brahmaputra (Bromhoputro). It curves southeast to join the Meghna River near Dhaka. The Padma and Meghna converge near Chandpur and flow out into the Bay of Bengal. This final part of the river is called Meghna.

The Brahmaputra enters the plains of Bangladesh after turning south around the Garo Hills below Dhuburi, India. After flowing past Chilmari, Bangladesh, it is joined on its right bank by the Tista River and then follows a 150-mi (240-km) course due south as the Jamuna River. (South of Gaibanda, the Old Brahmaputra leaves the left bank of the main stream and flows past Jamalpur and Mymensingh to join the Meghna River at Bhairab Bazar.) Before its confluence with the Ganges, the Jamuna receives the combined waters of the Baral, Atrai, and Hurasagar Rivers on its right bank and becomes the point of departure of the large Dhaleswari River on its left bank. A tributary of the Dhaleswari, the Buriganga ("Old Ganges"), flows past Dhaka, the capital of Bangladesh, and joins the Meghna River above Munshiganj.

The Jamuna joins with the Ganges north of Goalundo Ghat, below which, as the Padma, their combined waters flow to the southeast for a distance of about 75 mi (120 km). After several smaller channels branch off to feed the Ganges-Brahmaputra delta to the south, the main body of the Padma reaches its confluence with the Meghna River near Chandpur and then enters the Bay of Bengal through the Meghna estuary and lesser channels flowing through the delta. The growth of the Ganges-Brahmaputra Delta is dominated by tidal processes.

The Ganges Delta, fed by the waters of numerous rivers, including the Ganges and Brahmaputra, is 59, 570 square kilometres (23, 000 sq mi) the largest river deltas in the world.

The Ganges-Brahmaputra system has the third-greatest average discharge of the world's rivers—roughly 1, 086, 500 ft3 (30, 770 m3) per second; about 700, 000 ft3 (19, 800 m3) per second of the total is supplied by the Brahmaputra alone. The rivers' combined suspended sediment load of about 1. 84 billion tons per year is the world's highest. In the past, the course of the lower Brahmaputra was different and passed through the Jamalpur and Mymensingh districts. In a 7. 5 magnitude earthquake on 2 April 1762, the main channel of the Brahmaputra at Bhahadurabad point was switched southwards and opened as Jamuna due to the result of tectonic uplift of the Madhupur tract.

Flooding

During the spring season (June–October), floods are a very common occurrence. Deforestation in the Brahmaputra watershed has resulted in increased siltation levels, flash floods, and soil erosion in critical downstream habitat, such as the Kaziranga National Park in middle Assam. Occasionally, massive flooding causes huge losses to crops, life, and property. Periodic flooding is a natural phenomenon which is ecologically important because it helps maintain the lowland grasslands and associated wildlife. Periodic floods also deposit fresh alluvium, replenishing the fertile soil of the Brahmaputra River Valley. Thus flooding, agriculture, and agricultural practices are closely connected. Flood-control measures are taken by water resource department and the Brahmaputra Board, but until now the flood problem remains unsolved. At least a third of the land of Majuli island has been eroded by the river. Recently, it is suggested that a highway protected by concrete mat along the river and excavation of the river bed can curb this menace. This project, named the Brahmaputra River Restoration Project, is yet to be implemented by the government.

Indus River

The principal rivers of the Indus River system are snow-fed. Their flow varies greatly at different times of the year: the discharge is at a minimum during the winter months (December to February), there is a rise of water in spring and early summer (March to June), and floods occur in the rainy season (July to September). Occasionally, there are devastating flash floods. The Indus and its tributaries receive all their waters in the hilly upper parts of their catchments. Therefore, their flow is at a maximum where they emerge out of the foothills, The Indus river feeds the Indus submarine fan, which is the second largest sediment body on the Earth. It consists of around 5 million cubic kilometres of material eroded from the mountains. Studies of the sediment in the modern river indicate that the Karakoram Mountains in northern Pakistan and India are the single most important source of material, with the Himalayas providing the next largest contribution, mostly via the large rivers of the Punjab (Jhelum, Ravi, Chenab, Beas and Sutlej). Analysis of sediments from the Arabian Sea has demonstrated that prior to five million years ago the Indus was not connected to these Punjab rivers which instead flowed east into the Ganga and were captured after that time. Earlier work showed that sand and silt from western Tibet was reaching the Arabian Sea by 45 million years ago, implying the existence of an ancient Indus River by that time. The delta of this proto-Indus river has subsequently been found in the Katawaz Basin, on the Afghan-Pakistan border.

In the Nanga Parbat region, the massive amounts of erosion due to the Indus river following the capture and rerouting through that area is thought to bring middle and lower crustal rocks to the surface. In November 2011, satellite images showed that the Indus river had re-entered India, feeding Great Rann of Kutch, Little Rann of Kutch and a lake near Ahmedabad known as Nal Sarovar. Heavy rains had left the river basin along with the Lake Manchar, Lake Hemal and Kalri Lake (all in modern-day Pakistan) inundated. This happened two centuries after the Indus river shifted its course westwards following the 1819 Rann of Kutch earthquake. The Induan Age at start of the Triassic Period of geological time is named for the Indus region.

Ravi River

The Ravi is a trans-boundary river flowing through Northwestern India and eastern Pakistan. It is one of the six rivers of theIndus System in Punjab region (name of Punjab means "Five Rivers"). The waters of Ravi are allocated to India under Indus Water Treaty.

Under the Indus Waters Treaty of 1960, the waters of the Ravi and five other rivers are divided between India and Pakistan. Subsequently, the Indus Basin Project has been developed in Pakistan and many inter-basin water transfers, irrigation, hydropower and multipurpose projects have been built in India.

Geography

The Ravi River, a trans-boundary river of India and Pakistan, is an integral part of the Indus River Basin and forms the headwaters of the Indus basin. The waters of the Ravi River drain into the Indian Ocean through the Indus River in Pakistan. The river rises in the Bara Bhangal, District Kangra in Himachal Pradesh, India. The river drains a total catchment area of 14, 442 square kilometres (5, 576 sq mi) in India after flowing for a length of 720 kilometres (450 mi). Flowing westward, it is hemmed by the Pir Panjal and Dhauladhar ranges, forming a triangular zone.

Source reach

The Ravi River originates in the Himalayas in the Multhan tehsil of Kangra district of Himachal Pradesh, India. It follows a north-westerly course and is a perennial river. It is the smallest of the five Punjab rivers that rises from glacier fields at an elevation of 14, 000 feet (4, 300 m), on the southern side of the Mid Himalayas. It flows through Barabhangal, Bara Bansu and Chamba districts. It flows in rapids in its initial reaches with boulders seen scattered in the bed of the river. The Ravi River in this reach flows in a gorge with a river bed slope of 183 feet per mile (34. 7 m/km) and is mostly fed by snow melt, as this region lies in a rain shadow. Two of its major tributaries, the Budhil and Nai or Dhona join 64 kilometres (40 mi) downstream from its source. The Budhil River rises inLahul range of hills and is sourced from the Manimahesh Kailash Peak and the Manimahesh Lake, at an elevation of 4, 080 metres (13, 390 ft), and both are Hindu pilgrimage sites. The entire length of Budhil is 72 kilometres (45 mi) where it has a bed slope of 314 feet per mile (59. 5 m/km). It flows through the ancient capital of Bharmwar, now known as Bharmour in Himachal Pradesh. During 1858–1860, the Raja of Bharmour had considered the Budhil valley as an excellent source of Deodar trees for supply to the British Raj. However, a part of the forest surrounding the temple was considered sacred and declared a reserved area. The second tributary, the Nai, rises at Kali Debi pass, and flows for 48 kilometres (30 mi), with a bed slope of 366 feet per mile (69. 3 m/km), from its source at Trilokinath to its confluence with the Ravi. This valley was also exploited for its forest wealth during the British period.

Another major tributary that joins the Ravi River, just below Bharmour, the old capital of Chamba, is the Seul River from the northern direction. The valley formed by the river was also exploited for its rich timber trees. However, the valley has large terraces, which are very fertile and known as "the garden of Chamba". crops grown here supply grains to the capital region and to Dalhousie town and its surrounding areas. One more major tributary that joins the Ravi River near Bissoli is the Siawa. This river was also exploited for its forest resources, (controlled by the then Raja of Chamba) originating from the Jammu region. The valley is also formed by another major tributary that joins Seul River, the Baira-Nalla. Its sub-basin is in the Chamba district, located above Tissa. Baira drains the southern slopes of the Pir Panjal Range. The valley has an elevation variation between 5, 321 metres (17, 457 ft) and 2, 693 metres (8, 835 ft).

Tant Gari is another small tributary that rises from the subsidiary hill ranges of the Pir Panjal Range east of Bharmour. The valley formed by this stream is U-shaped with a river bed scattered with boulders and glacial morainic deposits.

Main Ravi River

The main Ravi River flows through the base of Dalhousie hill, past the Chamba town. It is at an elevation of 856 metres (2, 807 ft) (where a long wooden bridge existed to cross the Ravi River). It flows into the south-west, near Dalhousie, and then cuts a gorge in the Dhauladhar Range, before entering the Punjab plain near Madhopur and Pathankot. It then flows along the Indo–Pak border for 80 kilometres (50 mi) before entering Pakistan and joining the Chenab River. The total length of the river is about 725 kilometres (450 mi).

Ujh River is another major tributary of the Ravi River. Its source is in the Kailash mountains at an elevation of 4, 300 metres (14, 100 ft), close to the Bhaderwah Mountains in Jammu district. After flowing for 100 kilometres (62 mi), it joins Ravi at Nainkot in Pakistan. As the Ravi flows past Lahore in Pakistan (26 kilometres (16 mi) below Amritsar in India) it is called "The river of Lahore" since that city is on its eastern bank. After passing through Lahore the river takes a turn at Kamlia and then debouches into the Chenab River, south of the town of Ahmadpur Sial. On its western bank is the town of Shahdara Bagh with the tomb of Jahangir and the Tomb of Noor Jahan.

Change of river course

According to satellite imagery studies carried out over a period of 20 years (between 1972–1973 and 1991–1993), the river coursing along the India–Pakistan border meanders substantially in the alluvial plains of the Amritsar andGurdaspur districts of Punjab. This has resulted in successive damage in the Indian Territory as a result of the river changing its course towards India. The reason attributed to this change in the course of the river is massive river training structures/bunds constructed by Pakistan in its part of the river, close to the old course of the river. The shift in the course of the river is reported to be 4. 8 kilometres (3. 0 mi) towards India.

River water pollution

In the trans-boundary Ravi River flowing from India to Pakistan, in urban areas of Lahore the pollution levels in the river discharge are reportedly very high, which is attributed to careless disposal of large amount of industrial and agricultural waste water and faulty drainage systems in both countries. A 72-kilometre (45 mi) stretch of the Ravi River from Lahore Siphon to Baloki headworks indicates heavy contamination of the water and sediment with Cd, Cr, Co and Cu. The river sediments are highly contaminated and have become secondary source for pollution of the river water, even though some control over unauthorised discharges into the river have been checked. Hence, measures to check metal re-mobilization from sediments into the river flows needs attention. The worst affected drainage is the Hudiara drain, a tributary of the Ravi River. It is also a trans-border problem involving both India and Pakistan. A UNDP funded a special programme was launched in 2006 to address the issue in both countries.

Sutlej River

The Sutlej River is the longest of the five rivers that flow through the historic crossroads region of Punjab in northern India and Pakistan. It is located north of the Vindhya Range, south of the Hindu Kush segment of the Himalayas, and east of the Central Sulaiman Range in Pakistan. The Sutlej River is also known as Satadree.

The Sutlej is sometimes known as the Red River. It is the easternmost tributary of the Indus River. Its source is Lake Rakshastal in Tibet. From there, under the Tibetan name Langqên Zangbo (Elephant River), it flows at first west-northwest for about 260 kilometres (160 mi) to theShipki La pass, entering India in Himachal Pradesh state. It then turns slightly, heading west-southwest for about 360 kilometres (220 mi) to meet the Beas River near Makhu, Firozpur district, Punjab state.

Continuing west-southwest, the Sutlej enters Pakistan about 15 kilometres (9. 3 mi) east ofBhedian Kalan, Kasur District, Punjab province, continuing southwest to water the ancient and historical former Bahawalpur princely state.

About 17 kilometres (11 mi) north of Uch Sharif, the Sutlej unites with the Chenab River, forming the Panjnad River, which finally flows into the Indus river about 100 kilometres (62 mi) west of the city of Bahawalpur. The area to the southeast on the Pakistani side of the Indian border is called the Cholistan Desert and, on the Indian side, the Thar Desert. The Indus then flows through a gorge near Sukkur and the fertile plains region of Sindh, forming a large delta region between the border of Gujarat, India and Pakistan, finally terminating in theArabian Sea near the port city of Karachi, Pakistan.

Contrary to the claims of Punjab state in India, a small part of Panchkula district in Haryanastate is part of the Sutlej river basin area.

The waters of the Sutlej are allocated to India under the Indus Waters Treaty between India and Pakistan, and are mostly diverted to irrigation canals in India. There are several major hydroelectric projects on the Sutlej, including the 1, 000 MW Bhakra Dam, the 1, 000 MWKarcham Wangtoo Hydroelectric Plant, and the 1, 530 MW Nathpa Jhakri Dam. There has been a proposal to build a 214-kilometre (133 mi) long heavy freight canal, to be known as theSutlej-Yamuna Link (SYL), in India to connect the Sutlej and Yamuna rivers. The project is intended to connect the Ganges, which flows to the east coast of the subcontinent, with points west, via Pakistan. When completed, the SYL would enable inland shipping from India's east coast to its west coast (on the Arabian sea) without having to round the southern tip of India by sea, vastly shortening shipping distances, alleviating pressures on seaports, avoiding sea hazards, creating business opportunities along the route, raising real estate values, raising tax revenue, and establishing important commercial links and providing jobs for north-central India's large population. However, the proposal has met with obstacles and has been referred to the Supreme Court of India.

Geology

The Sutlej, along with all of the Punjab rivers, is thought to have drained east into the Ganges prior to 5 mya.

There is substantial geologic evidence to indicate that prior to 1700 BC, and perhaps much earlier, the Sutlej was an important tributary of theGhaggar-Hakra River (thought to be the legendary Sarasvati River) rather than the Indus, with various authors putting the redirection from 2500 to 2000 BC, from 5000 to 3000 BC, or before 8000 BC. Geologists believe that tectonic activity created elevation changes which redirected the flow of Sutlej from the southeast to the southwest. If the diversion of the river occurred recently (about 4000 years ago), it may have been responsible for the Ghaggar-Hakra (Saraswati) drying up, causing desertification of Cholistan and the eastern part of the modern state of Sindh, and the abandonment of Harappan settlements along the Ghaggar. However, the Sutlej may have already been captured by the Indus thousands of years earlier.

There is some evidence that the high rate of erosion caused by the modern Sutlej River has influenced the local faulting and rapidly exhumed rocks above Rampur.

This would be similar to, but on a much smaller scale than, the exhumation of rocks by the Indus River in Nanga Parbat, Pakistan. The Sutlej River also exposes a doubled inverted metamorphic gradient.

Beas River

The Beas River is a river in north India. The river rises in the Himalayas in central Himachal Pradesh, India, and flows for some 470 kilometres (290 mi) to the Sutlej River in the Indian state of Punjab. Its total length is 470 kilometres (290 mi) and its drainage basin is 20,303 square kilometres (7,839 sq mi) large. As of 2017 the river is home to a tiny isolated population of the Indus dolphin, the only one not located in Pakistan.

Course

The river rises 4,361 metres (14,308 ft) above sea-level on the southern face of Rohtang Pass in Kullu. It traverses the Mandi District and enters the Kangra District at Sandhol, 590 metres (1,940 ft) above sea-level. During its lower course the Beas is crossed by numerous ferries, many of which consist of inflated skins (darais). Near Reh in Kangra District it divides into three channels, which reunite after passing Mirthal, 300 metres (980 ft) above sea-level. On meeting the Sivalik Hills in Hoshiarpur, the river sweeps sharply northward, forming the boundary with Kangra District. Then bending round the base of the Sivalik Hills, it takes the southerly direction, separating the districts of Gurdaspur and Hoshiapur. After touching the Jalandhar district for a short distance, the river forms the boundary between Amritsar and Kapurthala. Finally the Beas joins the river Sutlej at the south-western boundary of Kapurthala district of Punjab after a total course of 470 kilometres (290 mi). The chief tributaries are Bain, Banganga, Luni and Uhal. The Sutlej continues into Pakistani Punjab and joins the Chenab River at Uch near Bahawalpur to form the Panjnad River; the latter in turn joins the Indus River at Mithankot. The water of the Beas river is allocated to India under the terms of the Indus Waters Treaty between India and Pakistan.

Chenab River

The Chenab River is a major river that flows in India and Pakistan, and is one of the 5 major rivers of the Punjab region. It forms in the upper Himalayas in the Lahaul and Spiti district of Himachal Pradesh state, India, and flows through the Kishtwar, Doda, Ramban, Reasi and Jammu districts of Jammu region in Indian state of Jammu and Kashmir into the plains of the Punjab, Pakistan, before flowing into the Indus River near the city of Uch Sharif. The waters of the Chenab were allocated to Pakistan under the terms of the Indus Waters Treaty.

The river is formed by the confluence of two rivers, Chandra and Bhaga, at Tandi, 8 km southwest of Kyelang, in the Lahaul and Spiti district in the Indian state of Himachal Pradesh. The Bhaga river originates from Surya taal lake, which is situated a few kilometers east of the Bara-lacha la pass in the in Himachal Pradesh. The Chandra river originates from glaciers east of the same pass. This pass also acts as a water-divide between these two rivers. The Chandra river transvers 115 km (71 mi) before the confluence. The Bhaga river transverses through narrow gorges a distance of 60 km (37 mi) before the confluence at Tandi.

The river was known to Indians in the Vedic period In 325 BC, Alexander the Great allegedly founded the town of Alexandria on the Indus (present day Uch Sharif or Mithankot or Chacharan in Pakistan) at the confluence of the Indus and the combined stream of Punjab rivers (currently known as the Panjnad River).

Gilgit River

The Gilgit River is a tributary of the Indus River, and flows past the town of Gilgit. It is located in the Gilgit–Baltistan, the northmost political entity of Pakistan. The Gilgit River starts from Shandur Lake, joins the Indus River at junction point of three mountains ranges near town of Juglot. It is also referred to as the Ghizer River.

Gomal River

Gomal River is a 400 km long river in Afghanistan and Pakistan. The river was described in Ramayana, as Gomati River (2.49.11). The river lends its name to the Gomal University in Dera Ismail Khan and the like-named Gomal District in Paktika Province of Afghanistan.

Gomal River's headwaters are located south-east of Ghazni. The springs which form the headwaters of the Gomal's main branch emerge above the fort at Babakarkol in Katawaz, a district inhabited by Ghilji Pashtuns from the Kharoti and Sulaimankhel clans, in Paktika Province. The Gomal's other branch, the "Second Gomal", joins the main channel about 14 miles below its source. The Gomal flows southeast through eastern Ghilji country for approximately 110 miles before it merges with the Zhob River, its major tributary, near Khajuri Kach. It is about 100 miles from the Zhob River to the Indus River.

Within Pakistan, the Gomal river surrounds the South Waziristan agency and forms the boundary between Khyber-Pakhtunkhwa and Balochistan. From South Waziristan, the river enters the Gomal Valley in the district of Tank, Pakistan at a place known as Girdavi, Murtuza which is inhabited by the Miani tribe. It is mainly here that the water of Gomal is used to cultivate the lands in Gomal Valley through Zam System (Rod Kohi). The river passes then through the Damaan plain in Kulachi Tehsil and later on through Dera Ismail Khan Tehsil, It then joins the Indus River 20 miles south of Dera Ismail Khan.

Hunza River

Hunza River is the principal river of Hunza in Gilgit–Baltistan, Pakistan It is formed by the confluence of the Chapursan and Khunjerab nalas (gorges) which are fed by glaciers. It is joined by the Gilgit River and the Naltar River, before it flows into the Indus River.

The river cuts through the Karakoram range, flowing from north to south. The Karakoram Highway (N-35) runs along the Hunza River valley, switching to the Khunjerab River valley at the point of confluence, eventually reaching the Khunjerab Pass at the border with China Xinjiang.

The Attabad landslide disaster in January 2010 completely blocked the Hunza Valley. A new lake — now called the Attabad Lake or Gojal Lake — which extends 30 kilometers and rose to a depth of 400 feet, was formed as the Hunza River backed-up. The landslide completely covered sections of the Karakoram Highway.

Jhelum River

The Jhelum River (/'d?e?l?m/),is a river of northwestern India and northern and eastern Pakistan. It is the westernmost of the five rivers of Punjab, and passes through Srinager District. It is a tributary of the Indus River and has a total length of about 725 kilometres (450 mi).

The river Jhelum is called Vitasta in the Rigveda and Hydaspes by the ancient Greeks. The Vitasta (Sanskrit: ?, fem., also, Vetasta) is mentioned as one of the major rivers by the holy scriptures — the Rigveda. It has been speculated that the Vitasta must have been one of the seven rivers (sapta-sindhu) mentioned so many times in the Rigveda. The name survives in the Kashmiri name for this river as Vyeth. According to the major religious work Srimad Bhagavatam, the Vitasta is one of the many transcendental rivers flowing through land of Bharata, or ancient India.

The river was regarded as a god by the ancient Greeks, as were most mountains and streams; the poet Nonnus in the Dionysiaca (section 26, line 350) makes the Hydaspes a titan-descended god, the son of the sea-god Thaumas and the cloud-goddess Elektra. He was the brother of Iris, the goddess of the rainbow, and half-brother to the Harpies, the snatching winds. Since the river is in a country foreign to the ancient Greeks, it is not clear whether they named the river after the god, or whether the god Hydaspes was named after the river. Alexander the Great and his army crossed the Jhelum in BC 326 at the Battle of the Hydaspes River where he defeated the Indian king, Porus. According to Arrian (Anabasis, 29), he built a city "on the spot whence he started to cross the river Hydaspes", which he named Bukephala (or Bucephala) to honour his famous horse Bukephalus or Bucephalus which was buried in Jalalpur Sharif. It is thought that ancient Bukephala was near the site of modern Jhelum City. According to a historian of Gujrat district, Mansoor

Behzad Butt, Bukephalus was buried in Jalalpur Sharif, but the people of Mandi Bahauddin, a district close to Jehlum, believed that their tehsil Phalia was named after Bucephalus, Alexander's dead horse. They say that the name Phalia was the distortion of the word Bucephala. The waters of the Jhelum are allocated to Pakistan under the terms of the Indus Waters Treaty. India is working on a hydropower project on a tributary of Jhelum river to establish first-use rights on the river water over Pakistan as per the Indus Waters Treaty.

The river Jhelum rises from Verinag Spring situated at the foot of the Pir Panjal in the south-eastern part of the valley of Kashmir. It's joined by its tributaries Lidder River at Mirgund Khannabal and Sind River at Shadipora in Kashmir Valley. It flows through Srinagar and the Wular lake before entering Pakistan through a deep narrow gorge. The Neelum River, the largest tributary of the Jhelum, joins it, at Domel Muzaffarabad, as does the next largest, the Kunhar River of the Kaghan valley. It also connects with rest of Pakistan and Pakistani Kashmir on Kohala Bridge east of Circle Bakote. It is then joined by the Poonch river, and flows into the Mangla Dam reservoir in the district of Mirpur. The Jhelum enters the Punjab in the Jhelum District. From there, it flows through the plains of Pakistan's Punjab, forming the boundary between the Chaj and Sindh Sagar Doabs. It ends in a confluence with the Chenab at Trimmu in District Jhang. The Chenab merges with the Sutlej to form the Panjnad River which joins the Indus River at Mithankot.

Kabul River

The Kabul River, the classical Cophes, is a 700-kilometre (430 mi) long river that emerges in the Sanglakh Range (Maidan Wardak Province) of the Hindu Kush mountains in Afghanistan and empties into the Indus River near Attock, Pakistan. It is the main river in eastern Afghanistan and is separated from the watershed of the Helmand by the Unai Pass. It is one of the most important right tributaries of Indus river.

Kunar River

The Kunar River, also called the Chitral River or the Kama River, is about 480 km long, located in northern Khyber Pakhtunkhwa, Pakistan, and eastern Afghanistan. The river system is fed by melting glaciers and snow of the Hindu Kush mountains. It is also part of the Indus watershed basin.

The river rises in the far north glaciated Hindu Kush mountains of Chitral, Khyber Pakhtunkhwa, Pakistan. Downstream as far as Mastuj it is known as the Mastuj River from there to its confluence with the Lotkoh River just north of the important regional centre of Chitral. It is then called the Chitral River, before flowing south into the upper Kunar Valley of Afghanistan. At the confluence in Asadabad, historically Chaga Sarai, it meets with Pech River and finally empties into the Kabul River just to the east of the city of Jalalabad in Afghanistan. The combined rivers then flow eastwards into Pakistan again, roughly following the Grand Trunk Road through the Khyber Pass, and joining the Indus River at the city of Attock.

Before the political division of the area divided the Kunar and Chitral Valley between the modern nation states of Afghanistan and Pakistan, it had formed an important trade route, being the easiest way to travel from the Pamir Mountains' passes to the plains of the subcontinent. While navigable in parts by expert kayakers, etc..., it is more precise to say its valley forms a trade route since, like nearly all of the rivers in Africa and Asia, it is not navigable for commerce or transport.

"At about six miles from Jellalabad, we quitted the valley of the Cabul river, and entered that of the Kama or Kooner river. I have not seen so fine a valley as this anywhere. The Kama is a large rapid stream, with about a mile of rich soil on either side, sometimes considerably more. The villages are large and well peopled." - Major - General Augustus Abbott, mid 19th century.

Kurram River

Kurram River is located in Paktia and Khost provinces of Afghanistan and Kurram Agency, North Waziristan Agency and Khyber Pakhtunkhwa, Pakistan. It drains the southern flanks of the Spin Ghar mountain range and is a right bank tributary of the Indus River. Its tributaries include the Kirman River and the Khurmana River.

Panjnad River

Panjnad River is a river at the extreme end of Bahawalpur district in Punjab, Pakistan. Panjnad River is formed by successive confluence or merger of the five rivers of the Punjab, namely Jhelum, Chenab, Ravi, Beas and Sutlej. Jhelum and Ravi join Chenab, Beas joins Sutlej, and then Sutlej and Chenab join to form Panjnad 10 miles north of Uch Sharif in Muzaffar Garh district. The combined stream runs southwest for approximately 45 miles and joins the Indus River at Mithankot. The Indus continues and then drains into the Arabian Sea. A barrage on Panjnad has been erected; it provides irrigation channels for Punjab and Sindh provinces south of the Sutlej and east of the Indus rivers.

Beyond the confluence of Indus and Panjnad rivers, the Indus river was known as Satnad (Sat = seven) carrying the waters of seven rivers including Indus river, which is believed to be in earlier times the Saraswati/Ghaggar/Hakra river which eventually dried and became a seasonal river due to seismic shifts in the glacial region of Himachal Pradesh where it originated and later on Kabul river and the five rivers of Punjab.

Shyok River

The Shyok River flows through northern Ladakh in India and the Ghangche District of Gilgit–Baltistan of Pakistan spanning some 550 km (340 mi). The Shyok River, a tributary of the Indus River, originates from the Rimo Glacier, one of the tongues of Siachen Glacier. The river widens at the confluence with the Nubra River. The alignment of the Shyok river is very unusual, originating from the Rimo glacier, it flows in a southeasterly direction and, joining the Pangong range, it takes a northwestern turn, flowing parallel to its previous path. The Shyok flows in a wide valley, suddenly entering a narrow gorge after Chalunka, continuing through Turtuk and Tyakshi before crossing into Pakistan. The Shyok joins the Indus at Keris, to the east of the town of Skardu.

The Nubra river, originating from the Siachen glacier, also behaves like the Shyok. Before Tirkit, the SE flowing river Nubra takes a NW turn on meeting the river Shyok. The similarity in the courses of these two important rivers probably indicates a series of paleo fault lines trending NW-SE in delimiting the upper courses of the rivers. The importance of the Indus and the Shyok rivers is in the deposition of the thick Quaternary sediments—a treasure trove for geology researchers.

Soan River

The Soan River, also referred to as the Swan, Sawan, or Sohan, is a river in Punjab, Pakistan. The Soan River is an important stream of the Pothohar region of Pakistan. It drains much of the water of Pothohar. It starts near the small village of Bun in the foothills of Patriata and Murree. It provides water to Simly Dam, which is the water reservoir for Islamabad. Near Pharwala Fort it cuts through a high mountain range, a wonderful natural phenomenon called Soan Cut. No stream can cut such a high mountain, which proves that the Soan was there before the formation of this range. And as the mountain rose through millions of years, the stream continued its path by cutting the rising mountain. Ling stream, following a relatively long course through Lehtrar and Kahuta falls in the Soan near Sihala on southern side of Village Gagri/Bhandar.

Islamabad Highway crosses this stream near Sihala where the famous Kak Pul bridge is constructed over it. The Ling Stream joins the Soan river just before the Kak Pul. Two other famous streams, the Korang River joins it just before the Soan Bridge and Lai stream, joins this stream after the Soan Bridge. After following a tortuous path along a big curve, the stream reaches Kalabagh proposed Dam Site close to Pirpiyahi where it falls into the Indus river. There is a famous railway station by the name of Sohan and a beautiful railway bridge very close to it. This relatively small stream is more than 250 kilometers long. Due to its mountainous course and shallow bed, it is hardly used for irrigation purposes. There are many species of fish in this river like China Rahu, Mahasher, Snakehead, Balm and Catfish are the main species of fish in this stream and Kingfisher birds hunt here too. In addition, there are many species of tortoises and turtles. The marine life of this river has been endangered due to drainage of chemicals of Sihala Industrial State and effluent water of Rawalpindi city into it. There was a time when peoples used to fish in this stream with nets and hooks and bring home handsome catch but now, not only the above species of fish have diminished but also peoples avoid to eat Soan River's fish as it feeds on contaminated water, which can cause different diseases. Ministry of Environment must take action and treat the effluent water of Rawalpindi city before releasing it into Soan River and also punish and fine those industrialists who are releasing chemicals and other industrial waste into the river and destroying the beautiful marine life of Soan River.

The river might be the Sushoma River of the Rig Veda. According to an important Hindu scripture, the Srimad Bhagavatam, the Sushoma is one of the many transcendental rivers flowing to the north of the land of Bharata.

Lakes

Lakes in India include Pulicat Lake, Kolleru Lake, Pangong Tso, Chilika Lake, Kuttanad Lake, Sambhar Salt Lake, and Pushkar Lake.

Wetlands

India is a signatory of the Ramsar Convention, an international treaty for the conservation and sustainable utilisation of wetlands

Water supply and sanitation

Water supply and sanitation in India continue to be inadequate, despite long-standing efforts by the various levels of government and communities at improving coverage. The level of investment in water and sanitation, albeit low by international standards, has increased during the 2000s. Access has also increased significantly. For example, in 1980 rural sanitation coverage was estimated at 1% and reached 21% in 2008. Also, the share of Indians with access to improved sources of water has increased significantly from 72% in 1990 to 88% in 2008. At the same time, local government institutions in charge of operating and maintaining the infrastructure are seen as weak and lack the financial resources to carry out their functions. In addition, no major city in India is known to have a continuous water supply and an estimated 72% of Indians still lack access to improved sanitation facilities.

In spite of adequate average rainfall in India, there is large area under the less water conditions/drought prone. There are lot of places, where the quality of groundwater is not good. Another issue lies in interstate distribution of rivers. Water supply of the 90% of India's territory is served by inter-state rivers. It has created growing number of conflicts across the states and to the whole country on water sharing issues.

A number of innovative approaches to improve water supply and sanitation have been tested in India, in particular in the early 2000s. These include demand-driven approaches in rural water supply since 1999, community-led total sanitation, a public-private partnerships to improve the continuity of urban water supply in Karnataka, and the use of micro-credit to women in order to improve access to water.

Water quality issues

When sufficient salt export is not taking place from a river basin to the sea in an attempt to harness the river water fully, it leads to river basin closer and the available water in downstream area of the river basin becomes saline and/ or alkaline water. Land irrigated with saline or alkaline water becomes gradually in to saline or alkali soils. The water percolation in alkali soils is very poor leading to waterlogging problems. Proliferation of alkali soils would compel the farmers to cultivate rice or grasses only as the soil productivity is poor with other crops and tree plantations. Cotton is the preferred crop in saline soils compared to many other crops as their yield is poor. In north eastern states high acidic nature of soils due to excessive rainfall is effecting the agriculture productivity. Interlinking water surplus rivers with water deficit rivers is needed for the long term sustainable productivity of the river basins and for mitigating the anthropogenic influences on the rivers by allowing adequate salt export to the sea in the form of environmental flows. Also baseflows in rivers are to be restored by stopping excessive ground water use and augmenting surface water by canals to achieve adequate salt export to the sea and preserve the water quality.

Water disputes

There is intense competition for the water available in the inter state rivers such as Kavery, Krishna, Godavari, Vamsadhara, Mandovi, Ravi-Beas-Sutlez, Narmada, Tapti, Mahanadi, etc. among the riparian states of India in the absence of water augmentation from the water surplus rivers such as Brahmaputra, Himalayan tributaries of Ganga and west flowing coastal rivers of western ghats.

Water pollution

Out of India's 3,119 towns and cities, just 209 have partial treatment facilities, and only 8 have full wastewater treatment facilities (WHO 1992)[needs update]. 114 cities dump untreated sewage and partially cremated bodies directly into the Ganges River. Downstream, the untreated water is used for drinking, bathing, and washing. This situation is typical of many rivers in India and river Ganga is less polluted comparatively.

Ganga

The Ganges River is the largest river in India. The extreme pollution of the Ganges affects 600 million people who live close to the river. The river water starts getting polluted when it enters the plain. The commercial exploitation of the river has risen in proportion to the rise of population. Gangotri and Uttarkashi are good examples too. Gangotri had only a few huts of Sadhus until the 1970s and the population of Uttrakashi has swelled in recent years.

Yamuna

Yamuna is one of the few sacred rivers in India which is worshipped by many Indians as a goddess. However, due to the exponentially rising amounts of tourists and pilgrims with addition to the rising population of the inhabitants of its banks, Yamuna has come under extreme duress. Due to this unprecedented rise, the river has become polluted. The river has become extremely polluted such that the Indian government has launched the Yamuna Action Plan to help the cause.

4

WATER POLLUTION

Water pollution is the contamination of water bodies (e.g. lakes, rivers, oceans, aquifers and groundwater). This form of environ-mental degradation occurs when pollutants are directly or indirectly discharged into water bodies without adequate treatment to remove harmful compounds. Water pollution affects the entire biosphere–plants and organisms living in these bodies of water. In almost all cases the effect is damaging not only to individual species and population, but also to the natural biological communities. Water pollution is a major global problem which requires ongoing evaluation and revision of water resource policy at all levels (international down to individual aquifers and wells). It has been suggested that water pollution is the leading worldwide cause of deaths and diseases, and that it accounts for the deaths of more than 14,000 people daily. An estimated 580 people in India die of water pollution related illness every day. Around 90% the water in the cities of China is polluted, and as of 2007, half a billion Chinese had no access to safe drinking water. In addition to the acute problems of water pollution in developing countries, developed countries also continue to struggle with pollution problems.

For example, in the most recent national report on water quality in the United States, 45 percent of assessed stream miles, 47% of assessed lake acres, and 32 percent of assessed bays and estuarine square miles were classified as polluted. The head of China's national development agency said in 2007 that one quarter the length of China's seven rivers were so poisoned the water harmed the skin. Water is typically referred to as polluted when it is impaired by anthropogenic contaminants and either does not support a human use, such as drinking water, or undergoes a marked shift in its ability to support its constituent biotic communities, such as fish. Natural phenomena such as volcanoes, algae blooms, storms, and earthquakes also cause major changes in water quality and the ecological status of water.

Categories

Although interrelated, surface water and groundwater have often been studied and managed as separate resources. Surface water seeps through the soil and becomes groundwater. Conversely, groundwater can also feed surface water sources. Sources of surface water pollution are generally grouped into two categories based on their origin.

Point sources

Point source water pollution refers to contaminants that enter a waterway from a single, identifiable source, such as a pipe or ditch. Examples of sources in this category include discharges from a sewage treatment plant, a factory, or a city storm drain. The U.S. Clean Water Act (CWA) defines point source for regulatory enforcement purposes. The CWA definition of point source was amended in 1987 to include municipal storm sewer systems, as well as industrial storm water, such as from construction sites.

Nonpoint sources

Nonpoint source pollution refers to diffuse contamination that does not originate from a single discrete source. NPS pollution is often the cumulative effect of small amounts of contaminants gathered from a large area. A common example is the leaching out of nitrogen compounds from fertilized agricultural lands. Nutrient runoff in storm water from "sheet flow" over an agricultural field or a forest are also cited as examples of NPS pollution. Contaminated storm water washed off of parking lots, roads and highways, called urban runoff, is sometimes included under the category of NPS pollution. However, because this runoff is typically channeled into storm drain systems and discharged through pipes to local surface waters, it becomes a point source.

Ground water pollution

Groundwater is the water located beneath Earth's surface in soil pore spaces and in the fractures of rock formations. A unit of rock or an unconsolidated deposit is called an aquifer when it can yield a usable quantity of water. The depth at which soil pore spaces or fractures and voids in rock become completely saturated with water is called the water table. Groundwater is recharged from, and eventually flows to, the surface naturally; natural discharge often occurs at springs and seeps, and can form oases or wetlands. Groundwater is also often withdrawn for agricultural, municipal, and industrial use by constructing and operating extraction wells. The study of the distribution and movement of groundwater is hydrogeology, also called groundwater hydrology. Typically, groundwater is thought of as liquid water flowing through shallow aquifers, but, in the technical sense, it can also include soil moisture, permafrost (frozen soil), immobile water in very low permeability bedrock, and deep geothermal or oil formation water. Groundwater is hypothesized to provide lubrication that can possibly influence the movement of faults. It is likely that much of Earth's subsurface contains some water, which may be mixed with other fluids in some instances. Groundwater may not be confined only to Earth. The formation of some of the landforms observed on Mars may have been influenced by groundwater. There is also evidence that liquid water may also exist in the subsurface of Jupiter's moon Europa.

An aquifer is a layer of porous substrate that contains and transmits groundwater. When water can flow directly between the surface and the saturated zone of an aquifer, the aquifer is unconfined. The deeper parts of unconfined aquifers are usually more saturated since gravity causes water to flow downward.

The upper level of this saturated layer of an unconfined aquifer is called the water table or phreatic surface. Below the water table, where in general all pore spaces are saturated with water, is the phreatic zone.

Substrate with low porosity that permits limited transmission of groundwater is known as an aquitard. An aquiclude is a substrate with porosity that is so low it is virtually impermeable to groundwater.

A confined aquifer is an aquifer that is overlain by a relatively impermeable layer of rock or substrate such as an aquiclude or aquitard. If a confined aquifer follows a downward grade from its recharge zone, groundwater can become pressurized as it flows. This can create artesian wells that flow freely without the need of a pump and rise to a higher elevation than the static water table at the above, unconfined, aquifer.

The characteristics of aquifers vary with the geology and structure of the substrate and topography in which they occur. In general, the more productive aquifers occur in sedimentary geologic formations. By comparison, weathered and fractured crystalline rocks yield smaller quantities of groundwater in many environ-ments. Unconsolidated to poorly cemented alluvial materials that have accumulated as valley-filling sediments in major river valleys and geologically subsiding structural basins are included among the most productive sources of groundwater.

The high specific heat capacity of water and the insulating effect of soil and rock can mitigate the effects of climate and maintain groundwater at a relatively steady temperature. In some places where groundwater temperatures are maintained by this effect at about 10 °C (50 °F), groundwater can be used for controlling the temperature inside structures at the surface. For example, during hot weather relatively cool groundwater can be pumped through radiators in a home and then returned to the ground in another well. During cold seasons, because it is relatively warm, the water can be used in the same way as a source of heat for heat pumps that is much more efficient than using air.

Groundwater is often cheaper, more convenient and less vulnerable to pollution than surface water. Therefore, it is commonly used for public water supplies. Groundwater provides the largest source of usable water storage in the United States. Underground reservoirs contain far more water than the capacity of all surface reservoirs and lakes, including the Great Lakes. Many municipal water supplies are derived solely from groundwater.

The volume of groundwater in an aquifer can be estimated by measuring water levels in local wells and by examining geologic records from well-drilling to determine the extent, depth and thickness of water-bearing sediments and rocks. Before an investment is made in production wells, test wells may be drilled to measure the depths at which water is encountered and collect samples of soils, rock and water for laboratory analyses. Pumping tests can be performed in test wells to determine flow characteristics of the aquifer.

Polluted ground water is less visible, but more difficult to clean up, than pollution in rivers and lakes. Ground water pollution most often results from improper disposal of wastes on land. Major sources include industrial and household chemicals and garbage landfills, industrial waste lagoons, tailings and process wastewater from mines, oil field brine pits, leaking underground oil storage tanks and pipelines, sewage sludge and septic systems. Polluted groundwater is mapped by sampling soils and groundwater near suspected or known sources of pollution, to determine the extent of the pollution, and to aid in the design of groundwater remediation systems. Preventing groundwater pollution near potential sources such as landfills requires lining the bottom of a landfill with watertight materials, collecting any leachate with drains, and keeping rainwater off any potential contaminants, along with regular monitoring of nearby groundwater to verify that contaminants have not leaked into the groundwater.

The danger of pollution of municipal supplies is minimized by locating wells in areas of deep ground water and impermeable soils, and careful testing and monitoring of the aquifer and nearby potential pollution sources.

Issues

Overview

Certain problems have beset the use of groundwater around the world. Just as river waters have been over-used and polluted in many parts of the world, so too have aquifers. The big difference is that aquifers are out of sight. The other major problem is that water management agencies, when calculating the "sustainable yield" of aquifer and river water, have often counted the same water twice, once in the aquifer, and once in its connected river. This problem, although understood for centuries, has persisted, partly through inertia within government agencies. In Australia, for example, prior to the statutory reforms initiated by the Council of Australian Governments water reform framework in the 1990s, many Australian states managed groundwater and surface water through separate government agencies, an approach beset by rivalry and poor communication.

In general, the time lags inherent in the dynamic response of groundwater to development have been ignored by water management agencies, decades after scientific understanding of the issue was consolidated. In brief, the effects of groundwater overdraft (although undeniably real) may take decades or centuries to manifest themselves. In a classic study in 1982, Bredehoeft and colleagues modeled a situation where groundwater extraction in an intermontane basin withdrew the entire annual recharge, leaving 'nothing' for the natural groundwater-dependent vegetation community. Even when the borefield was situated close to the vegetation, 30% of the original vegetation demand could still be met by the lag inherent in the system after 100 years. By year 500, this had reduced to 0%, signalling complete death of the groundwater-dependent vegetation. The science has been available to make these calculations for decades; however, in general water management agencies have ignored effects that will appear outside the rough timeframe of political elections (3 to 5 years). Marios Sophocleous argued strongly that management agencies must define and use appropriate timeframes in groundwater planning. This will mean calculating groundwater withdrawal permits based on predicted effects decades, sometimes centuries in the future.

As water moves through the landscape, it collects soluble salts, mainly sodium chloride. Where such water enters the atmosphere through evapotranspiration, these salts are left behind. In irrigation districts, poor drainage of soils and surface aquifers can result in water tables' coming to the surface in low-lying areas. Major land degradation problems of soil salinity and waterlogging result, combined with increasing levels of salt in surface waters. As a consequence, major damage has occurred to local economies and environments.

Four important effects are worthy of brief mention. First, flood mitigation schemes, intended to protect infrastructure built on floodplains, have had the unintended consequence of reducing aquifer recharge associated with natural flooding. Second, prolonged depletion of groundwater in extensive aquifers can result in land subsidence, with associated infrastructure damage – as well as, third, saline intrusion. Fourth, draining acid sulphate soils, often found in low-lying coastal plains, can result in acidification and pollution of formerly freshwater and estuarine streams.

Another cause for concern is that groundwater drawdown from over-allocated aquifers has the potential to cause severe damage to both terrestrial and aquatic ecosystems – in some cases very conspicuously but in others quite imperceptibly because of the extended period over which the damage occurs.

Overdrafting

Groundwater is a highly useful and often abundant resource. However, over-use, or overdraft, can cause major problems to human users and to the environment. The most evident problem (as far as human groundwater use is concerned) is a lowering of the water table beyond the reach of existing wells. As a consequence, wells must be drilled deeper to reach the groundwater; in some places (e.g., California, Texas, and India) the water table has dropped hundreds of feet because of extensive well pumping. In the Punjab region of India, for example, groundwater levels have dropped 10 meters since 1979, and the rate of depletion is accelerating. A lowered water table may, in turn, cause other problems such as groundwater-related subsidence and saltwater intrusion.

Groundwater is also ecologically important. The importance of groundwater to ecosystems is often overlooked, even by freshwater biologists and ecologists. Groundwaters sustain rivers, wetlands, and lakes, as well as subterranean ecosystems within karst or alluvial aquifers.

Not all ecosystems need groundwater, of course. Some terrestrial ecosystems – for example, those of the open deserts and similar arid environments – exist on irregular rainfall and the moisture it delivers to the soil, supplemented by moisture in the air. While there are other terrestrial ecosystems in more hospitable environments where groundwater plays no central role, ground water is in fact fundamental to many of the world's major ecosystems. Water flows between groundwaters and surface waters. Most rivers, lakes, and wetlands are fed by, and (at other places or times) feed groundwater, to varying degrees. Groundwater feeds soil moisture through percolation, and many terrestrial vegetation communities depend directly on either groundwater or the percolated soil moisture above the aquifer for at least part of each year. Hyporheic zones (the mixing zone of streamwater and groundwater) and riparian zones are examples of ecotones largely or totally dependent on groundwater.

Overdrafting is the process of extracting groundwater beyond the safe yield or equilibrium yield of the aquifer.

Since every groundwater basin recharges at a different rate depending upon precipitation, vegetative cover and soil conservation practises, the quantity of groundwater that can be safely pumped varies greatly among regions of the world and even within provinces. Some aquifers require a very long time to recharge and thus the process of overdrafting can have consequences of effectively drying up certain sub-surface water supplies. Subsidence occurs when excessive groundwater is extracted from rocks that support more weight when saturated. This can lead to a capacity reduction in the aquifer.

Effects on climate

Aquifer drawdown or overdrafting and the pumping of fossil water may be a contributing factor to sea-level rise. By increasing the amount of moisture available to fall as precipitation, severe weather events are more likely to occur. To some extent moisture in the atmosphere accelerates the probability of a global warming event. The correlation coefficient is not yet scientifically determined.

Accelerated decline in subterranean reservoirs

According to a 2013 report by research hydrologist, Leonard F. Konikow, at the United States Geological Survey (USGC), the depletion of the Ogallala Aquifer between 2001–2008, inclusive, is about 32 percent of the cumulative depletion during the entire 20th century (Konikow 2013:22)." In the United States, the biggest users of water from aquifers include agricultural irrigation and oil and coal extraction."Cumulative total groundwater depletion in the United States accelerated in the late 1940s and continued at an almost steady linear rate through the end of the century. In addition to widely recognized environmental consequences, groundwater depletion also adversely impacts the long-term sustainability of groundwater supplies to help meet the Nation's water needs."

According to another USGS study of withdrawals from 66 major US aquifers, the three greatest uses of water extracted from aquifers were agriculture (68%), public water supply (19%), and self-supplied industrial (4%). The remaining about 8% of groundwater withdrawals were for "self-supplied domestic, aquaculture, livestock, mining, and thermoelectric power uses."

Socio-economic effects

Scores of countries are overpumping aquifers as they struggle to satisfy their growing water needs, including each of the big three grain producers—China, India, and the United States. These three, along with a number of other countries where water tables are falling, are home to more than half the world's people.

Water is intrinsic to growth, and overdraft limits its available supply. According to Liebig's law of the minimum, growth is therefore impeded. Deeper wells must be drilled as the water table drops, which can become expensive. In addition, the energy needed to extract a given volume of water increases with the amount the aquifer has been depleted. Saltwater intrusion is another consequence of overdrafting, leading to a reduction in water quality.Subsidence

Subsidence occurs when too much water is pumped out from underground, deflating the space below the above-surface, and thus causing the ground to collapse. The result can look like craters on plots of land. This occurs because, in its natural equilibrium state, the hydraulic pressure of groundwater in the pore spaces of the aquifer and the aquitard supports some of the weight of the overlying sediments. When groundwater is removed from aquifers by excessive pumping, pore pressures in the aquifer drop and compression of the aquifer may occur. This compression may be partially recoverable if pressures rebound, but much of it is not. When the aquifer gets compressed, it may cause land subsidence, a drop in the ground surface. The city of New Orleans, Louisiana is actually below sea level today, and its subsidence is partly caused by removal of groundwater from the various aquifer/aquitard systems beneath it. In the first half of the 20th century, the city of San Jose, California dropped 13 feet from land subsidence caused by overpumping; this subsidence has been halted with improved groundwater management.

Seawater intrusion

In general, in very humid or undeveloped regions, the shape of the water table mimics the slope of the surface. The recharge zone of an aquifer near the seacoast is likely to be inland, often at considerable distance. In these coastal areas, a lowered water table may induce sea water to reverse the flow toward the land. Sea water moving inland is called a saltwater intrusion. In alternative fashion, salt from mineral beds may leach into the groundwater of its own accord.

Mechanism

Water pollution of groundwater, from pollutants released to the ground that can work their way down into groundwater, can create a contaminant plume within an aquifer. Movement of water and dispersion within the aquifer spreads the pollutant over a wider area, its advancing boundary often called a plume edge, which can then intersect with groundwater wells or daylight into surface water such as seeps and springs, making the water supplies unsafe for humans and wildlife. Different mechanism have influence on the transport of pollutants, e.g. diffusion, adsorption, precipitation, decay, in the groundwater. The interaction of groundwater contamination with surface waters is analyzed by use of hydrology transport models.

The stratigraphy of the area plays an important role in the transport of these pollutants. An area can have layers of sandy soil, fractured bedrock, clay, or hardpan. Areas of karst topography on limestone bedrock are sometimes vulnerable to surface pollution from groundwater. Earthquake faults can also be entry routes for downward contaminant entry. Water table conditions are of great importance for drinking water supplies, agricultural irrigation, waste disposal (including nuclear waste), wildlife habitat, and other ecological issues.

Pollution from landfills

Love Canal was one of the most widely known examples of groundwater pollution. In 1978, residents of the Love Canal neighborhood in upstate New York noticed high rates of cancer and an alarming number of birth defects. This was eventually traced to organic solvents and dioxins from an industrial landfill that the neighborhood had been built over and around, which had then infiltrated into the water supply and evaporated in basements to further contaminate the air. Eight hundred families were reimbursed for their homes and moved, after extensive legal battles and media coverage.

Pollution with arsenic

Another example of widespread groundwater pollution is in the Ganges Plain of northern India and Bangladesh where severe contamination of groundwater by naturally occurring arsenic affects 25% of water wells in the shallower of two regional aquifers. The pollution occurs because aquifer sediments contain organic matter that generates anaerobic conditions in the aquifer. These conditions result in the microbial dissolution of iron oxides in the sediment and, thus, the release of the arsenic, normally strongly bound to iron oxides, into the water. As a consequence, arsenic-rich groundwater is often iron-rich, although secondary processes often obscure the association of dissolved arsenic and dissolved iron.

Pollution from on-site sanitation systems

Groundwater pollution with pathogens and nitrate can also occur from the liquids infiltrating into the ground from on-site sanitation systems such as pit latrines and septic tanks, depending on the population density and the hydrogeological conditions. As liquid leaches from the pit it first passes the unsaturated soil zone (which is not completely filled with water), then it enters the groundwater and is transported with the flow, all too often towards a nearby supply well which pumps groundwater. During this passage, pathogens can die off or be adsorbed significantly, mostly depending on the travel time between the pit and the well. Most, but not all pathogens die within 50 days of travel through the subsurface. The degree of pathogen removal strongly varies with soil type, aquifer type, distance and other environmental factors. A well-researched example of this is the strongly karstified system in areas of Lusaka, Zambia where groundwater pollution from pit latrines is a major public health threat.

Pollution from other point sources

Groundwater pollution can also occur from leaking sewers which has been observed for example in Germany.

Government regulations

In the United States, laws regarding ownership and use of groundwater are generally state laws; however, regulation of groundwater to minimize pollution of groundwater is by both states and the federal-level Environmental Protection Agency. Ownership and use rights to groundwater typically follow one of three main systems:

Rule of Capture

The Rule of Capture provides each landowner the ability to capture as much groundwater as they can put to a beneficial use, but they are not guaranteed any set amount of water. As a result, well-owners are not liable to other landowners for taking water from beneath their land. State laws or regulations will often define "beneficial use", and sometimes place other limits, such as disallowing groundwater extraction which causes subsidence on neighboring property.

Riparian Rights

Limited private ownership rights similar to riparian rights in a surface stream. The amount of groundwater right is based on the size of the surface area where each landowner gets a corresponding amount of the available water. Once adjudicated, the maximum amount of the water right is set, but the right can be decreased if the total amount of available water decreases as is likely during a drought. Landowners may sue others for encroaching upon their groundwater rights, and water pumped for use on the overlying land takes preference over water pumped for use off the land.

Reasonable Use Rule (American Rule)

This rule does not guarantee the landowner a set amount of water, but allows unlimited extraction as long as the result does not unreasonably damage other wells or the aquifer system. Usually this rule gives great weight to historical uses and prevents new uses that interfere with the prior use.

Environmental protection of groundwater

In November 2006, the Environmental Protection Agency published the Ground Water Rule in the United States Federal Register. The EPA was worried that the ground water system would be vulnerable to contamination from fecal matter. The point of the rule was to keep microbial pathogens out of public water sources. The 2006 Ground Water Rule was an amendment of the 1996 Safe Drinking Water Act.

Groundwater scrutiny upon real estate property transactions in the US

In the US, upon commercial real estate property transactions both groundwater and soil are the subjects of scrutiny, with a Phase I Environmental Site Assessment normally being prepared to investigate and disclose potential pollution issues. In the San Fernando Valley of California, real estate contracts for property transfer below the Santa Susana Field Laboratory (SSFL) and eastward have clauses releasing the seller from liability for groundwater contamination consequences from existing or future pollution of the Valley Aquifer.

Interactions between groundwater and surface water are complex. Consequently, groundwater pollution, sometimes referred to as groundwater contamination, is not as easily classified as surface water pollution. By its very nature, groundwater aquifers are susceptible to contamination from sources that may not directly affect surface water bodies, and the distinction of point vs. non-point source may be irrelevant. A spill or ongoing release of chemical or radionuclide contaminants into soil (located away from a surface water body) may not create point or non-point source pollution but can contaminate the aquifer below, creating a toxic plume. The movement of the plume, called a plume front, may be analyzed through a hydrological transport model or groundwater model. Analysis of groundwater contamination may focus on soil characteristics and site geology, hydrogeology, hydrology, and the nature of the contaminants.

Causes

The specific contaminants leading to pollution in water include a wide spectrum of chemicals, pathogens, and physical changes such as elevated temperature and discoloration. While many of the chemicals and substances that are regulated may be naturally occurring (calcium, sodium, iron, manganese, etc.) the concentration is often the key in determining what is a natural component of water and what is a contaminant. High concentrations of naturally occurring substances can have negative impacts on aquatic flora and fauna.

Oxygen-depleting substances may be natural materials such as plant matter (e.g. leaves and grass) as well as man-made chemicals. Other natural and anthropogenic substances may cause turbidity (cloudiness) which blocks light and disrupts plant growth, and clogs the gills of some fish species.

Many of the chemical substances are toxic. Pathogens can produce waterborne diseases in either human or animal hosts. Alteration of water's physical chemistry includes acidity (change in pH), electrical conductivity, temperature, and eutrophication. Eutrophication is an increase in the concentration of chemical nutrients in an ecosystem to an extent that increases in the primary productivity of the ecosystem. Depending on the degree of eutrophication, subsequent negative environmental effects such as anoxia (oxygen depletion) and severe reductions in water quality may occur, affecting fish and other animal populations

NONPOINT SOURCE POLLUTION

Muddy river

Nonpoint source (NPS) pollution refers to both water and air pollution from diffuse sources. Nonpoint source water pollution affects a water body from sources such as polluted runoff from agricultural areas draining into a river, or wind-borne debris blowing out to sea. Nonpoint source air pollution affects air quality from sources such as smokestacks or car tailpipes. Although these pollutants have originated from a point source, the long-range transport ability and multiple sources of the pollutant make it a nonpoint source of pollution. Nonpoint source pollution can be contrasted with point source pollution, where discharges occur to a body of water or into the atmosphere at a single location.

NPS may derive from many different sources with no specific solution may change to rectify the problem, making it difficult to regulate. Non point source water pollution is difficult to control because it comes from the everyday activities of many different people, such as fertilizing a lawn, using a pesticide, or constructing a road or building. It is the leading cause of water pollution in the United States today, with polluted runoff from agriculture the primary cause.

Other significant sources of runoff include hydrological and habitat modification, and silviculture (forestry).

Contaminated stormwater washed off parking lots, roads and highways, and lawns (often containing fertilizers and pesticides) is called urban runoff. This runoff is often classified as a type of NPS pollution. Some people may also consider it a point source because many times it is channeled into municipal storm drain systems and discharged through pipes to nearby surface waters. However, not all urban runoff flows through storm drain systems before entering water bodies. Some may flow directly into water bodies, especially in developing and suburban areas. Also, unlike other types of point sources, such as industrial discharges, sewage treatment plants and other operations, pollution in urban runoff cannot be attributed to one activity or even group of activities. Therefore, because it is not caused by an easily identified and regulated activity, urban runoff pollution sources are also often treated as true nonpoint sources as municipalities work to abate them.

Principal types

Sediment

Sediment (loose soil) includes silt (fine particles) and suspended solids (larger particles). Sediment may enter surface waters from eroding stream banks, and fromsurface runoff due to improper plant cover on urban and rural land Sediment creates turbidity (cloudiness) in water bodies, reducing the amount of light reaching lower depths, which can inhibit growth of submerged aquatic plants and consequently affect species which are dependent on them, such as fish andshellfish. High turbidity levels also inhibit drinking water purification systems.

Sediment can also be discharged from multiple different sources. Sources includeconstruction sites (although these are point sources, which can be managed witherosion controls and sediment controls), agricultural fields, stream banks, and highly disturbed area

Nutrients

Nutrients mainly refers to inorganic matter from runoff, landfills, livestock operations and crop lands. The two primary nutrients of concern are phosphorus and nitrogen.

Phosphorus is a nutrient that occurs in many forms that are bioavailable. It is notoriously over-abundant in human sewag-esludge. It is a main ingredient in many fertilizers used for agriculture as well as on residential and commercial properties, and may become a limiting nutrient in freshwater systems and some estuaries. Phosphorus is most often transported to water bodies via soil erosion because many forms of phosphorus tend to be adsorbed on to soil particles. Excess amounts of phosphorus in aquatic systems (particularly freshwater lakes, reservoirs, and ponds) leads to proliferation of microscopic algae called phytoplankton. The increase of organic matter supply due to the excessive growth of the phytoplankton is called eutrophication. A common symptom of eutrophication is algae blooms that can produce unsightly surface scums, shade out beneficial types of plants, and poison the water due to toxins produced by the algae. These toxins are a particular problem in systems used for drinking water because some toxins can cause human illness and removal of the toxins is difficult and expensive. Bacterial decomposition of algal blooms consumes dissolved oxygen in the water, generating hypoxia with detrimental consequences for fish and aquatic invertebrates.

Nitrogen is the other key ingredient in fertilizers, and it generally becomes a pollutant in saltwater or brackish estuarine systems where nitrogen is a limiting nutrient. Similar to phosphorus in freshwaters, excess amounts of bioavailable nitrogen in marine systems lead to eutrophication and algae blooms. Hypoxia is an increasingly common result of eutrophication in marine systems and can impact large areas of estuaries, bays, and near shore coastal waters. Each summer, hypoxic conditions form in bottom waters where the Mississippi River enters the Gulf of Mexico. During recent summers, the aereal extent of this "dead zone" is comparable to the area of New Jersey and has major detrimental consequences for fisheries in the region. Nitrogen is most often transported by water as nitrate (NO3). The nitrogen is usually added to a watershed as organic-N oram-monia (NH3), so nitrogen stays attached to the soil until oxidation converts it into nitrate. Since the nitrate is generally already incorporated into the soil, the water traveling through the soil (i.e., interflow and tile drainage) is the most likely to transport it, rather than surface runoff.

Toxic contaminants and chemicals

Compounds including heavy metals like lead, mercury, zinc, and cadmium, organics like polychlorinated biphenyls (PCBs) and polycyclic aromatic hydrocarbons (PAHs), fire retardants, and other substances are resistant to breakdown. These contaminants can come from a variety of sources including human sewage sludge, mining operations, vehicle emissions,fossil fuel combustion, urban runoff, industrial operations and landfills.

Toxic chemicals mainly include organic compounds and inorganic compounds. These compounds include pesticides likeDDT, acids, and salts that have severe effects to the ecosystem and waterbodies. These compounds can threaten the health of both humans and aquatic species while being resistant to environmental breakdown, thus allowing them to persist in the environment. These toxic chemicals could come from croplands, nurseries, orchards, building sites, gardens, lawns and landfills.

Acids and salts mainly are inorganic pollutants from irrigated lands, mining operations, urban runoff, industrial sites and landfills.

Pathogens

Pathogens are bacteria and viruses that can be found in water and cause diseases in humans. Typically, pathogens cause disease when they are present in public drinking water supplies. Pathogens found in contaminated runoff may include:

- Cryptosporidium parvum
- Giardia lamblia
- Salmonella
- Novovirus and other viruses
- Parasitic worms (helminths).

Coliform bacteria and fecal matter may also be detected in runoff. These bacteria are a commonly used indicator of water pollution, but not an actual cause of disease.

Pathogens may contaminate runoff due to poorly managed livestock operations, faulty septic systems, improper handling of pet waste, the over application of human sewage sludge, contaminated storm sewers, and sanitary sewer overflows.

Principal sources

Urban and suburban areas

Urban and suburban areas are a main sources of nonpoint source pollution due to the amount of runoff that is produced due to the large amount of paved surfaces. Paved surfaces, such as asphalt and concrete are impervious to water penetrating them. Any water that is on contact with these surfaces will run off and be absorbed by the surrounding environment. These surfaces make it easier for stormwater to carry pollutants into the surrounding soil. Construction sites tend to have disturbed soil that is easily eroded by precipitation like rain, snow, and hail. Additionally, discarded debris on the site can be carried away by runoff waters and enter the aquatic environment. Typically, in suburban areas, chemicals are used for lawn care. These chemicals can end up in runoff and enter the surrounding environment via storm drains in the city. Since the water in storm drains is not treated before flowing into surrounding water bodies, the chemicals enter the water directly.

Agricultural operations

Agricultural operations account for a large percentage of all nonpoint source pollution in the United States. When large tracts of land are plowed to grow crops, it exposes and loosens soil that was once buried. This makes the exposed soil more vulnerable to erosion during rainstorms. It also can increase the amount of fertilizer and pesticides carried into nearby bodies of water.

Atmospheric inputs

Atmospheric inputs of pollution into the air can come from multiple sources. Typically, industrial facilities, like factories, emit air pollution via a smokestack. Although this is a point source, due to the distributional nature, long-range transport, and multiple sources of the pollution, it is considered a nonpoint source. Additionally, atmospheric pollution can become water pollution, by being washed out of the atmosphere in the form of rain or snow.

Highway Runoff

Highway runoff accounts for a small but widespread percen-tage of all nonpoint source pollution. Harned (1988) estimated that runoff loads were composed of atmospheric fallout (9%), vehicle deposition (25%) and highway maintenance materials (67%) he also estimated that about 9 percent of these loads were reentrained in the atmosphere.

Forestry and mining operations

Forestry and mining operations can have significant inputs to nonpoint source pollution.

Forestry

Forestry operations reduce the number of trees in a given area, thus reducing the soil stability in that area as well. This action, coupled with the heavy machinery rolling over the soil increases the risk of erosion.

Mining

Active mining operations are considered point sources, however runoff from abandoned mining operations contribute to nonpoint source pollution. In strip mining operations, the top of the mountain is removed to expose the desired ore. If this area is not properly reclaimed once the mining has finished, soil erosion can occur. Additionally, there can be chemical reactions with the air and newly exposed rock to create acidic runoff. Water that seeps out of abandoned subsurface minescan also be highly acidic. This can seep into the nearest body of water and change the pH in the aquatic environment.

MARINAS AND BOATING ACTIVITIES

Chemicals used for boat maintenance, like paint, solvents, and oils find their way into water through runoff. Additionally, spilling fuels or leaking fuels directly into the water from boats contribute to nonpoint source pollution. Nutrient and bacteria levels are increased by poorly maintained sanitary waste receptacles on the boat and pump-out stations.

Control

Urban and suburban areas

To control nonpoint source pollution, many different approaches can be undertaken in both urban and suburban areas. Buffer strips provide a barrier of grass in between impervious paving material like parking lots and roads, and the closest body of water. This allows the soil to absorb any pollution before it enters the local aquatic system. Retention ponds can be built in drainage areas to create an aquatic buffer between runoff pollution and the aquatic environment. Runoff and storm water drain into the retention pond allowing for the contaminates to settle out and become trapped in the pond. The use of porous pavement allows for rain and storm water to drain into the ground beneath the pavement, reducing the amount of runoff that drains directly into the water body. Restoration methods such as constructing wetlands are also used to slow runoff as well as absorb contamination.

Construction sites typically implement simple measures to reduce pollution and runoff. Firstly, sediment or silt fences are erected around construction sites to reduce the amount of sediment and large material draining into the nearby water body. Secondly, laying grass or straw along the border of construction sites also work to reduce nonpoint source pollution. In areas served by single-home septic systems, local government regulations can force septic system maintenance to ensure compliance with water quality standards. In Washington (state), a novel approach to was developed through creation of a "shellfish protection district" when either a commercial or recreational shellfish bed is downgraded because of ongoing nonpoint source pollution. The shellfish protection district is a geographic area designated by a county to protect water quality and tideland resources, and provides a mechanism to generate local funds for water quality services to control nonpoint sources of pollution. At least two shellfish protection districts in south Puget Sound have instituted septic system operation and maintenance requirements with program fees tied directly to property taxes.

Agricultural operations

To control sediment and runoff, farmers may utilize erosion controls to reduce runoff flows and retain soil on their fields. Common techniques include contour plowing, crop mulching, crop rotation, planting perennial crops and installing riparian buffers. Conservation tillage is a concept used to reduce runoff while planting a new crop. The farmer leaves some crop reside from the previous planting in the ground to help prevent runoff during the planting process.

Nutrients are typically applied to farmland as commercial fertilizer; animal manure; or spraying of municipal or industrial wastewater (effluent) or sludge. Nutrients may also enter runoff from crop residues, irrigation water, wildlife, andatmospheric deposition. Farmers can develop and implement nutrient management plans to reduce excess application of nutrients.

To minimize pesticide impacts, farmers may use Integrated Pest Management (IPM) techniques (which can includebiological pest control) to maintain control over pests, reduce reliance on chemical pesticides, and protect water quality.

Forestry operations

With well planned placement of both logging trails, also called skid trails, can reduce the amount of sediment generated. By planning the trails location as far away from the logging activity as possible as well as contouring the trails with the land, it can reduce the amount of loose sediment in the runoff. Additionally, by replanting trees on the land after logging, it provides a structure for the soil to regain stability as well as replaces the logged environment.

Marinas

Installing shut off valves on fuel pumps at a marina dock can help reduce the amount of spillover into the water. Additionally, pump-out stations that are easily accessible to boaters in a marina can provide a clean place in which to dispose of sanitary waste without dumping it directly into the water. Finally, something as simple as having trash containers around a marina can prevent larger objects entering the water.

Ater pollution is the contamination of natural water bodies by chemical, physical, radioactive orpathogenic microbial substances. Adverse alteration of water quality presently produces large scale illness and deaths, accounting for approximately 50 million deaths per year worldwide, most of these deaths occurring in Africa and Asia. In China, for example, about 75 percent of the population (or 1.1 billion people) are without access to unpolluted drinking water, according to China's own standards. Widespread consequences of water pollution upon ecosystems include species mortality, biodiversity reduction and loss of ecosystem services. Some consider that water pollution may occur from natural causes such as sedimentation from severe rainfall events; however, natural causes, including volcanic eruptions and algae blooms from natural causes constitute a minute amount of the instances of world water pollution. The most problematic of water pollutants are microbes that induce disease, since their sources may be construed as natural, but a preponderance of these instances result from human intervention in the environment or human overpopulation phenomena.

Classes of water pollutants

Chemical water pollutants are generally atoms or molecules, which have been discharged into natural water bodies, usually by activities of humans. Common examples of such chemical water pollutants are mercury emanating from mining activity, certain nitrogen compounds used in agriculture, chlorinated organic molecules arising from sewage or water treatment plants or various acids which are the externalities of various manufacturing activities.

Physical water pollutants are either (a) much larger particles or (b) physical factors such as temperature change, both of which while not typically toxic, cause a variety of harmful effects. The most obvious of physical pollutants are (a) excessive sediment load, mostly arising from over-intense land use practices and (b) rubbish discarded from human manufacturing activity (e.g. plastic bags, bottles). While these materials are not so harmful to human health as chemicals or pathogens, they comprise the majority of visual impact of water pollution. In the case of thermal pollution, these point source discharges typically affect the metabolism of aquatic fauna in adverse ways.

Radioactive substances are really merely a special sub-class of chemical pollutants, and by mass represent the smallest of the contributors to water pollution; however, their potential for harm allows recognition as a separate class. In fact, most discharge of radioactivity is not from the negligible escape from nuclear power plants, but rather arises from agricultural practices such as tobacco farming, where radioactive contamination of phosphate fertilizer is a common method of introduction of radioactive materials into the environment.

Common pathogenic microbes introduced into natural water bodies are pathogens from untreated sewage or surface runoff from intensive livestock grazing. One of the most common disease agents is Giardia lamblia, a parasitic protozoan common in fecal material of many fauna including humans; this microbe is particularly insidious, due to its resistance to conventional sewage treatment. This and other protozoans and bacteria are important causes of illness and mortality in developing countries where population density, water scarcity and inadequate sewage treatment combine to occasion widespread parasitic andbacterial disease.

Sources

Water pollutant sources can be grouped into two super categories: (a) point sources which can be attributed to discrete discharge from a factory or sewage outfall and (b) non-point sources that include agricultural run off, urban stormwater runoff and other area wide sources.

Many of the common inorganic chemical water pollutants are produced by non-point sources, chiefly relating to intensive agriculture and high-density urban areas. Specific inorganic chemicals and their major sources are: monopotassium phosphate, ammonium nitrate and a host of related phosphate and nitrogen compounds used in agricultural fertilizers; heavy metals (present in urban runoff and mine tailings area runoff). However, some inorganics such as chlorine and related derivatives are produced chiefly from point sources, ironically employed in water treatment facilities. Moreover, some of the large dischargers of heavy metals to aquatic media are fixed point industrial plants.

Improper storage and use of automotive fluids produce common organic chemicals causing water pollution are: methanol and ethanol (present in wiper flluid); gasoline and oil compounds such as octane, nonane (overfilling of gasoline tanks); most of these foregoing discharges are considered non-point sources since their pathway to watercourses is mainly overland flow. However, leaking underground and above ground storage tanks can be considered point sources for some of these same chemicals, and even more toxic organics such as perchloroethylene. Grease and fats (higher chain length carbon molecules such as present in auto lubrication and restaurant effluent can be either point or non-point sources depending upon whether the restaurant releases grease into the wastewater collection system (point source) or disposes of such organics on the exterior ground surface or transports to large landfills, both of which last two cases lead to non-point release to water systems.

The most significant physical pollutant is excess sediment in runoff from agricultural plots, clearcut forests, improperly graded slopes, urban streets and other poorly managed lands, especially when steep slopes or lands near streams are involved. Other physical pollutants include a variety of plastic refuse products such as packaging materials; the most pernicious of these items are ring shaped objects that can trap or strangle fish and other aquatic fauna. Other common physical objects are timber slash debris, waste paper and cardboard. Finally power plants and other industrial facilities that use natural water bodies for cooling are the main sources of thermal pollution.

Common pathogenic microbes, in addition to *G. lamblia*, are: species of the genus Salmonella (which variously cause typhoid fever and food-borne illnesses); species in the genus Cryptos-poridium, which are fecal-oral route parasites often transmitted as water pollutants and are associated with inadequate sanitation; parasitic worms that live inside faunal digestive systems for part of their life cycle (This widespread syndrome is spread partially as water pollutants, with an estimated three billion people currently affected). Hepatitis A is a viral disease, one of whose pathways of transmission is water-borne.

Historic trends

While it is not possible to reconstruct water pollution conditions throughout prehistory, certain facts are clear. Modern prevalence of chemical and radioactive water pollutants are clearly correlated with the population explosion and resource use of modern humans. The trends in chemical water pollution increasing from the early Holocene to the 1960s is relatively clear worldwide; starting with the advent of the Australia in the USA, a turnaround in most aspects of water quality began in the early 1970s for most of North America; similar trends in much of Europe as well as India and Madagascar began slightly later. Only in the developing countries including China and rainforest, has chemical water pollution failed to reach a peak, due to high population growth coupled with the priority of economic development above environmental protection in many cases.

Sediment loading of surface waters is a clear long term increasing problem, due to the intensification of agriculture (both for crops and livestock) and increased runoff from urbanization. In extreme cases, such as the north central highlands of coral reef massive topsoil loss has followed extreme slash-and-burn Holocene destruction from the 1970s onward. Most of this loss has resulted in ongoing heavy sediment loads to the central Madagascar river system as well as many of the nearby coastal waters in the Indian Ocean.

A more interesting situation arises with pathogenic microbes. Even though there is evidence that many of the present day water-borne microbes existed earlier in the Quaternary for humans and even as early asCretaceous Period times for other fauna, a case may be made that the rate of incidence of microbial pathogenic infection may be at an all time high for humans, given the overcrowding and inability to supply fresh drinking water to a large percentage of the human population in its present level of 2010. These effects are exacerbated by the likelihood that modern diseases may be mutating at a more rapid rate than historic, given the abrupt man-induced alterations of the chemical, physical and biological environment; human adaptation is not likely to be able to keep up with the pace of such disease mutation velocities; therefore, it is likely that present day water-borne pathogenic disease is at a higher rate of occurrence with present day human populations, in all world regions except for those where water pollution control, medical care and prevention are at the highest levels (e.g. USA, New Zealand, Australia, Sweden, Denmark).

Water Pollution Control

Non-point source control relates chiefly to land management practices in the fields of agriculture, silviculture, mining and urban design and sanitation. Agricultural practices leading to the greatest improvement of sediment control include: contour grading, avoidance of bare soils in rainy and windy seasons, polyculture farming resulting in greater vegetative cover, and increasing fallow periods. Minimization of fertilizer, pesticide and herbicide runoff is best accomplished by reducing the quantities of these materials, as well as using application times removed from periods of high precipitation. Other techniques include avoidance of highly water soluble pesticide and herbicide compounds, and use of materials that have the most rapid decay times to benign substances.

The chief water pollutants associated with mines and quarries are aqueous slurries of minute rock particles, which result from rainfall scouring exposed soils and haul roads and also from rock washing and grading activities. Runoff from metal mines and ore recovery plants is typically contaminated by the minerals present in the native rock formations. Control of this runoff is chiefly derived by controlling rapid runoff and designing mining operations to avoid tailings either on steep slopes or near streams.

In the case of urban stormwater control, the most important methods are achieved in urban planning by use of minimal net surface runoff of impermeable surfaces. This is not merely a simply geometric design issue of avoiding sprawl and minimizing paved surfaces, but also a strategy of incorporating holding ponds into landscaping and use of bioswales and permeable pavers. At an operational level, the use of native plant and xeriscape techniques reduces water use and water runoff and also minimizes need for pesticides and nutrients.

5

WATER CONSERVATION

Water conservation is a big thing, but every little bit helps, so don't think that what you do doesn't matter. A whole lot of people doing a little bit adds up to a whole lot. We must all make changes in our lifestyles that will change the course of our water and its quality. Water conservation needs to be a way of life, not just something we think about once in a while. If we all do our part in conserving Washington's water, we can make a huge difference for the environment.

Water conservation means using our water wisely and caring for it properly. Since each of us depends on water for life, it is our responsibility to learn more about water conservation and how we can help keep our water pure and safe for generations to come. Since we all enjoy the benefits of having pure, clean water, we must help conserve water so that we may continue to enjoy these benefits.

Water conservation is not a job that is just for the technician, soil scientist, hydrologist, forester, wildlife manager, plant scientist, city planner, park manager, farmer, rancher, or mine owner alone. It is a job for the every day person who just likes to have access to the life sustaining resource of water. We must all recognize that water conservation really is our personal responsibility and not just leave it up to other people. We all enjoy the many ways that we use water, so why not do our part in caring for our water?

Our water supply is finite, which means that we do not have an endless supply. We only have the water that we have now. Ninety - seven percent of all the water on the earth is salt water which is not suitable for drinking. Only three percent of all the water is fresh water, and only one percent is available for drinking water. The other two percent is locked in ice caps and glaciers.

With all the people on Earth relying on such a small percentage of all the water on Earth, it only makes sense that we must preserve and conserve our water. We must not pollute our water because it is the only water we will ever have. Some people do not realize the importance of water, and they are continually polluting it. Only about ten percent of waste water is disposed of properly.

We must learn to save water now for the future. The quality of our water is very important. We have the same amount of water now as there was when the earth was created. This is the water we have, and we must preserve its quality. Water is the foundation of food and life. Next to air, water is our most precious resource. We cannot live without water. Saving water helps to preserve our environment. It reduces the energy required to process and deliver water, which helps in reducing pollution and in conserving fuel resources. Saving water now means having water available in the future for recreational purposes, too. Conserving the water we have minimizes the effects of water shortages and helps build a better defense against future drought years. If we save water now, we are helping to ensure a water supply adequate for future generations. Saving water saves money. $It makes cents¢!

We must save water today so that it will be available to us in the future. We need to think of future generations, people who will not have a sufficient supply of water unless we become more concerned with how we use our water today. There are many things that we can do to help preserve Washington's water. These are not things that are hard. All it takes is a little bit of extra effort, and soon it will be second nature. Save the water, its all we've got!

Water Conservation is the most important step in the direction of finding solutions to issues of water and environmental conservation is to change people's attitudes and habits--this includes each one of us. Conserve water because it is the right thing to do. A step to conserve water is the step to secure the future. The most essential among all the natural resources on earth is water. A drop of water is worth more than a sack of gold for the thirsty man. If each one of us make efforts to save water today, it will save us later. Water conservation is the most effective and environmentally sound method to fight global warming. Water conservation is what that can reduce the scarcity of water. It aims to improve the efficiency of use of water, and reduce losses and waste. We can follow some of the simple things that have been listed below and contribute to water conservation.

- Try to do one thing each day that will result in saving water. Don't worry if the savings are minimal¾every drop counts! You can make a difference.
- Remember to use only the amount you actually need.
- Form a group of water-conscious people and encourage your friends and neighbours to be part of this group. Promote water conservation in

community newsletters and on bulletin boards. Encourage your friends, neighbours and co-workers to also contribute.

- Encourage your family to keep looking for new ways to conserve water in and around your home.
- Make sure that your home is leak-free. Many homes have leaking pipes that go unnoticed.
- Do not leave the tap running while you are brushing your teeth or soaping your face.
- See that there are no leaks in the toilet tank. You can check this by adding colour to the tank. If there is a leak, colour will appear in the toilet bowl within 30 minutes. (Flush as soon as the test is done, since food colouring may stain the tank.)
- Avoid flushing the toilet unnecessarily. Put a brick or any other device that occupies space to cut down on the amount of water needed for each flush.
- When washing the car, use water from a bucket and not a hosepipe.
- Do not throw away water that has been used for washing vegetables, rice or dals--use it to water plants or to clean the floors, etc
- You can store water in a variety of ways. A simple method is to place a drum on a raised platform directly under the rainwater collection source. You can also collect water in a bucket during the rainy season.

Our ancient religious texts and epics give a good insight into the water storage and conservation systems that prevailed in those days.

Over the years rising populations, growing industrialization, and expanding agriculture have pushed up the demand for water. Efforts have been made to collect water by building dams and reservoirs and digging wells; some countries have also tried to recycle and desalinate (remove salts) water. Water conservation has become the need of the day. The idea of ground water recharging by harvesting rainwater is gaining importance in many cities.

In the forests, water seeps gently into the ground as vegetation breaks the fall. This groundwater in turn feeds wells, lakes, and rivers. Protecting forests means protecting water 'catchments'. In ancient India, people believed that forests were the 'mothers' of rivers and worshipped the sources of these water bodies.

Some Ancient Indian Methods of Water Conservation

The Indus Valley Civilization, that flourished along the banks of the river Indus and other parts of western and northern India about 5,000 years ago, had one of the most sophisticated urban water supply and sewage systems in the world. The fact that the people were well acquainted with hygiene can be seen from the covered drains running beneath the streets of the ruins at both Mohenjodaro and Harappa. Another very good example is the well-planned city of Dholavira, on Khadir Bet, a low plateau in the Rann in Gujarat. One of the oldest water harvesting systems is found about 130 km from Pune along Naneghat in the Western Ghats. A large number of tanks were cut in the rocks to provide drinking water to tradesmen who used to travel along this ancient trade route. Each fort in the area had its own water harvesting and storage system in the form of rock-cut cisterns, ponds, tanks and wells that are still in use today. A large number of forts like Raigad had tanks that supplied water.

In ancient times, houses in parts of western Rajasthan were built so that each had a rooftop water harvesting system. Rainwater from these rooftops was directed into underground tanks. This system can be seen even today in all the forts, palaces and houses of the region.

Underground baked earthen pipes and tunnels to maintain the flow of water and to transport it to distant places, are still functional at Burhanpur in Madhya Pradesh, Golkunda and Bijapur in Karnataka, and Aurangabad in Maharashtra.

Water Situation in India: An Overview

India as a nation uses 500 billion cubic meters (BCM) of water every year for various purposes, most important of which are irrigation and domestic use. This equates with a per capita use of around 470 cu meters per year, which is far short of the amount used in a developed country like United Stated where per capita water use is around 1600 cu meters. But then, United States has a larger territory, and a far smaller population. The lower per capita use also denotes a scarcity that is likely to become much worse by year 2025 due to population growth and increasing urbanisation.

India receives a rainfall of around 4000 BCM every year, actually enough to take care of India's water needs, but most of this water is received within 4 months of 'monsoon' season between June and September, or be more precise, during less than 100 hours of precipitation spread over these four months. Most of this water is lost. A small part of it goes into soil and rechargesground water, but most of it evaporates and is lost.

Traditionally, till the middle of twentieth century, India largely managed its water needs by its major rivers, and rainfall, supplemented by traditional water bodies called by different names like 'johad', 'talab', bavdi, or 'jheel' - most of which were nothing but natural or man-made ponds that collected water during rainfall that was used by surrounding villagers for various purposes. After the great demographic divide of 1919, after which the population of India started exploding, these sources were unable to satisfy the needs. As a result more and more people resorted to drilling for groundwater. Today, the groundwater levels are falling everywhere throughout the country. In some cities they are falling at an alarming rate of almost 10 meters every year, indicating a crisis. The realisation that ground water will not last forever has resulted in a strong demand for finding solutions to water scarcity that threatens the country. Though there is a realisation for water conservation policy, a comprehensive policy in this regard is yet to emerge.

Water Conservation Policy as a Solution to Water Scarcity

With a rainfall that is adequate to take care of its needs, there is a realisation that water conservation is the only answer to India's problems. Some cities have made rules making it compulsory to have water harvesting systems in all new buildings. Many non-governmental organisations (NGOs) have taken up projects related to water conservation in rural and semi-arid areas with excellent results. One such NGO, 'Tarun Bharat Sangh' has been involved with people and by way of self-help, it has created 3000 water conservation structures in 650 villages. However, a lot more needs to be done in this regard.

Different experts and stakeholders have suggested different methods, but there are some that everyone agrees as a solution.

(i) Local Water Bodies - This used to be one of the main local sources of water earlier, but gradual filling of those bodies and their pollution by people and industries have made most of them dysfunctional. Now their renovation is looked upon as one of the means to store and conserve rain water.

(ii) Urban Water Harvesting - As India gradually shifts from rural to urban areas, the role of water conservation in urban areas is increasingly recognised and accepted as a solution towater problem.

(iii) Afforestation - Planting of trees has been accepted as a single biggest means of holding water in a territory, and is increasingly resorted to as a means especially in the semi-arid regions. However most of the efforts are not well coordinated till now in this regard.

(iv) Improving Technology & Water Use Methods - Most Indian farmers are still not fully acquainted with efficient water usage methods for irrigation. Methods like drip irrigation are still unknown in most parts even where water is scarce. Awareness in this regard needs to grow to allow efficient use of water.

(v) Water Pricing As Means To Minimize Wastage - So far, the governments have generally desisted from pricing water at its real cost. The unrealistically low prices tend to encourage overuse and wastage. The World Bank has been advocating pricing as a means to avoid wastage, but the democratic governments have so far not been inclined to accept this advice.

Need to Plan for the Impending Crisis

India stands today at a point where it faces a real danger of a water crisis if it is unable to plan for the solutions and start implementing them very soon. The most feasible solution is toconserve water by all the different means possible, and a comprehensive water conservation policy with necessary legal sanction for enforcement, and enough propaganda to create public awareness is an acute need of the hour.

Water conservation refers to reducing the usage of water and recycling of waste water for different purposes such as cleaning, manufacturing, and agricultural irrigation. Water conservation can be defined as:

1. Any beneficial reduction in water loss, use or waste as well as the preservation of water quality.
2. A reduction in water use accomplished by implementation of water conservation or water efficiency measures; or,
3. Improved water management practices that reduce or enhance the beneficial use of water.

Goals

The goals of water conservation efforts include as follows:

- Sustainability. To ensure availability for future generations, the withdrawal of fresh water from an ecosystem should not exceed its natural replacement rate.
- Energy conservation. Water pumping, delivery, and wastewater treatment facilities consume a significant amount of energy. In some regions of the world over 15% of total electricity consumption is devoted to water management.
- Habitat conservation. Minimizing human water use helps to preserve fresh water habitats for local wildlife and migrating waterfowl, as well as reducing the need to build new dams and other water diversion infrastructures.
- Reduce water consumption per capita.

Important Points to Follow for Water Conservation

The most important step in the direction of finding solutions to problems of water and environment conservation is to change people's attitudes and habits which includes each one of us. Conserve water because it is the right thing to do. We can follow some of the simple things that have been listed below and contributed to water conservation.

I. Try to do one thing each day that will result in saving water. Don't worry if the savings are minimal because every drop counts ! You can make a difference.

II. Remember to use only the amount you actually need.

III. Form a group of water -conscious people and encouarage your friends and neighbours to be part of this group. Promote water conservation in community news letters and on bulletin boars. Encourage your friends, neighbours and co-workers to also contribute.

IV. Encourage your family to keep looking for new ways to conserve water in and around your home.

V. Make sure that your home is leak free. Many homes have leaking pipes that goods unnoticed.

VI. Do not leave the tap running while you are brushing your teeth or soaping your face.

VII. See that there are no leaks in the toilet tank. You can check this by adding colour to the tank. If there is a leak, color will appear in the toilet bowl within 30 minutes. (Flush as soon as the test is done, since food colouring may stain the tank.)

IX. Avoid flushing the toilet unnecessarily. Put a brick or any other device that occupies space to cut down on the amount of water needed in each flush.

X. When washing the car, use water from bucket and not hosepice.

XI. Do not throw away water that has been used for washing vegetable, rice or dals. Use it to water plants or to clean the floors, etc.

XII. You can store water in a variety of ways. A simple method is to collection source. You can also collect water in a bucket during the rainy season.

SOCIAL SOLUTIONS

Water conservation programs are typically initiated at the local level, by either municipal water utilities or regional governments. Common strategies include public outreach campaigns, tiered water rates (charging progressively higher prices as water use increases), or restrictions on outdoor water use such as lawn watering and car washing. Cities in dry climates often require or encourage the installation of xeriscaping or natural landscaping in new homes to reduce outdoor water usage.

One fundamental conservation goal is universal metering. The prevalence of residential water metering varies significantly worldwide. Recent studies have estimated that water supplies are metered in less than 30% of UK households, and about 61% of urban Canadian homes (as of 2001). Although individual water meters have often been considered impractical in homes with private wells or in multifamily buildings, the U.S. Environmental Protection Agency estimates that metering alone can reduce consumption by 20 to 40 percent. In addition to raising consumer awareness of their water use, metering is also an important way to identify and localize water leakage. Water metering would benefit society in the long run it is proven that water metering increases the efficiency of the entire water system, as well as help unnecessary expenses for individuals for years to come. One would be unable to waste water unless they are willing to pay the extra charges, this way the water department would be able to monitor water usage by public, domestic and manufacturing services.

Some researchers have suggested that water conservation efforts should be primarily directed at farmers, in light of the fact that crop irrigation accounts for 70% of the world's fresh water use. The agricultural sector of most countries is important both economically and politically, and water subsidies are common. Conservation advocates have urged removal of all subsidies to force farmers to grow more water-efficient crops and adopt less wasteful irrigation techniques.

New technology poses a few new options for consumers, features such and full flush and half flush when using a toilet are trying to make a difference in water consumption and waste. Also available in our modern world is shower heads that help reduce wasting water, old shower heads are said to use 5-10 gallons per minute. All new fixtures available are said to use 2.5 gallons per minute and offer equal water coverage.

Household Applications

Water-saving technology for the home includes:

- Low-flow shower heads sometimes called energy-efficient shower heads as they also use less energy,
- Low-flush toilets and composting toilets. These have a dramatic impact in the developed world, as conventional Western toilets use large volumes of water.
- Dual flush toilets created by Caroma includes two buttons or handles to flush different levels of water. Dual flush toilets use up to 67% less water than conventional toilets.
- water (sea water) or rain water can be used for flushing toilets.
- Faucet aerators, which break water flow into fine droplets to maintain "wetting effectiveness" while using less water. An additional benefit is that they reduce splashing while washing hands and dishes.
- Wastewater reuse or recycling systems, allowing:
- Reuse of graywater for flushing toilets or watering gardens
- Recycling of wastewater through purification at a water treatment plant. See also Wastewater - Reuse
- Rainwater harvesting

- High-efficiency clothes washers
- Weather-based irrigation controllers
- Garden hose nozzles that shut off water when it is not being used, instead of letting a hose run.
- Using low flow taps in wash basins
- Swimming pool covers that reduce evaporation and can warm pool water to reduce water, energy and chemical costs.
- Automatic faucet is a water conservation faucet that eliminates water waste at the faucet. It automates the use of faucets without the use of hands.
- Water can also be conserved by landscaping with native plants and by changing behavior, such as shortening showers and not running the faucet while brushing teeth.
- Use waste water for growth of plants

Commercial Applications

Many water-saving devices (such as low-flush toilets) that are useful in homes can also be useful for business water saving. Other water-saving technology for businesses includes:

- Waterless urinals
- Waterless car washes
- Infrared or foot-operated taps, which can save water by using short bursts of water for rinsing in a kitchen or bathroom
- Pressurized water brooms, which can be used instead of a hose to clean sidewalks
- X-ray film processor re-circulation systems
- Cooling tower conductivity controllers
- Water-saving steam sterilizers, for use in hospitals and health care facilities.
- Rain water harvesting.

Agricultural Applications

For crop irrigation, optimal water efficiency means minimizing losses due to evaporation,runoff or subsurface drainage while maximizing production. An evaporation pan in combination with specific crop correction factors can be used to determine how much water is needed to satisfy plant requirements. Flood irrigation, the oldest and most common type, is often very uneven in distribution, as parts of a field may receive excess water in order to deliver sufficient quantities to other parts. Overhead irrigation, using center-pivot or lateral-moving sprinklers, has the potential for a much more equal and controlled distribution pattern. Drip irrigation is the most expensive and least-used type, but offers the ability to deliver water to plant roots with minimal losses. However, drip irrigation is increasingly affordable, especially for the home gardener and in light of rising water rates. There are also cheap effective methods similar to drip irrigation such as the use of soaking hoses that can even be submerged in the growing medium to eliminate evaporation.

As changing irrigation systems can be a costly undertaking, conservation efforts often concentrate on maximizing the efficiency of the existing system. This may include chiseling compacted soils, creating furrow dikes to prevent runoff, and using soil moisture and rainfall sensors to optimize irrigation schedules. Usually large gains in efficiency are possible through measurement and more effective management of the existing irrigation system.

Minimum Water Network Target and Design

The cost effective minimum water network is a holistic framework/guide for water conservation that helps in determining the minimum amount of freshwater and wastewater target for an industrial or urban system based on the water management hierarchy i.e. it considers all conceivable methods to save water. The technique ensure that the designer desired payback period is satisfied using Systematic Hierarchical Approach for Resilient Process Screening (SHARPS) technique.

Few suggestions for improving conservation practices at the community level. These include:

- Encouraging participation of beneficiaries: Placing more emphasis at the community level and encouraging participation of people, particularly women and making the community as a whole realise the importance of efficient use of water.
- Saving irrigation water: The agricultural sector is the largest consumer of water and, therefore, needs to be focused for water conservation practices
- Saving industrial water: The industries should be encouraged to purify their water and to reutilise the once used water and to dispose off effluent safely, avoiding natural water resources.
- Limiting evaporation: Evaporation from open water bodies and reservoirs constitutes a major constraint in the efficient use of water. Suitable evaporation retardation or suppression practices must be adopted for limiting evaporation
- Augmenting aquifer recharge: Efforts at large scale artificial recharge can have the potential for storing water underground with minimum evaporation and surface storage facility.

- Use of non-conventional water sources: For example, treated municipal wastewater, agriculture drainage and brackish/ saline ground water etc., can be encouraged for supplemental use in agricultural irrigation after treatment.
- Capacity building: To develop skilled manpower, encourage field level training and live demonstrations of rational water use through modern irrigation methods, advice on soil moisture & crop water requirement etc.

Tips to save water :

- Avoid leakage of water from the taps.
- Turn the tap off when not in use especially when you brush your teeth or wash clothes.
- Rainwater harvesting is the another method to conserve water.
- The water supply should be limited in those areas which enjoys the unlimited water supplies.
- Check the leakage of water in the toilets. Also get check the hidden water leaks.
- Educate the mind of the people in the rural areas to save the water.
- Promote the conservation of water through media and wall posters.
- Never throw the water unnecessary on roads which can be used for gardening and cleaning.
- Avoid unnecessary flushing the toilets. Dispose off the tissues, cigarettes and other waste into the bin instead of toilets.
- Use minimum amount of water to bath.
- Water Waste restrictions.
- Improvement in the water distribution system.
- Water your lawn only when it is needed.
- Use a broom instead of hose to clean the sidewalks or to wash the car.
- Capture the water that is leaking and repair it as soon as possible.
- You can use washing machine to wash clothes that does not consume much water.
- Donot leave the tap running while washing the dishes in the kitchen.
- Install small shower heads to reduce the flow of water.

Benefits to Conserve water :

- If you save water it can save your money bills.
- Reduction in interior water use cuts waste water flows, especially overflowing of gutters which contaminates the environment.

- Environment benefits include eco system and habitat protection.
- Water conservation helps in improving the quality of your drinking water.

Importance of Water to Plants

The importance of water to plants, cannot be underestimated. In fact, without water, the plants cannot survive as their physiological activities will come to a standstill. Some examples are cited here.

1. The plants take their food nutrients in solution form. In the absence of water in the soil, nutrients become unavailable to plants.
2. Water in the plant cells help in the proper functioning of the cells.
3. With the movement of water inside the plant, food materials are also carried to different parts of the cells.
4. Water is used as a raw material in photosynthesis wherein the plant manufactures carbohydrate in the presence of light using water and carbon dioxide as raw materials.
5. Water influences the chemical, physical and biological activities of soil.
6. Plants transpire water and keep the temperature normal. Thus water controls the temperature of plant.
7. In winter, it prevents quick freezing of plants and protects them from frost.
8. Hormone and enzymes move inside the plant by dissolving in water.
9. Water is an integral constituent of the cell protoplasm. It keeps the cell turgid which is necessary for all physiological processes to operate.

Meaning and Scope of Water Conservation

Water conservation is a broad term involving all the measures which aim at arresting maximum amount of run-off water, storing or collecting it in safe, artificially raised structures and recycling the stored water again to the field in times of need. The ultimate aim of water conservation is to produce maximum from the limited availability of water, by using it most efficiently. An efficient system of collecting, storing and reusing run-off water is sometimes also referred to as water harvesting.

From the point of dry land agriculture, water harvesting indicates not only storing rain water in irrigation tanks/ponds but also employing mechanical and cultural manipulations of soil to improve its rainfall absorption and moisture retention capacity. However, the measures adopted are location specific.

To decide on the suitable measures to be taken a preliminary inventory of soil water and other resources, a study of the land use pattern, a dialogue with the target group and a study of socio-political, economic and cultural setting are a must. This implies a scientific approach to water conservation. The need for educating, and organizing people to make them accept necessary reforms for effective implementation of integrated water conservation programmes need not be over-emphasized.

The chief agent of soil erosion is water. It is estimated that about 5333 million tonnes of soil (amounting to about 16 tonnes/ha) is eroded from Indian territory. Nearly 29% of this is carried into the sea and nearly 10% is deposited in surface reservoirs resulting in loss of storage capacity of reservoirs and occurrence of floods. Conservation of water virtually reduces the soil erosion. Thus the scope of conservation is very wide, encompassing all the measures of soil conservation, moisture retention and nutrient preservation in the soil.

Importance of Water Conservation

Water is very important content. Without water there is no life in earth. We use water in so many ways in our daily life. Water is a substance which covers ¾ part of the world. It means water is occupying more portion compared to land. But this water is becoming more polluted because of the environmental changes. So it is necessary to conserve the water. It is every one duty to protect the water because everyone uses this water and we the people are making water polluted. Now we may get the doubt how to conserve the water. A few simple changes in home about the water consumption make a vast contribution to water conservation.

The basic changes we have to do is stop leaking of water. We have to see whether there is a leakage in our taps. Because of the water leakage taps we can't protect the water. We can also protect the water by installing Low-Flow Showerheads. With these we can protect the water because it flows lower. And we won't use much water with these Low-Flow Showerheads. When we are brushing our teeth or when we are shaving we have to turn off faucets. After completing the work we have to use the water. By this change also we can preserve the water. Using sprinklers for yard is also a good remedy for water conservation. Turn off the water in the shower when shampooing the hair Water will be saved. It is better to replace the new toilets in place of old toilets. These are high efficiency and will use less water. To collect a rain water use a rain barrel at the bottom of gutter down spouts. It may use for watering plants. These are some of the things to follow to conserve the water.

We all use the water we must also see how we are using this. It is important to note that there are many reasons why we waste this natural resource; it has become necessary to waste the resources for innovating new technologies. However we have now proved it is not necessary to waste. It is easy now to conserve the water. By conserving water we can also save the money. There are hundreds of ways to conserve the water. But may not follow all those ways. We can follow only those ways which are easier to us. The easier things we can follow are the things which are mentioned above. They are the things we follow easily in our home.

It is the responsibility of every one to educate our children. With this we are indirectly conserving the water in future also. The children who are educating to save the water will save the water in future also. Now if we cut the usage of water it definitely helps the future generations. Water is mainly related to environment. Green also helps for the environment. Growing greenery everywhere and encouraging forest causes rain fall. So green is also conserving the water

Need of Water Conservation

Water is absolutely essential for utilization of all other natural resources including land. Agriculture depends on the availability and supply of water either from rainfall or from surface or underground fresh water resources. Most of the water is received through monsoon rains and is obtained from July to September. Only in some rivers of North India, water is received through the melting of snow in summer.

Because of high intensity of monsoon rains coupled with poor infiltration rates of soils, there is a heavy run-off of water. Run-off occurs when amount of precipitation is more than infiltration of water into the soil. Run-off is very heavy particularly in first rains. As a consequence of heavy run-off, the discharge in rivers increases during rainy period. The experience of last 2-3 decades, shows that every passing year the amount of run-off water discharged into the rivers is in creasing. It indicates that the amount of water which was conserved earlier in the site of rainfall (catchment area) has been reduced.

Our country receives more annual rainfall than the global average of annual rainfall. Therefore, water conservation has to be adopted as a national objective, not only for the sake of water availability but also for protecting the soil which is a precious resource.

Technical methods to conserve water

Rainwater Harvesting : Rainwater harvesting is the gathering and collection of water from the rooftop. The traditional method of rain water harvesting is the most effective and simple way to conserve the water. It means utilization of rain water for the domestic as well as agricultural purposes. There are three technical methods of rain water harvesting such as Catchment, Conveyance and storage.

Historical Water Bodies : There are many traditional water bodies which have been in disuse for the longer time. These bodies can be reused as the recharging points.

Ponds : Steps should be taken to avoid dumping of sewage into the village ponds. Efforts need to be made to deepen these ponds with the dragline machines. Garbage and other waste should not be dumped into the ponds.

Water Erosion Problem in India

Out of the total 328 m ha of geographical area of the country, about 143 m ha is cultivated and about 65 m ha is barren uncultivable waste and cultural wastelands. By and large these lands are more or less devoid of protective vegetation and physical barriers and hence are subjected to very serious water erosion. Nature takes 500 to 1000 years to develop 2.5 cm top soil but gets eroded in a course of a year or two. As a result it becomes the most harmful single factor in the deterioration of productive land. Along with the loss of run-off water a colossal amount of soil is also lost.

Overgrazing, deforestation, faulty cultivation, shifting cultivation and carelessly built roads in catchment areas have led to devastating down streams. These include gullying and floods leading to destruction of farm lands and villages, drop in flow during the dry season and consequent loss of crops, and siltation of reservoirs and canals. The problem has been further aggravated due to high rate of population growth: both human and livestock, resulting in indiscriminate exploitation of natural sources, for meeting the ever-increasing demand for fodder, fuel, fibre and fertilizers. Thus continuous degradation of production base and imbalance in land - water -plant-man-animal system is leading to ecological imbalance and economic insecurity through severe soil erosion both by water and wind and threat to the quality of our life and cultivation.

Even though no systematic survey for accelerating the extent of water and soil loss in different areas of the country, has been carried out, it. is estimated that about 5334 mt of soil is detached annually from its original place. The country's rivers carry 2052 mt of this, nearly 1572 mt is carried away by the rivers into the sea every year and 480 mt is being deposited in various reservoirs.

Soil erosion by water is mainly of four types:

1. Sheet erosion

About 4-10 tonnes/ha/year in red soil, 17-43 tonne/ha/year in black soil and 4-14 tonne/ha/year in alluvial soil.

2. Gully erosion

   About 33 tonnes/ha/year in ravine regions

3. Hill side erosion

   More than 80 tonnes/ha/year in landslide, mine spoil areas etc.

4. Stream bank erosion

Along with soil, about 2.5 mt of nitrogen, 1.5 mt of phosphorus, 2.2 mt of potash, large amount of organic matter and micronutrients are also lost. The loss of nutrients is valued at about Rs.15000 million per annum. Mohanjodaro and Harappa are the examples of this fact. Glaring evidence of neglect of rainwater can be seen in the denuded hill tops all around and the extensive network of raville in the states of Gujarat, Madhya Pradesh, Uttar Pradesh and Rajasthan. Once a thick forest, Rajasthan is now a desert, Conservation of water is prerequisite to conserve soil which is a natural reservoir of plant nutrients. Taking into consideration the magnitude of problems, there is a great scope to streamline the entire water conservation programme more scientifically on micro watershed basis in cluster instead of sporadic efforts. Watershed management programme to conserve water and soil is a new approach and it is getting popularity very rapidly. The programme involves an appropriate combination of agronomic, mechanical, agrostological and afforestation measures, fitting to soil-climatic and topographical conditions of area and socio-economic conditions of the target group.

Types of Water Losses

Water losses with special reference to soil erosion aloe de- scribed below:

A. Raindrop or splash loss

Raindrops fall on the ground and form a splash along with suspended soil patlicles. Water and soil move away from their original place.

B. Sheet erosion

When the intensity of rainfall is bit higher, a fairly uniform layer of soil is lost from the soil along with the run-off water. This type of soil and water loss is serious on slightly sloppy lands. The existence of this kind of soil and water erosion is seldom known to farmers as it does not leave behind any visible change on soil topography. Though the field appears the same as before but the cream of the soil i.e. highly fertile upper most layer, is skimmed off along with moving water. The estimates reveal that sheet loss of water erode away many tonnes of soil and has damaged millions of hectares of land in our country.

C. Rill erosion

The speed of run-off water is sometimes high enough to create rills on the soil surface. Rills are very shallow finger sized channels which can be smoothened out by ploughing. Rill erosion is the intensified form of sheet erosion and more distinctly noticeable. The width of rills keeps on widening with every passing rainy season.

D. Gully erosion

Gully erosion of water and soil is the aggravated form of rill loss. When rills are not smoothened out, water keeps on drawing through them and these shallow rills go deeper and deeper with every passing rainy season. The depth and size of rills go so high that they cannot be smoothened out with ploughing, they are called gullies. The size of gullies may be many metres wide and deep. They provide a big route to run-off water and a huge amount of water is lost through them.

Factors Affecting Water Losses

Water in motion removes the soil and causes erosion of soil. Water erosion is due to dispersive action and transporting power of water. So it descends in the form of rain and leaves the land in the form of run-off. The intensity and speed of run-off are directly correlated to the amount of soil' and water loss. Some of the factors discussed below greatly influence the extent of soil and water loss.

A. Climate

The major climatic factors affecting run-off and erosion at'e rainfall, temperature and wind. Of these, rainfall is the most important and the knowledge of rainfall characteristics is important in erosion studies. Raindrop impact on a soil surface has a pronounced effect on erosion. Raindrops cause soil splash, detach soil particles and make them available for transport. The amount of soil detached by the rain depends on the intensity of rain, its duration and the protective value of any cover present.

Erosive capacity of rain drops results from three factors, namely (1) amount and intensity of rainfall, (2) the diameter of raindrops, and (3) the velocity of drops as they strike the soil. The amount of soil transported by raindrops is governed by the transporting capacity of the raindrop energy; the case with which detached particles are called; and the resistance afforded by the vegetation and other obstructions.

The amount, intensity, frequency and duration of rainfall have very definite effects on the amount and rate of resultant run-off. As large total. rainfall may not cause excessive erosion if the intensity is low. Likewise, an intensive rain of short duration may not produce sufficient run-off to cause erosion. When both amount and intensity factors are high in a given storm, run-off erosion will be serious.

B. Topography

Sloppy soils frequently permit run-off, resulting in the loss of water and soil, while on flat lands erosion is comparatively negligible. The degree of slope and length of slope are the main features of topography affecting erosion. Velocity of run-off water is influenced mainly by the degree of slope. If the land slope is increased four times, the velocity of water flowing over it is approximately doubled. When the velocity is doubled the erosion capacity as represented by the kinetic energy of flowing water is increased about four times, tile amount of material of a given size that can be carried is increased by about 32 times, and the size of particles that can be transported by pushing or rolling is increased by about 64 times. There is a definite limit to the amount of suspended material that can be carried by given stream of water. When this limit is reached, more material cannot be picked up without increasing either the velocity or the depth. Any reduction in velocity on the other hand, will result in deposition of the suspended soil.

C. Vegetation

Vegetation is one of tile most efficient means to conserve water. A thick growth of grass or a dense forest may negate the effect of all those factors which accelerate water erosion. Vegetation reduces water erosion through the ways mentioned below.

1. Interception of rainfall

The foliage of vegetation intercepts tile raindrops, retains a portion of water and evaporates it back into the atmosphere. However, the plant absorbs some of this water through foliage and utilizes it for metabolic purposes. The amount directly evaporated does not contribute to the run-off water.

2. Decreasing run-off velocity

Vegetative cover serves as a hindrance to run-off water. A weIl distributed, close-growing vegetation not only slows down the rate at which water travels down the slope, but also tends to prevent " the concentration of water in a particular area. These two effects greatly reduce the erosive capacity of the run-off water.

3. Root effect

Roots of grasses and other vegetation make the soil porous and permeable to water. The roots die and decay in the soil leaving behind numerous small cavities which increase the pore space and consequently the absorption of water.

4. Biological effect

Remnants of vegetation serve as the food material for numerous organisms inhabiting the soil. These organisms range from large sized earthworms to microscopic bacteria, fungi, algae and actionmycetes. Macroorganisms work the soil and make it porous and permeable. Microorganisms decompose organic matter of the soil, and humus is formed as a decomposition product. Humus is very efficient in retaining water. It can retain water as much as 100 times of its own weight.

D. Soil

The amount of water left on the soil surface for run-off very much depends on soil properties. If the infiltration capacity of the soil is high, a large amount of water is soaked in by the soil and there is little run-off. In case soil is very compact or hardpan lies in the subsoil, a large amount of water is lost by run-off. The infiltration capacity or permeability of soil depends on number of factors such as soil structure, texture, organic matter content, moisture content and slope percentage.

Avenues of Water Loss From Soil

Factors which influence behavior of water in soil in the context of soil erosion have been discussed above. Here, it is useful to briefly highlight the avenues through which water disappears from the soil.

A. Infiltration

The downward entry or movement of water into the soil is called infiltration. When the water is introduced to the soil either by irrigation or rain, it penetrates the surface of the soil. The speed of movement or the depth covered in a unit time is termed as infiltration rate. It is usually measured in centimeters per hour. The infiltration rate is greatly influenced by the moisture content of the soil. The soil takes in or absorbs water rapidly, when water is just applied on the field surface. If water supply continues, surface becomes saturated and infiltration rate decreases until it becomes merely constant.

The soil layer with the lowest transmission rate, either at the surface or below it, usually determines the infiltration rate. Surface sealing which is the formation of a thin compact layer on the ground surface rapidly reduces the rate of infiltration. The surface seal results from a breakdown in soil structure due to the beating action of raindrops and the action of the water flowing over the surface whereby, the fine particles are filtered around the larger particles to form a relatively impervious seal. Very sandy soils may have infiltration rates of over 25 cm per hour, whereas clay soils may have rates approaching zero. Poor soil structure, salinity and alkalinity reduce infiltration rate.

B. Run-off

Run-off may be defined as the portion of rain or irrigation water which leaves the field either as surface flow or subsurface flow. As long as the rate of rainfall (or irrigation water) reaching the soil surface is less than the infiltration capacity, all the water is absorbed into the soil. When the rate of rain- fall exceeds the rate of infiltration, shallow depressions are filled to overflow level, water begins to flow over the land towards streams. The water required to fill depressions prior to the beginning of surface run-off is called initial detention or 'depression storage'.

Overland run-off usually reaches a tiny rivulet or channel which connects to the big rivers and ultimately to ocean. The time of travel from raindrop to stream channels depends on distance, slope, surface condition and depth of flow. Surface run-off is the run-off which travels over the ground surface to the channel of a watershed. The channels may range from rills to major streams. Subsurface flow is that portion of infiltrated water which penetrates to a shallow depth, travels laterally and is intercepted by channels or emerges to the surface. Appreciable amount of subsurface flow, relative to total run-off, is likely to occur where permeable surface soils lie over slowly permeable or impervious horizons at shallow depths.

C. Evaporation from the soil

Evaporation is the process of vaporization by which liquid water is transferred into the atmosphere. The rate of evaporation of water depends on atmospheric temperature, soil humidity, organic matter content and wind velocity. Dry air has a greater capacity for absorbing moisture than moist air, hence evaporation increases under conditions of low humidity. High evaporation occurs at places of high temperature, low atmospheric humidity and high winds.

D. Transpiration by plants

Transpiration is the discharge of water as vapour into the atmosphere by the leaves and stems of living plants. The amount and frequency with which irrigation water must be applied will be governed in part by the rate at which crops transpire. This rate of transpiration varies from practically nothing at young seedling stage during cold cloudy weather to a maximum during hot windy weather at a time when crops are growing most luxuriantly. Rapidly growing crops of a leafy nature may use about 5 cm of water per week during the peak growing season. The volume of water transpired by the plants depends mainly on the water at their disposal and also on temperature and humidity of the air, wind, intensity and duration of sunlight, stage of development of the plant, species of the plant and nature of the leaves.

E. Deep percolations

When the amount of water entering the soil becomes greater than its water holding capacity, deep percolation occurs. This is an intensified form of infiltration in which a great loss of water takes place. Water goes beyond the exploration capacity of plant roots. This water moves due to the force of gravity.

Uses of Water

There are many, many ways that we use our water, and that is partly why it is so important that we conserve our water. Water is our most precious resource. Water is vital to life. Humans, plants, and animals are made up of mostly water. All living things would die if it weren't for water. We use water for drinking, washing, cleaning, cooking, and growing our food as well as many, many other things. The average American uses around 150-250 gallons of water daily. Even more water is used by industries to generate electricity, manufacture things, and transport people and goods.

Common household uses consume a lot of water. It may take between 30 and 40 gallons for one bath. The average toilet uses about 5 gallons of water per flush. It takes 20-40 gallons of water for one shower. Washing machines use an average of 25 gallons per load. The kitchen sink takes roughly 20 gallons per day for preparing food and washing dishes. The bathroom sink, used for washing hands, shaving and brushing teeth, requires about 15 gallons per day. These numbers are estimated for the average household in America.

Much of our fresh water is also used outdoors for watering lawns, flower beds, and vegetable gardens, as well as washing cars and filling swimming pools. We must be careful not to pollute the water that we use outside. Many people use chemicals on their lawns and gardens and then water them with pure water. The water will wash the chemicals off of your plants and then run down a storm drain and go straight to the rivers and streams where fish make their homes. This kind of polluted water can kill fish and wildlife. We should always be mindful of whether or not we are polluting the water around us. We need to take care of our water, not abuse it.

Every town and city, whether small or big, uses water. Cities use water for fire fighting, street cleaning, and watering public areas such as parks, grass, trees, shrubs, and flowers. Water is also used to fill public drinking fountains, including those at schools and libraries. All of the different businesses in your community also use water. Think about all the water that is used by restaurants, hospitals, laundries, dry cleaners, golf courses, hotels, car washes, beauty shops, barber shops, gas stations, and health clubs as well as all of the other businesses in town. These all add up to quite a big demand on Washington's water supply. We need to try to think about how many people need to use water and conserve our water so that there is enough to go around.

The amount of water needed to run a farm is tremendous. When we think of water on a farm, we think of watering crops; but the amount of water needed on a dairy farm is a huge amount also. Chickens, pigs, sheep, and all the other animals in a farmyard need drinking water to stay alive. Food must be grown for them to eat, and water is also required in the cooling systems used to keep production meat fresh. Vegetable and grain crops also require water. Water is used in spreading fertilizers, herbicides, and pesticides, which produce a greater crop yield, but can also contaminate the water. Most of the water used on farms is used for irrigation.

Studies show that by using drip irrigation, farmers can conserve up to 60% of the water that it would normally take to irrigate their crops using other systems of irrigation.

It takes about 26 gallons of water to produce one ear of corn while it takes about 2000-2500 gallons of water to produce one pound of beef. Around 120 gallons of water is required to produce one egg. About 300 gallons of water is needed to produce one loaf of bread, and it takes about 12,000 gallons of water to grow a bushel of wheat. Believe it or not, about 1,400 gallons of water are used during the final production of one fast-food meal including a burger, fries, and a soft drink.

Hydroelectric plants are the largest users of water. Hydroelectric plants capture the kinetic energy of falling water to make electricity. They do this with a dam. The dam forces the water level to go up so that the water will have more power when falling. The force of the falling water pressing against the turbines' blades cause them to spin. The spinning turbines transmit the kinetic energy of the falling water to generators. The generators spin when the turbines spin generating electricity that will be transmitted on the power lines to homes and businesses.

Of all the electricity in the world, about 20% is generated by hydropower. About 10% of all the electricity in the United States is provided by hydropower. Hydropower generating prevents a lot of pollution. Hydropower generating is clean and does not leave any waste. Because of the electricity generated by hydropower, the amount of oil and coal needed to produce enough electricity is reduced. It prevents the need to burn about 22 billion gallons of oil or 120 million tons of coal each year.

The amount of electricity that a hydroelectric plant produces depends on two things: how far the water falls and the quantity of water falling. The higher the dam, the further the water falls and the more electric power produced. If the water falls twice as far, there will be twice as much electricity generated. The quantity of water that falls also affects the amount of power produced. The more water that flows through the turbines making them spin, the more electric power produced.

Water is also essential in industry. It is heated and the steam is used to run machinery. Water is used to cool hot metal such as in the production of steel. Water is also used to cool the air. It is an important element in many products like chemicals, drugs, lotions, shampoos, cosmetics, cleaners, and also beverages. Water is used in processing food and in innumerable factories and industrial processes including the manufacturing of paper. Water used in processing foods and beverages must be absolutely clean, while other industries such as a manufacturing plant may use a lower quality of water. "In the early 1900s, American industry used about 10 to 15 billion gallons of water a day. With the huge growth in industry following World War II, the industrial use of water also grew. By 1980, industry was using about 150-200 billion gallons each day." - Water: A Resource in Crisis by Eileen Lucas

Recreation is another way that we use and enjoy Washington state's water. Many people enjoy fishing, boating, sailing, canoeing, rafting, and swimming, as well as many other recreational activities that depend on water. Most of us enjoy at least one of these activities and would be disappointed if we could never do any of them again. If our water is continually polluted though, then it could become dangerous to go swimming and all of the fish could die so that no one could go fishing either. No one would enjoy boating, sailing, canoeing, or rafting in polluted water either. We need to remember that it is important to guard our waters from becoming polluted.

Transportation is yet another way that we use water. Many people use boats and ferries to commute to and from work every day. People also enjoy going on cruise ships or just going sailing. Water transportation has been a channel for growth in America, but that growth has not come without cost. The destruction caused by shipping includes fuel spills, which coat the water with oil and coal dust, and erosion of the shoreline which causes damage to the fish and other inhabitants of the water.

Role of People

It's 85 degrees and your throat is parched. As sweat drips down your neck, you envision a tall glass of cool water. You enter the kitchen, grab a glass, and turn on the faucet. Nothing happens. You stare, incredulous. Your mouth believes it's filled with cotton as you sprint to the bathroom, twist the faucet and wait for that nice clear water to rush out. Nothing happens. What's going on?? You stare at your empty glass and reevaluate your dog, who laps happily out of the toilet bowl.

Does that make you think twice about whether or not you should help conserve water? It is our responsibility to help conserve and preserve Washington's water so that we will always have plenty of pure, clean water to drink and use for other purposes as well. There are many places in the world where there is not enough pure, clean water for the people, but people cannot live without drinking.

Every 8 seconds, a young person in a third world country dies from insufficient water supply or lack of pure water. Right now most of us have plenty of water and by just turning on the faucet, it is at our disposal. What we have plenty of with such little effort, we often take for granted. We don't usually think twice about the water that goes down our drain. When we turn on the faucet and water comes pouring out right down the drain, we don't even think about the fact that we are wasting water. The average American uses between 150-250 gallons of water daily. We could easily get by with a lot less if we just took the time to think about little things like turning off the faucet while brushing our teeth or shaving. We need to think when we use water, pay attention to what we are using it for, and we will soon realize just how much water we have been wasting.

The demands on our fresh water are ever increasing in today's modern society. New uses for this precious resource are being discovered constantly. Of all the water used in the United States, about 47% is used in agriculture, 44% inindustry, and only about 9% is used in households. Of the 9% that is used in households, only a very small portion is actually used as drinking water.

Americans use more water per household than any other country in the world. The results of the United States Geological Survey record that about 450 billion gallons of water are used in this country per day. That amounts to somewhere around 2,000 gallons per person per day. I know, you don't think that you use anywhere near that much water every day. You are right, you don't use that much water every day. That figure includes water that is used on farms and in factories to produce the food you eat; to cool your house, school, stores, and work places; to generate energy for running lights and appliances; and to produce manufactured goods. Even if you don't use 2,000 gallons of water every day, you probably use a lot more than you think. It is estimated that the average American uses between 100 and 200 gallons of water every day in the home.

People have always relied on lakes and rivers to provide food. Commercial fishing has been a prominent industry, as well as a vital source of food. Laws have been made about how many fish may be caught when and where so that people do not over-fish any body of water. When our water becomes polluted, many fish die and this creates a problem not only in the food chain, but also in the job line. Many people make their living fishing, and if there are not enough fish to keep all of the fisherman busy, then there could be many people without jobs.

The human body is made up of about 60-70% water. Your body uses water to digest food, maintain body temperature, remove wastes and toxins, cleanse the eyes, lubricate joints, and so much more. You probably drink a couple of glasses of water a day, but you also take in water from the food you eat. It is vital to your health that your body have enough water. You can probably live for several weeks without eating any food, but you would last only a few days without any water. With a loss of only 15% of your body's water, you are likely to die.

Fish and Wildlife

Animals rely on lakes and rivers for food and water. There are many animals that rely on the lakes and rivers and other bodies of water to supply them with life sustaining water, and for many of them food, also. Bears and birds are examples of animals that rely on fish as a main part of their diet. There are other animals that also eat a lot of fish. These animals would not know what to do if there were not enough fish for them to eat. Fish have always been a main part of their diet. Every part of nature depends on the other parts to help keep everything working right. Just as bears depend on salmon as a major part of their diet, and salmon depend on the water as their habitat, so every other thing in nature has things that it depends on. Some people call this the web of life, it is also known as an ecosystem.

Wetlands and Watersheds are also areas where numerable plants and animals other than fish make their home. When these areas are threatened by water pollution, the plants and animals are also threatened. Once a plant or animal becomes extinct, it is gone forever. Whales are one example of a species that has been in danger of becoming extinct. People have hunted whale for many years, and the water that they live in has also been polluted by oil and sewage. The ocean covers four-fifths of our planet. It is alive with activity - from tiny plants to gigantic whales. We must look after our water and preserve it, for without water there would be no life.

Nature

Plants are the source of all of our food. These plants receive energy from sunlight and store it in the food that they produce. This process called photosynthesis, is a process of the plant converting carbon dioxide and water to oxygen. This process is accelerated by a green pigment called clorophyll which is the pigment that makes the leaves green. The oxygen which is released during this process is what we breathe. This oxygen is also a life sustaining resource without which we could not live. Only green plants are capable of carrying out this amazing process of photosynthesis. The earth's entire source of food and most of its oxygen come from these plants.

Because plants are the source of all of our food and most of our oxygen, we need to make sure that they have enough water to carry out the process of making food and oxygen. Plants need water. They absorb the water and then let it pass through to their leaves where it is evaporated. Without water, there would be no life and there would be no life without plants either. It is necessary that we recognize the importance of plants and realize that we must conserve water so that we will continue to have a good supply of food and oxygen which we need to sustain life.

Forests provide a natural habitat for wildlife in Washington. Forests slow the water flowing down slopes, which reduces the amount of sediment runoff that contaminates rivers and streams. This keeps the water purer and helps give fish a healthier place to live. The forests also provide the earth with healthy topsoil. In the fall when the leaves and pine needles drop to the ground, they disintegrate and leave very healthy topsoil.

We need to protect the forests. They are an important part of our state. The trees purify the air that we breathe just as green plants do. Trees depend on water to live just as all other living things do. We should never cut down a tree within 100 feet of water. Doing this would contaminate the water and hurt the fish and other life in the water. When we cut down trees, we should plant new trees. This way there will always be trees available to meet our needs.

If we abuse our natural resources now, we will ruin them, and they will not be available for people in the future. This means that our children and grandchildren may not have enough clean, pure water to drink if we don't conserve it for them and ourselves. We have a responsibility to the future. We must protect and conserve our state's natural resource of water!

Pollution

Scientists have specified eight general categories for the major kinds of pollutants. These categories include sediment, sewage and organic wastes, plant nutrients, infectious agents, organic chemicals, salts and mineral substances, radioactive substances, and thermal (heat) pollution. We need to do our best to help keep these and other pollutants out of our streams, rivers, and lakes.

There are 8 categories for the major types of pollution: sediment, plant nutrients, infectious agents, organic chemicals, sewage and organic wastes, salts and mineral substances, and thermal (heat) pollution. It is important that we not only help conserve water, but that we help preserve water by trying not to pollute it.

Sediment runoff is when the loose, rich top soil erodes, and is washed into rivers, streams and lakes. Much of the sediment runoff comes from poorly managed land in rural areas. A good portion of sediment run off also comes from unprotected areas in or near cities. The solution to preventing water pollution by sediment runoff is to control erosion.

Plant nutrients is another category of water pollution that creates special problems. There are three main plant nutrients that cause problems: nitrogen, phosphorus, and potassium. The plant nutrients are the main problem mostly because of the large amounts of them that into surface waters. Infectious Agents such as microorganisms and bacteria are especially significant pollutants of water because they cause diseases. The solution to plant nutrient pollution as well as infectious agent pollution is to properly treat all sewage and waste.

Organic chemicals include a wide variety of materials that pollute water including pesticides and herbicides used to get rid of undesirable plants and insects. These chemicals get into the water by being washed into rivers, streams, and lakes, and are usually carried in the sediment runoff. It is vital that we do as much as possible to keep these chemicals out of our rivers, streams, and lakes by disposing of the harmful products properly.

Sewage and Organic Wastes are extremely significant pollutants because of the quantity involved. In addition to the wastes being carried from homes, large amounts of organic wastes result from food processing and industrial operations. Salts and Mineral Substances are pollutants that can be very troublesome. As water moves across the land, it picks up natural salts and minerals. Salt in water is a major problem especially in irrigation. Mining waste has caused severe damage to some streams and rivers. There have been laws made to regulate mining which have helped to reduce the pollution of water from erosion and drainage of mines.

Thermal pollution is when the temperature of water is raised, like when water is used to heat or cool something. When water is heated and then put back into rivers, streams, and lakes, it causes a lot of problems for plants and animals that depend on that water. Plants and animals cannot survive for long when the temperature is much higher than normal. Some industries are using cooling towers and other things to help reduce the temperature of water they have used so that they can reuse it for cooling purposes.

These different kinds of pollutants are only a few of the many things that pollute water. You may have noticed that these pollutants are each related to the ways people use water. This means that if we are going to have clean, pure water we are all going to have to work together to see that our water is not damaged by the necessary ways we use it.

Importance of Conservation of Water

With only 1% of water available for human consumption, doesn't it make sense that we should treat our water supply with more respect?

Conservation should not be considered an option any longer. Current circumstances require our full attention if we hope to thrive as a civilization.

If these statements sound dramatic, it is because much of the world is currently suffering due to a lack of clean water.

Importance of Conservation of Water / Stats

Statistics around the reveal that our fresh water supply is practically nonexistent.

That is why it is so important to seek out, find and start using all the innovative water conservation solutions and methods that are available today.

Whether you live in Australia or China or the US, it is time to wake up and take responsibility. It is easy to practice water conservation in the home, but there is more to be done. Our world needs help on a commercial level as well so that our waste can be controlled in such areas as agriculture and irrigation.

Water Conservation ideas are cropping up in exciting legislation. That's why our participation in the voting process is so important. Let's act now, so our kids will have a brighter, greener future. Demands are increasing every year for water while resources are becoming more and more limited. Since many individuals are unaware (or, sadly, just don't care) that this issue needs attention, it is up to more informed and proactive individuals and companies to take up the slack. A 40% increase is expected in water demand over the period of next two decades. The increase in water demand is a contribution of various factors including growing population, increased agricultural needs, industrial use of water and water needed for electricity production.

The problem of water waste is severe in countries where people are using the same inefficient methods for irrigation of agricultural land. Water needs are increasing every year and the proven fact is that clean water is not available to 1 out of 5 people on earth. Solutions are clearly needed.

The increased demand in water combined with the pollution of water has had many adverse effects on the environment, growth and economy of many countries.

Tankless Water Heaters Save Money, Energy & Water >>more here

Importance of Conservation of Water / Consequences of Not Conserving

- o Increased expenditures on pipeline development, sewage and other infrastructure to supply the water to the households and industries
- o Investment to build more dams to fulfill the water demands and increased electric supply
- o Huge amounts of money going out every year for recycling, cleaning and purification of the water
- o Large amounts of industrial pollutants dumped in the rivers which are responsible for destroying and upsetting the ecosystem of the whole planet
- o Rising instances of erosion of land because of increased demand of agricultural land and irrigation

Importance of Conservation of Water

The problem of water is local as well as global. The resources of water on earth are limited and are reducing every year. Due to the effects of global warming the frequency of rain has become unpredictable.

In some parts of world there is rain throughout the year, while in other parts individuals are suffering the consequences. Widespread dry spells are occurring over much of Europe and Asia, Canada, western and southern Africa, and eastern Australia.

And in parts of the world where rainwater is available, major problems are still prevalent because of a lack of proper systems to clean the water.

Pollutants like oil, cigarette butts and other impurities make the percentage of usable water practically nonexistent. Methods of water conservation must be implemented in these areas if our planet is to continue to sustain itself.

Importance of Conservation of Water / The Future

The United Nations has predicted that by the next two decades there will be 17% more demand of water compared to resources available.

The unpleasant fact is that the most people remain careless and do not recognize the importance of water conservation. Huge campaigns are conducted every year by many government organizations and NGOs to spread awareness of the importance of water conservation. As citizens of this planet it is our individual responsibility to take an active role in water conservation. The simple tasks we perform every day like gardening, tending to laundry, washing our cars, etc. are opportunities for us to make a difference. Change begins with us, and if we lead by example, others are more likely to follow.

6

WATER RECYCLING PROCESS

Recycled water, is former wastewater (sewage) that is treated to remove solids and certain impurities, and used in sustainable landscaping irrigation or to recharge groundwater aquifers. The purpose of these processes is sustainability and water conservation, rather than discharging the treated water to surface waters such as rivers and oceans.

The definition of reclaimed water, as defined by Levine and Asano, is "The end product of wastewater reclamation that meets water quality requirements for biodegradable materials, suspended matter and pathogens." In more recent conventional use, the term refers to water that is not treated as highly in order to offer a way to conserve drinking water. This water is thusly given to uses such as agriculture and sundry industry uses.

Cycled repeatedly through the planetary hydrosphere, all water on Earth is recycled water. But, typically when we hear the term "recycled water" or "reclaimed water" it means wastewater that is sent from our home or business through a pipeline system to a treatment facility where is treated to a level consistent with its intended use. It is then routed directly to a recycled water system for uses such as irrigation or industrial cooling.

The recycling and recharging is often done by using the treated wastewater for designated municipal sustainable gardening irrigation applications. In most locations, it is intended to only be used for nonpotable uses, such as irrigation, dust control, and fire suppression. There are examples of communities that have safely used recycled water for many years. Los Angeles County's sanitation districts have provided treated wastewater for landscape irrigation in parks and golf courses since

1929. The first reclaimed water facility in California was built at San Francisco's Golden Gate Park in 1932. The Irvine Ranch Water District (IRWD) was the first water district in California to receive an unrestricted use permit from the state for its recycled water; such a permit means that water can be used for any purpose except drinking. IRWD maintains one of the largest recycled water systems in the nation with more than 400 miles serving more than 4,500 metered connections. The Irvine Ranch Water District and Orange County Water District in Southern California are established leaders in recycled water. Further, the Orange County Water District, located in in Orange County, and in other locations throughout the world such as Singapore, water is given more advanced treatments and is used indirectly for drinking.

In spite of quite simple methods that incorporate the principles of water-sensitive urban design (WSUD) for easy recovery of stormwater runoff, there remains a common perception that reclaimed water must involve sophisticated and technically complex treatment systems, attempting to recover the most complex and degraded types of sewage. As this effort is supposedly driven by sustainability factors, this type of implementation should inherently be associated with point source solutions, where it is most economical to achieve the expected outcomes. Harvesting of stormwater or rainwater can be an extremely simple to comparatively complex, as well as energy and chemical intensive, recovery of more contaminated sewage.

HISTORY

Storm and sanitary sewers were necessarily developed along with the growth of cities. By the 1840s the luxury of indoor plumbing, which mixes human waste with water and flushes it away, eliminated the need for cesspools. Odor was considered the big problem in waste disposal and to address it, sewage could be drained to a lagoon, or "settled" and the solids removed, to be disposed of separately. This process is now called "primary treatment" and the settled solids are called "sludge."

At the end of the 19th century, since primary treatment still left odor problems, it was discovered that bad odors could be prevented by introducing oxygen into the decomposing sewage. This was the beginning of the biological aerobic and anaerobic treatments which are fundamental to waste water processes. By the 1920s, it became necessary to further control the pollution caused by the large quantities of human and industrial liquid wastes which were being piped into rivers and oceans, and modern treatment plants were being built in the US and other industrialized nations by the 1930s.

Designed to make water safe for fishing and recreation, the Clean Water Act of 1972 mandated elimination of the discharge of untreated waste from municipal and industrial sources, and the US federal government provided billions of dollars in grants for building sewage treatment plants around the country. Modern treatment plants, usually using oxidation and/or chlorination in addition to primary and secondary treatment, were required to meet certain standards.

Current treatment improves the quality of separated wastewater solids or sludge. The separated water is given further treatment considered adequate for non potable use by local agencies, and discharged into bodies of water, or reused as reclaimed water. In places like Florida, where it is necessary to avoid nutrient overload of sensitive receiving water, reuse of treated or reclaimed water can be more economically feasible than meeting the higher standards for surface water disposal mandated by the Clean Water Act

Maximum water recovery

To determine maximum water recovery there are various techniques that have been developed by researchers; for maximum water reuse/reclamation/recovery strategies such as water pinch analysis. The techniques help a user to target the minimum freshwater consumption and wastewater target. It also helps in designing the network that achieves the target. This provides a benchmark to be used by users in improving their water systems.

Benefits

The cost of reclaimed water exceeds that of potable water in many regions of the world, where a fresh water supply is plentiful. However, reclaimed water is usually sold to citizens at a cheaper rate to encourage its use. As fresh water supplies become limited from distribution costs, increased population demands, or climate change reducing sources, the cost ratios will evolve also.

Using reclaimed water for non-potable uses saves potable water for drinking, since less potable water will be used for non-potable uses.

It sometimes contains higher levels of nutrients such as nitrogen, phosphorus and oxygen which may somewhat help fertilize garden and agricultural plants when used for irrigation.

The usage of water reclamation decreases the pollution sent to sensitive environments. It can also enhance wetlands, which benefits the wildlife depending on that eco-system. For instance, The San Jose/Santa Clara Water Pollution Control Plant instituted a water recycling program to protect the San Francisco Bay area's natural salt water marshes.

Concerns

Reclaimed water is highly engineered for safety and reliability so that the quality of reclaimed water is more predictable than many existing surface and groundwater sources. Reclaimed water is considered safe when appropriately used. Reclaimed water planned for use in recharging aquifers or augmenting surface water receives adequate and reliable treatment before mixing with naturally occurring water and undergoing natural restoration processes. Some of this water eventually becomes part of drinking water supplies.

A water quality study published in 2009 compared the water quality differences of reclaimed/recycled water, surface water, and groundwater. Results indicate that reclaimed water, surface water, and groundwater are more similar than dissimilar with regard to constituents. The researchers tested for 244 representative constituents typically found in water. When detected, most constituents were in the parts per billion and parts per trillion range. DEET (a bug repellant), and Caffeine were found in all water types and virtually in all samples. Triclosan (in anti-bacterial soap & toothpaste) was found in all water types, but detected in higher levels (parts per trillion) in reclaimed water than in surface or groundwater. Very few hormones/steroids were detected in samples, and when detected were at very low levels. Haloacetic acids (a disinfection by-product) were found in all types of samples, even groundwater. The largest difference between reclaimed water and the other waters appears to be that reclaimed water has been disinfected and thus has disinfection by-products (due to chlorine use).

A 2005 study titled "Irrigation of Parks, Playgrounds, and Schoolyards with Reclaimed Water" found that there had been no incidences of illness or disease from either microbial pathogens or chemicals, and the risks of using using reclaimed water for irrigation are not measurably different from irrigation using potable water. Studies by the National Academies of Science, the Monterey Regional Water Pollution Control Agency, and others have found reclaimed water to be safe for agricultural use.

There is debate about possible health and environmental effects. To address these concerns, A Risk Assessment Study of potential health risks of recycled water and comparisons to conventional Pharmaceuticals and Personal Care Product (PPCP) exposures was conducted by the WateReuse Research Foundation. For each of four scenarios in which people come into contact with recycled water used for irrigation - children on the playground, golfers, and landscape, and agricultural workers - the findings from the study indicate that it could take anywhere from a few years to millions of years of exposure to nonpotable recycled water to reach the same exposure to PPCPs that we get in a single day through routine activities.

Distribution and Demand

Reclaimed water is often distributed with a dual piping network that keeps reclaimed water pipes completely separate from potable water pipes. In the United States, reclaimed water is always distributed in lavender (light purple) pipes to distinguish it from potable water. The use of the color purple for pipes carrying recycled water was pioneered by the Irvine Ranch Water District in Irvine, California.

In many cities using reclaimed water, it is now in such demand that consumers are only allowed to use it on assigned days. Some cities that previously offered unlimited reclaimed water at a flat rate are now beginning to charge citizens by the amount they use.

Testing standards

Reclaimed water is not regulated by the Environmental Protection Agency (EPA) but by the states, using standards formulated decades ago. Newer information shows serious public health concerns about pathogens in the water. Many pathogens cannot be detected by currently used tests.

Recent literature also questions the validity of testing for "indicator organisms" instead of pathogens. Nor do present standards consider interactions of heavy metals and pharmaceuticals which may foster the development of drug resistant pathogens in waters derived from sewage.

Potable uses

In most locations, reclaimed water is not directly mixed with potable (drinking) water for several reasons:

- Utilities providing reclaimed water for nonpotable uses do not treat the water to drinking water standards.
- Varying amounts of pathogens, pharmaceutical chemicals (e.g., hormones from female hormonal contraception) and other trace chemicals are able to pass through the treatment and filtering process, potentially causing danger to humans. Modern technologies such as reverse osmosis may help to somewhat overcome this problem. An experiment by the University of New South Wales reportedly showed a reverse osmosis system removed ethinylestradiol and paracetamol from the wastewater, even at 1000 times the expected concentration.
- Drinking water standards were developed for natural ground water, and are not appropriate for identifying contaminants in reclaimed water. In addition to pathogens, and organic and endocrine disrupting chemicals, a large number of compounds may be present in reclaimed water. They cannot all be tested for, and there is a paucity of toxicity information on many of the compounds.

Because of this, state regulatory agencies do not allow reclaimed water to be used for drinking, bathing, or filling swimming pools. They also warn those who use reclaimed water for irrigation to place a sign on their property warning people not to drink from the irrigation system, and to not use it directly on fruits or vegetables.

Aboard the International Space Station, astronauts have been able to drink recycled urine due to the introduction of the ECLSS system. The system cost $250 million and has been working since May 2009. The system recycles wastewater and urine back into potable water used for drinking, food preparation, and oxygen generation. This cuts back on the need for resupplying the space station so often.

Indirect potable reuse

Some municipalities are now investigating Indirect Potable Use (IPU) of reclaimed water. For example, reclaimed water may be pumped into (subsurface recharge) or percolated down to (surface recharge) groundwater aquifers, pumped out, treated again, and finally used as drinking water. This technique may also be referred to as groundwater recharging.

UNPLANNED INDIRECT POTABLE REUSE

Unplanned Indirect Potable Use has existed even before the introduction of reclaimed water. Many cities already use water from rivers that contain effluent discharged from upstream sewage treatment plants. There are many large towns on the River Thamesupstream of London (Oxford, Reading, Swindon, Bracknell) that discharge their treated sewage into the river, which is used to supply London with water downstream.

This phenomenon is also observed in the United States, where the Mississippi River serves as both the destination of sewage treatment plant effluent and the source of potable water. Research conducted in the 1960s by the London Metropolitan Water Board demonstrated that the maximum extent of recycling water is about 11 times before the taste of water induces nausea in sensitive individuals. This is caused by the build up of inorganic ions such as Cl, SO4, K and Na, which are not removed by conventional sewage treatment.

Space Travel

Wastewater reclamation can be especially important in relation to human spaceflight.

- In 1998, NASA announced it had built a human waste reclamation bioreactor designed for use in the International Space Station and a manned Mars mission. Human urine and feces are input into one end of the reactor and pure oxygen, pure water, and compost(humanure) are output from the other end. The soil could be used for growing vegetables, and the bioreactor also produceselectricity.

Treatment Improvements

As world populations require both more clean water and better ways to dispose of wastewater, the demand for water reclamation will increase. Future success in water reuse will depend on whether this can be done without adverse effects on human health and the environment.

In the United States, reclaimed waste water is generally treated to secondary level when used for irrigation, but there are questions about the adequacy of that treatment. Some leading scientists in the main water society, AWWA, have long believed that secondary treatment is insufficient to protect people against pathogens, and recommend adding at least membrane filtration, reverse osmosis, ozonation, or other advanced treatments for irrigation water.

Recent Advances in Reverse Osmosis have been in different countries, but have consistently produced very high quality water all the same. In Singapore, reclaimed water, also known as NEWater has become cleaner than the government issue tap water. Also, according to Bartels, the Bedok Demonstration Plant, which uses RO membranes, has successfully run for the past 3 years, producing high quality wastewater all the while.

Seepage of nitrogen and phosphorus into ground and surface water is also becoming a serious problem, and will probably lead to at least tertiary treatment of reclaimed to remove nutrients in future. Even using secondary treatment, water quality can be improved. Testing for pathogens using Polymerase Chain Reaction (PCR) instead of older culturing techniques, and changing the discredited fecal coloform "indicator organism" standard would be improvements.

In a large study treatment plants showed that they could significantly reduce the numbers of parasites in effluent, just by making adjustments to the currently used process. But, even using the best of current technology, risk of spreading drug resistance in the environment through wastewater effluent, would remain.

Some scientists have suggested that there need to be basic changes in treatment, such as using bacteria to degrade waste based on nitrogen (urine) and not just carbonaceous (fecal) waste, saying that this would greatly improve effectiveness of treatment. Currently designed plants do not deal well with contaminants in solution (e.g. pharmaceuticals). "Dewatering" solids is a major problem. Some wastes could be disposed of without mixing them with water to begin with. In an interesting innovation, solids (sludge) could be removed before entering digesters and burned into a gas that could be used to run engines.

Emerging disinfection technologies include ultrasound, pulse arc electrohydrolic discharge, and bank filtration. Another issue is concern about weakened mandates for pretreatment of industrial wastes before they are made part of the municipal waste stream. Some also believe that hospitals should treat their own wastes. The safety of drinking reclaimed water which has been given advanced treatment and blended with other waters, remains controversial.

Other Alternatives

In urban areas where climate change has threatened long-term water security and reduced rainfall over catchment areas, using reclaimed water for indirect potable use may be superior to other water supply augmentation methods. One other commonly used option is seawater desalination. Recycling wastewater and desalinating seawater may have many of the same disadvantages, including high costs of water treatment, infrastructure construction, transportation, and waste disposal problems. Although the best option varies from region to region, desalination is often superior economically, as reclaimed water usually requires a dual piping network, often with additional storage tanks, when used for nonpotable use.

A less elaborate alternative to reclaimed water is a greywater system. Greywater is wastewater that has been used in sinks, baths, showers, or washing machines, but does not contain sewage (see blackwater) and has not been treated at the same levels as recycled water. In a home system, treated or untreated greywater may be used to flush toilets or for irrigation. Some systems now exist which directly use greywater from a sink to flush a toilet or even combine the two into one piece of furniture.

Perhaps the simplest option is a rainwater harvesting system. Although there are concerns about the quality of rainwater in urban areas, due to air pollution and acid rain, many systems exist now to use untreated rainwater for nonpotable uses or treated rainwater for direct potable use. Urban design systems which incorporate rainwater harvesting and reduce runoff are known as Water Sensitive Urban Design (WSUD) in Australia, Low Impact Development (LID) in United states and Sustainable urban drainage systems (SUDS) in United Kingdom. There are also concerns about rainwater harvesting systems reducing the amount of run-off entering natural bodies of water.

Worldwide applications and acceptance

As of 2010, Israel leads the world in the proportion of water it recycles. Israel treats 80% of its sewage (400 billion liters a year), and 100% of the sewage from the Tel Aviv metropolitan area is treated and reused as irrigation water for agriculture and public works. The remaining sludge is currently pumped into the Mediterranean, however a new bill has passed stating a conversion to treating the sludge to be used as manure. Only 20% of the treated water is lost (due to evaporation, leaks, overflows and seeping). The recycled water allows farmers to plan ahead and not be limited by water shortages. There are many levels of treatment, and many different ways of treating the water-which leads to a big difference in the quality of the end product. The best quality of reclaimed sewage water comes from adding a gravitational filtering step, after the chemical and biological cleansing. This method uses small ponds in which the water seeps through the sand into the aquifer in about 400 days, then is pumped out as clear purified water. This is nearly the same process used in the space station water recycling system, which turns urine and feces into purified drinking water, oxygen and manure. To add to the efficiency of the Israeli system - the reclaimed sewage water may be mixed with reclaimed sea water (Plans are in action to increase the desalinization program up to 50% of the countries usage by 2013 - 600 billion liters of drinkable sea water a year), along with aquifer water and fresh sweet lake water - monitored by computer to account for the nationwide needs and input. This action reduced the outdated risk of salt and mineral percentages in the water. Plans to implement this overall usage of reclaimed water for drinking are discouraged by the psychological preconception of the public for the quality of reclaimed water, and the fear of its origin. As of today, all the reclaimed sewage water in Israel is used for agricultural and land improvement purposes.

The second largest waste reclamation program in the world is in Spain, where 12% of the nation's waste is treated.

The leaders in use of reclaimed water in the U.S. are Florida and California, with Irvine Ranch Water District as one of the leading developers. They were the first district to approve the use of reclaimed water for in-building piping and use in flushing toilets.

In a January 2012 U.S. National Research Council report, a committee of independent experts found that expanding the reuse of municipal wastewater for irrigation, industrial uses, and drinking water augmentation could significantly increase the United States' total available water resources. The committee noted that a portfolio of treatment options is available to mitigate water quality issues in reclaimed water. The report also includes a risk analysis that suggests the risk of exposure to certain microbial and chemical contaminants from drinking reclaimed water is not any higher than the risk from drinking water from current water treatment systems-and in some cases, may be orders of magnitude lower. The report concludes that adjustments to the federal regulatory framework could enhance public health protection and increase public confidence in water reuse.

As Australia continues to battle the 7-10-year drought, nationwide, reclaimed effluent is becoming a popular option. Two major capital cities in Australia, Adelaide and Brisbane, have already committed to adding reclaimed effluent to their dwindling dams. Brisbane has been seen as a leader in this trend, and other cities and towns will review the Western Corridor Recycled Water Project once completed. Goulbourn, Canberra, Newcastle, and Regional Victoria, Australia are already considering building a reclaimed effluent process.

According to an EU-funded study, "Europe and the Mediterranean countries are lagging behind" California, Japan, and Australia "in the extent to which reuse is being taken up." According to the study "the concept (of reuse) is difficult for the regulators and wider public to understand and accept."

Water efficiency

Water efficiency can be defined as:

1. The accomplishment of a function, task, process, or result with the minimal amount of water feasible;
2. An indicator of the relationship between the amount of water required for a particular purpose and the amount of water used or delivered.

Though the two are often used interchangeably, there is a difference between water conservation and water efficiency. Water efficiency differs from water conservation in that it focuses on reducing waste. A proposition is that the key for efficiency is reducing waste, not restricting use. It also emphasises the influence consumers can have in water efficiency by making small behavioural changes to reduce water wastage and by choosing more water efficient products. Examples of water efficient steps includes fixing leaking taps, taking showers rather than baths, installing displacements devices inside toilet cisterns, and using dishwashers and washing machines with full loads. These are things that fall under the definition of water efficiency, as their purpose is to obtain the desired result or level of service with the least necessary water.

According to the Second UN World Water Development Report, if present levels of consumption continue, two-thirds of the global population will live in areas of water stress by 2025. Increasing human demand for water coupled with the effects of climate change mean that the future of our water supply is not secure. As of now, 2.6 billion people do not have safe drinking water. Added to this, are the changes in climate, population growth and lifestyles. The changes in human lifestyle and activities require more water per capita. This tightens the competition for water amongst agricultural, industrial, and human consumption.

Residential

According to Waterwise, here are some simple ways to be more water efficient at home:

- Turning off the tap while brushing teeth- a running tap can waste over six liters per minute.

- Putting a "hippo" or other displacement device into the toilet cistern.
- Fixing dripping taps. A dripping tap wastes thousands of litres of water a year.
- Using a full load in the dishwasher and washing machine. A person should be sure to buy a water efficient model when buying a new machine.
- Having a short shower instead of a bath.
- Washing fruits and vegetables in a bowl rather than under a running tap.
- Using the leftover water to feed houseplants.
- Using a watering can or a hosepipe with a trigger nozzle instead of a sprinkler.
- Using a bucket and sponge when washing the car rather than a running hosepipe.
- Using little amount of water at every time
- We should wash our cloths in washing machine rather than washing with hands

Manufacturing

According to Save water!, these are solutions useful to manufacturers:

- Identifying and eliminating wastage (such as leaks) and inefficient processes (such as continual spray devices on stop-start production lines). This may be the most low cost area for water savings, as it involves minimal capital outlay. Savings can be made through implementing procedural changes, such as cleaning plant areas with brooms rather than water.
- Changing processes and plant machinery. A retrofit of key plant equipment may increase efficiency. Alternatively, upgrades to more efficient models can be factored in to planned maintenance and replacement schedules.
- Reusing wastewater. As well as saving on mains water, this option may improve the reliability of supply, whilst reducing trade waste charges and associated environmental risks.

Waterless products

- Using waterless car wash products to wash cars, boats, motorcycles and bicycles. This could save up to 150 gallons of water per wash.

Utilities

According to US EPA, here are some ideas for communities and utilities:

- Implementing a water-loss management program (e.g. locate and repair leaks).

- Utilities should strive for universal metering.
- Ensuring that fire hydrants are tamper proof.
- Equipment changes - Setting a good example by using water efficient equipment.
- Installing faucet aerators and low flow shower heads in municipal buildings.
- Replace worn out plumbing fixtures, appliances and equipment with water-saving models.
- Minimizing the water used in space cooling equipment in accordance with manufacturer's recommendations. Shut off cooling units when not needed.

Utilities can also modify their billing software to track customers who have taken advantage of various utility sponsored water conservation initiatives (toilet rebates, irrigation rebates, etc.) to see which initiatives provide the greatest water savings for the least cost.

7

WATERSHED

A watershed is an area of land that drains all the streams and rainfall to a common outlet such as the outflow of a reservoir, mouth of a bay, or any point along a stream channel. The word watershed is sometimes used interchangeably with drainage basin or catchment. Ridges and hills that separate two watersheds are called the drainage divide. The watershed consists of surface water—lakes, streams, reservoirs, and wetlands—and all the underlying ground water. Larger watersheds contain many smaller watersheds. It all depends on the outflow point; all of the land that drains water to the outflow point is the watershed for that outflow location. Watersheds are important because the streamflow and the water quality of a river are affected by things, human-induced or not, happening in the land area "above" the river-outflow point.

Most of the precipitation that falls within the drainage area of a stream's monitoring site collects in the stream and eventually flows by the monitoring site. Many factors, some listed below, determine how much of the streamflow will flow by the monitoring site. Imagine that the whole basin is covered with a big (and strong) plastic sheet. Then if it rained one inch, all of that rain would fall on the plastic, run downslope into gulleys and small creeks and then drain into main stream. Ignoring evaporation and any other losses, and using a 1-square mile example watershed, then all of the approximately 17,378,560 gallons of water that fell (you can use our interactive rainfall calculator to find out how many gallons of water fall during a storm) as rainfall would eventually flow by the watershed-outflow point.

An integrated watershed management approach needs to be adopted and the soil and water conservation technologies and approaches need to be applied in field situations by the officer-trainees. The Indian Institute of Soil and Water

Conservation (IISWC) director Pravin Shinagare said this while addressing a multi-disciplinary team of 20 officers from Odisha at the conclusion of a five-day training programme on soil and water conservation training-cum-exposure visit at the institute on Friday. The trainees were sponsored by the Institute on Management of Agricultural Extension, Bhubaneswar.

Mishra spoke on the concept, philosophy, importance of conserving natural resources through integrated watershed management. He advised the participants to give more thought on this mechanism. Speaking on the occasion, Plant Science head OP Chaturvedi stressed on importance of work in close collaboration with people and different agencies and not in isolation. He said that common property resources should be protected, conserved and utilised with community participation for their common cause and development. The HRD and Social Science division head Lakhan Singh motivated the officers to use past experiences and relate with existing agro-ecosystems. He highlighted the various factors promoting and inhibiting people's participation in watershed management. He said the commitment of villagers and officers towards watershed goal will make a difference in socio-economic transformation of people.

Watersheds sustain life, in more ways than one. According to the Environmental Protection Agency, more than $450 billion in foods, fiber, manufactured goods and tourism depend on clean, healthy watersheds. That is why proper watershed protection is necessary to you and your community. Watershed protection is a means of protecting a lake, river, or stream by managing the entire watershed that drains into it. Clean, healthy watersheds depend on an informed public to make the right decisions when it comes to the environment and actions made by the community.

CONTROLLING POLLUTION

In agricultural systems, common practices include the use of buffer strips, grassed waterways, the reestablishment of wetlands, and forms of sustainable agriculture practices such as conservation tillage, crop rotation and intercropping. After certain practices are installed, it is important to continually monitor these systems to ensure that they are working properly in terms of improving environmental quality.

In urban settings, managing areas to prevent soil loss and control stormwater flow are a few of the areas that receive attention. A few practices that are used to manage stormwater before it reaches a channel are retention ponds, filtering systems and wetlands. It is important that stormwater is given an opportunity to infiltrate so that the soil and vegetation can act as a "filter" before the water reaches nearby streams or lakes. In the case of soil erosion prevention, a few common practices include the use of silt fences, landscape fabric with grass seed and hydroseeding. The main objective in all cases is to slow water movement to prevent soil transport.

Governance

The 2nd World Water Forum held in The Hague in March 2000 raised some controversies that exposed the multilateral nature and imbalance the demand and supply management of freshwater. While donor organisations, private and government institutions backed by the World Bank, believe that freshwater should be governed as an economic good by appropriate pricing, NGOs however, held that freshwater resources should be seen as a social good. The concept of network governance where all stakeholders form partnerships and voluntarily share ideas towards forging a common vision can be used to resolve this clash of opinion in freshwater management. Also, the implementation of any common vision presents a new role for NGOs because of their unique capabilities in local community coordination, thus making them a valuable partner in network governance. Watersheds replicate this multilateral terrain with private industries and local communities interconnected by a common watershed. Although these groups share a common ecological space that could transcend state borders, their interests, knowledge and use of resources within the watershed are mostly disproportionate and divergent, resulting to the activities of a specific group adversely impacting on other groups. Examples being the Minamata Bay poisoning that occurred from 1932 to 1968, killing over 1,784 individuals and the Wabigoon River incidence of 1962. Furthermore, while some knowledgeable groups are shifting from efficient water resource exploitation to efficient utilization, net gain for the watershed ecology could be lost when other groups seizes the opportunity to exploit more resources. This gap in cooperative communication among multilateral stakeholders within an interconnected watershed, even with the likely presence of the usually reactive and political boundary-constraint state regulations, makes it necessary for the institutionalisation of an ecological-scale cooperative network of stakeholders. This concept supports an integrated management style for interconnected natural resources; resonating strongly with the Integrated Water Resources Management system proposed by Global Water Partnership.

Moreover, the need to create partnerships between donor organisations, private and government institutions and community representatives like NGOs in watersheds is to enhance an "organisational society" among stakeholders. This posits a type of public-private partnership, commonly referred to as Type II partnership, which essentially brings together stakeholders that share a common watershed under a voluntary, idea sharing and collectively agreed vision aimed at granting mutual benefits to all stakeholders. Also, it explicates the concept of network governance, which is "the only alternative for collective action", requiring government to rescale its role in decision making and collaborate with other stakeholders on a level playing field rather than in an administrative or hierarchical manner.

Several riparian states have adopted this concept in managing the increasingly scarce resources of watersheds. These include, the nine Rhine states, with a common vision of pollution control, the Lake Chad and river Nile Basins, whose common vision is to ensure environmental sustainability. As a partner in the commonly shared vision, NGOs has adopted a new role in operationalising the implementation of regional watershed management policies at the local level. For instance, essential local coordination and education are areas where the services of NGOs have been effective. This makes NGOs the "nuclei" for successful watershed management.

Watershed management overview

Every body of water (e.g., rivers, lakes, ponds, streams, and estuaries) has a watershed. The watershed is the area of land that drains or sheds water into a specific receiving waterbody, such as a lake or a river. As rainwater or melted snow runs downhill in the watershed, it collects and transports sediment and other materials and deposits them into the receiving waterbody.

What is Watershed Management?

Watershed management is a term used to describe the process of implementing land use practices and water management practices to protect and improve the quality of the water and other natural resources within a watershed by managing the use of those land and water resources in a comprehensive manner.

What is Watershed Management Planning?

Watershed management planning is a process that results in a plan or a blueprint of how to best protect and improve the water quality and other natural resources in a watershed. Very often, watershed boundaries extend over political boundaries into adjacent municipalities and/or states. That is why a comprehensive planning process that involves all affected municipalities located in the watershed is essential to successful watershed management.

Why is watershed management important?

Runoff from rainwater or snowmelt can contribute significant amounts of pollution into the lake or river. Watershed management helps to control pollution of the water and other natural resources in the watershed by identifying the different kinds of pollution present in the watershed and how those pollutants are transported, and recommending ways to reduce or eliminate those pollution sources.

All activities that occur within a watershed will somehow affect that watershed's natural resources and water quality. New land development, runoff from already-developed areas, agricultural activities, and household activities such as gardening/lawn care, septic system use/maintenance, water diversion and car maintenance all can affect the quality of the resources within a watershed. Watershed management planning comprehensively identifies those activities that affect the health of the watershed and makes recommendations to properly address them so that adverse impacts from pollution are reduced.

Watershed management is also important because the planning process results in a partnership among all affected parties in the watershed. That partnership is essential to the successful management of the land and water resources in the watershed since all partners have a stake in the health of the watershed. It is also an efficient way to prioritize the implementation of watershed management plans in times when resources may be limited.

Because watershed boundaries do not coincide with political boundaries, the actions of adjacent municipalities upstream can have as much of an impact on the downstream municipality's land and water resources as those actions carried out locally. Impacts from upstream sources can sometimes undermine the efforts of downstream municipalities to control pollution. Comprehensive planning for the resources within the entire watershed, with participation and commitment from all municipalities in the watershed, is critical to protecting the health of the watershed's resources.

What are some key steps in watershed management?

Familiarize Yourself with Your Watershed

Comprehensive watershed plans should first identify the characteristics of the watershed and inventory the watershed's natural resources. It is important to establish a baseline of the overall nature and quality of the watershed in order to plan properly for the improvement of the resources in the watershed and to actually measure those improvements.

The first steps in watershed management planning are to:

- Delineate and map the watershed's boundaries and the smaller drainage basins within the watershed;
- Inventory and map the resources in the watershed;
- Inventory and map the natural and manmade drainage systems in the watershed;
- Inventory and map land use and land cover;
- Inventory and map soils;

- Identify areas of erosion, including stream banks and construction sites;
- the quality of water resources in the watershed as a baseline; and
- Inventory and map pollution sources, both point sources (such as industrial discharge pipes) and nonpoint sources (such as municipal stormwater systems, failing septic systems, illicit discharges).

Much of this information may already be compiled and available through the DEP, the Natural Resources Conservation Service of the U. S. Department of Agriculture, and municipal offices such as planning and zoning, inland wetlands, and public works. Additional information specific to the watershed can be gathered during volunteer stream walks which allow for on the ground study of the general conditions of the receiving waters and the adjacent watershed areas.

Build Local Partnerships

Watershed planning should also identify and include the partners, or "stakeholders, " in the watershed. Development of local partnerships can also lead to greater awareness and support from the general public. Once individuals become aware of and interested in their watershed, they often become more involved in decision-making as well as hands-on protection and restoration efforts. Through such involvement, watershed management builds a sense of community, helps reduce conflicts, increases commitment to the actions necessary to meet environmental goals, and ultimately, improves the likelihood of success for the watershed management plan.

Local partnerships can include:

- Residents;
- Landowners;
- Federal, state, and municipal government officials;
- Watershed associations and other environmental and civic groups;
- Local business and industry leaders;
- Agricultural users;
- Developers;
- Teachers; and
- Recreational users.

Determine Priorities for Action

Watershed management planning should also determine what the opportunities are to reduce pollution or address other pressing environmental issues, prioritize those opportunities, and identify a time frame for accomplishing pollution reduction and resource and habitat improvements. Those issues that pose the greatest risk to human health or particular resources, or to desired uses of resources (i. e., swimming beaches), might be given highest priority for control and reduction. Watershed plans should establish clear goals, visions, and actions to be taken.

Examples of opportunities to reduce pollution and address other wide-ranging environmental issues include:

- Infrastructure improvements. More frequent maintenance of municipal stormwater systems or improving or replacing inadequate stormwater treatment systems, identifying and eliminating illicit (i. e., non-stormwater) connections to municipal stormwater systems;
- paved areas and other impervious cover, especially adjacent to waterbodies and wetlands. Zoning and subdivision regulations can be revised to address issues such as reducing lot coverage/impervious cover, reducing roadway widths, encouraging cluster and low impact development, limiting land disturbance such as grading and clearing, and increasing development setbacks from resources;
- Identifying appropriate areas for open space acquisition, greenways planning, and the establishment of vegetated buffers along waterbodies and wetland areas;
- Establishing sewer avoidance areas to limit development;
- Increasing inspections and maintenance of existing septic system and encouraging repairs to failing systems;
- Identifying other appropriate housekeeping practices for homeowners and landowners (encouraging the use of vegetated buffers adjacent to waterbodies and wetlands, reducing lawn areas and the amount of fertilizers and chemicals applied to them, recommending washing cars over lawns instead of driveways so rinse water can drain into the lawn and not run-off into storm drains, etc.);
- Identifying resource and wildlife habitat restoration priorities;
- Increasing and promoting public access and greenways and identifying areas where it is appropriate to do so; and
- Identifying and evaluating opportunities for nonstructural flood protection efforts;
- Improving waste management, pollution prevention, and recycling efforts at municipal facilities and businesses within the watershed.

CONDUCT EDUCATIONAL PROGRAMS

The degree of public education and participation in the planning process can greatly influence the success of watershed management. There are many ways to involve and educate the public in watershed management. The formation of citizen review groups and advisory committees can gain public support from the watershed and are an essential component to a successful, community-based, and locally led effort. These community-based groups and committees can also provide the means to keep the project going once the plan has been finalized to make sure that recommended actions are taken. It might also be helpful to identify a watershed coordinator to help in this effort.

- Outreach and education efforts can include:
- Periodic informational meetings;
- Stream walk assessments;
- Organized storm drain stenciling projects;
- Watershed clean-up days and riparian planting/habitat restoration days;
- Coordination with school systems within the watershed;
- Information kiosks and websites;
- Videos; and
- Newsletters and other printed materials to provide status and progress reports.

Ensure Implementation and Follow-up

It is important to establish a schedule with milestones and some sort of committee to ensure that projects proceed in a timely manner. A monitoring program should also be established to measure success through data gathering. It is also important to identify ways in which landowners can be assisted with undertaking necessary improvements, such as low interest loans or technical outreach information. Finally, it is important to ensure that the recommendations contained in the watershed plan, especially design standards, are integrated into municipal land use regulations (zoning, subdivision, inland wetlands).

Where can a municipality get additional information?

If you are interested in watershed planning, please contact the Department of Energy and Environmental Protection's Watershed Management and Coordination Program at 860-424-3020 or Office of Long Island Sound Programs at 860-424-3034.

DRAINAGE BASIN

A drainage basin or catchment basin is an extent or an area of land where all surface water from rain, melting snow, or ice converges to a single point at a lower elevation, usually the exit of the basin, where the waters join another body of water, such as a river, lake, reservoir, estuary, wetland, sea, or ocean. Thus if a tributary stream joins a brook that in turn joins a small river which is a tributary of a larger river, there is a series of successively larger (and lower elevation) drainage basins. For instance, the Missouri and Ohio rivers are within their own drainage basins and also within the drainage basin of the Mississippi River.

Other terms used to describe drainage basins are catchment, catchment area, drainage area, river basin and water basin. In North America, the term watershed is commonly used to mean a drainage basin, though in other English-speaking countries, it is used only in its original sense, to mean a drainage divide, the former meaning an area, the latter the high elevation perimeter of that area. Drainage basins drain into other drainage basins in a hierarchical pattern, with smaller sub-drainage basins combining into larger drainage basins.

In closed ("endorheic") drainage basins the water converges to a single point inside the basin, known as a sink, which may be a permanent lake, a dry lake, or a point where surface water is lost underground. The drainage basin includes all the streams and rivers that convey the water towards the sink, as well as the land surfaces from which water drains into those channels. The drainage basin acts as a funnel by collecting all the water within the area covered by the basin and channelling it to a single point. Each drainage basin is separated topographically from adjacent basins by a perimeter, the drainage divide, making up a succession of higher geographical features (such as a ridge, hill or mountains) forming a barrier.

Drainage basins are similar but not identical to hydrologic units, which are drainage areas delineated so as to nest into a multi-level hierarchical drainage system. Hydrologic units are defined to allow multiple inlets, outlets, or sinks. In a strict sense, all drainage basins are hydrologic units but not all hydrologic units are drainage basins.

Major drainage basins of the world

The following is a list of the major ocean basins:

- About 48.7% of the world's land drains to the Atlantic Ocean. In North America, surface water drains to the Atlantic via the Saint Lawrence River and Great Lakes basins, the Eastern Seaboard of the United States, the Canadian Maritimes, and most of Newfoundland and Labrador. Nearly all of South America east of the Andes also drains to the Atlantic, as does most of Western and Central Europe and the greatest portion of western Sub-Saharan Africa, as well as Western Sahara and part of Morocco. The two major mediterranean seas of the world also flow to the Atlantic:
- The Caribbean Sea and Gulf of Mexico basin includes most of the U. S. interior between the Appalachian and Rocky Mountains, a small part of the Canadian provinces of Alberta and Saskatchewan, eastern Central

America, the islands of the Caribbean and the Gulf, and a small part of northern South America.

- The Mediterranean Sea basin includes much of North Africa, east-central Africa (through the Nile River), Southern, Central, and Eastern Europe, Turkey, and the coastal areas of Israel, Lebanon, and Syria.
- The Arctic Ocean drains most of Western and Northern Canada east of the Continental Divide, northern Alaska and parts of North Dakota, South Dakota, Minnesota, and Montana in the United States, the north shore of the Scandinavian peninsula in Europe, and much of central and northern Russia, and parts of Kazakhstan and Mongolia in Asia, which totals to about 17% of the world's land.
- Just over 13% of the land in the world drains to the Pacific Ocean. Its basin includes much of China, eastern and southeastern Russia, Japan, the Korean Peninsula, most of Indochina, Indonesia and Malaysia, the Philippines, all of the Pacific Islands, the northeast coast of Australia, and Canada and the United States west of the Continental Divide (including most of Alaska), as well as western Central America and South America west of the Andes.
- The Indian Ocean's drainage basin also comprises about 13% of Earth's land. It drains the eastern coast of Africa, the coasts of the Red Sea and the Persian Gulf, the Indian subcontinent, Burma, and most of Australia.
- The Southern Ocean drains Antarctica. Antarctica comprises approximately eight percent of the Earth's land.

Largest river basins

The five largest river basins (by area), from largest to smallest, are the basins of the Amazon, the Río de la Plata, the Congo, the Nile, and the Mississippi. The three rivers that drain the most water, from most to least, are the Amazon, Ganga, and Congo rivers.

Endorheic drainage basins

Endorheic drainage basins are inland basins that do not drain to an ocean. Around 18% of all land drains to endorheic lakes or seas or sinks. The largest of these consists of much of the interior of Asia, which drains into the Caspian Sea, the Aral Sea, and numerous smaller lakes. Other endorheic regions include the Great Basin in the United States, much of the Sahara Desert, the drainage basin of the Okavango River (Kalahari Basin), highlands near the African Great Lakes, the interiors of Australia and the Arabian Peninsula, and parts in Mexico and the Andes. Some of these, such as the Great Basin, are not single drainage basins but collections of separate, adjacent closed basins.

In endorheic bodies of standing water where evaporation is the primary means of water loss, the water is typically more saline than the oceans. An extreme example of this is the Dead Sea.

Importance of drainage basins

Geopolitical boundaries

Drainage basins have been historically important for determining territorial boundaries, particularly in regions where trade by water has been important. For example, the English crown gave the Hudson's Bay Company a monopoly on the fur trade in the entire Hudson Bay basin, an area called Rupert's Land. Bioregional political organization today includes agreements of states (e.g., international treaties and, within the U.S.A., interstate compacts) or other political entities in a particular drainage basin to manage the body or bodies of water into which it drains. Examples of such interstate compacts are the Great Lakes Commission and the Tahoe Regional Planning Agency.

Hydrology

In hydrology, the drainage basin is a logical unit of focus for studying the movement of water within the hydrological cycle, because the majority of water that discharges from the basin outlet originated as precipitation falling on the basin. A portion of the water that enters the groundwater system beneath the drainage basin may flow towards the outlet of another drainage basin because groundwater flow directions do not always match those of their overlying drainage network. Measurement of the discharge of water from a basin may be made by a stream gauge located at the basin's outlet.

Rain gauge data is used to measure total precipitation over a drainage basin, and there are different ways to interpret that data. If the gauges are many and evenly distributed over an area of uniform precipitation, using the arithmetic mean method will give good results. In the Thiessen polygon method, the drainage basin is divided into polygons with the rain gauge in the middle of each polygon assumed to be representative for the rainfall on the area of land included in its polygon. These polygons are made by drawing lines between gauges, then making perpendicular bisectors of those lines form the polygons. The isohyetal method involves contours of equal precipitation are drawn over the gauges on a map. Calculating the area between these curves and adding up the volume of water is time consuming. Isochrone maps can be used to show the time taken for runoff water within a drainage basin to reach a lake, reservoir or outlet, assuming constant and uniform effective rainfall.

Geomorphology

Drainage basins are the principal hydrologic unit considered in fluvial geomorphology. A drainage basin is the source for water and sediment that moves from higher elevation through the river system to lower elevations as they reshape the channel forms.

Ecology

Drainage basins are important in ecology. As water flows over the ground and along rivers it can pick up nutrients, sediment, and pollutants. With the water, they are transported towards the outlet of the basin, and can affect the ecological processes along the way as well as in the receiving water source. Modern use of artificial fertilizers, containing nitrogen, phosphorus, and potassium, has affected the mouths of drainage basins. The minerals are carried by the drainage basin to the mouth, and may accumulate there, disturbing the natural mineral balance. This can cause eutrophication where plant growth is accelerated by the additional material.

Resource management

Because drainage basins are coherent entities in a hydro-logical sense, it has become common to manage water resources on the basis of individual basins. In the U. S. state of Minnesota, governmental entities that perform this function are called "watershed districts". In New Zealand, they are called catchment boards. Comparable community groups based in Ontario, Canada, are called conservation authorities. In North America, this function is referred to as "watershed management". In Brazil, the National Policy of Water Resources, regulated by Act 9. 433 of 1997, establishes the drainage basin as the territorial division of Brazilian water management. When a river basin crosses at least one political border, either a border within a nation or an international boundary, it is identified as a transboundary river. Management of such basins becomes the responsibility of the countries sharing it. Nile Basin Initiative, OMVS for Senegal River, Mekong River Commission are a few examples of arrangements involving management of shared river basins.

Management of shared drainage basins is also seen as a way to build lasting peaceful relationships among countries.

Catchment factors

The catchment is the most significant factor determining the amount or likelihood of flooding. Catchment factors are: topography, shape, size, soil type, and land use (paved or roofed areas). Catchment topography and shape determine the time taken for rain to reach the river, while catchment size, soil type, and development determine the amount of water to reach the river.

Topography

Generally, topography plays a big part in how fast runoff will reach a river. Rain that falls in steep mountainous areas will reach the primary river in the drainage basin faster than flat or lightly sloping areas (e.g., > 1% gradient).

Shape

Shape will contribute to the speed with which the runoff reaches a river. A long thin catchment will take longer to drain than a circular catchment.

Size

Size will help determine the amount of water reaching the river, as the larger the catchment the greater the potential for flooding. It also determined on the basis of length and width of the drainage basin.

Soil type

Soil type will help determine how much water reaches the river. Certain soil types such as sandy soils are very free-draining, and rainfall on sandy soil is likely to be absorbed by the ground. However, soils containing clay can be almost impermeable and therefore rainfall on clay soils will run off and contribute to flood volumes. After prolonged rainfall even free-draining soils can become saturated, meaning that any further rainfall will reach the river rather than being absorbed by the ground. If the surface is impermeable the precipitation will create surface run-off which will lead to higher risk of flooding; if the ground is permeable, the precipitation will infiltrate the soil.

Land use

Land use can contribute to the volume of water reaching the river, in a similar way to clay soils. For example, rainfall on roofs, pavements, and roads will be collected by rivers with almost no absorption into the groundwater.

Environmental law

Environmental laws often dictate the planning and actions that agencies take to manage watersheds. Some laws require that planning be done, others can be used to make a plan legally enforceable and others set out the ground rules for what can and cannot be done in development and planning. Most countries and states have their own laws regarding watershed management. Those concerned about aquatic habitat protection have a right to participate in the laws and planning processes that affect aquatic habitats. By having a clear understanding of whom to speak to and how to present the case for keeping our waterways clean a member of the public can become an effective watershed protection advocate.

Watershed Management Principles and Practices

Watershed management is the study of the relevant characteristics of a watershed aimed at the sustainable distribution of its resources and the process of creating and implementing plans, programs, and projects to sustain and enhance watershed functions that affect the plant, animal, and human communities within a watershed boundary. Features of a watershed that agencies seek to manage include water supply, water quality, drainage, stormwater runoff, water rights, and the overall planning and utilization of watersheds. Landowners, land use agencies, stormwater management experts, environmental specialists, water use surveyors and communities all play an integral part in watershed management.

Objectives of watershed management

The different objectives of watershed management programs are:

- To protect, conserve and improve the land of watershed for more efficient and sustained production.
- To protect and enhance the water resource originating in the watershed.
- To check soil erosion and to reduce the effect of sediment yield on the watershed.
- To rehabilitate the deteriorating lands.
- To moderate the floods peaks at down stream areas.
- To increase infiltration of rainwater.
- To improve and increase the production of timbers, fodder and wild life resource.
- To enhance the ground water recharge, wherever applicable.
- To reduce the occurrence of floods and the resultant damage by adopting strategies for flood management.

Need of watershed management

An integrated watershed management approach needs to be adopted and the soil and water conservation technologies and approaches need to be applied in field situations by the officer-trainees. The Indian Institute of Soil and Water Conservation (IISWC) director Pravin Shinagare said this while addressing a multi-disciplinary team of 20 officers from Odisha at the conclusion of a five-day training programme on soil and water conservation training-cum-exposure visit at the institute on Friday. The trainees were sponsored by the Institute on Management of Agricultural Extension, Bhubaneswar.

Mishra spoke on the concept, philosophy, importance of conserving natural resources through integrated watershed management. He advised the participants to give more thought on this mechanism. Speaking on the occasion, Plant Science head OP Chaturvedi stressed on importance of work in close collaboration with people and different agencies and not in isolation. He said that common property resources should be protected, conserved and utilised with community participation for their common cause and development. The HRD and Social Science division head Lakhan Singh motivated the officers to use past experiences and relate with existing agro-ecosystems. He highlighted the various factors promoting and inhibiting people's participation in watershed management. He said the commitment of villagers and officers towards watershed goal will make a difference in socio-economic transformation of people.

Watersheds sustain life, in more ways than one. According to the Environmental Protection Agency, more than $450 billion in foods, fiber, manufactured goods and tourism depend on clean, healthy watersheds. That is why proper watershed protection is necessary to you and your community. Watershed protection is a means of protecting a lake, river, or stream by managing the entire watershed that drains into it. Clean, healthy watersheds depend on an informed public to make the right decisions when it comes to the environment and actions made by the community.

Water Management Practices

Watershed management practices in terms of purpose

- To increase infiltration
- To increase water holding capacity
- To prevent soil erosion
- Method and accomplishment

In brief various control measures are:

- Vegetative measures (Agronomical measures)
- Strip cropping
- Pasture cropping
- Grass land farming
- Wood lands
- Engineering measures (Structural practices)
- Contour bunding
- contour trenching
- Terracing
- Construction of earthen embankment
- Construction of check dams
- Construction of farm ponds

- Construction of diversion
- Gully controlling structure
- Rock dam
- Establishment of permanent grass and vegetation
- Providing vegetative and stone barriers
- Construction of silt tanks dentension

Influence of soil conservation measures and vegetation cover on erosion, runoff and nutrient loss.

Principles of Watershed Management

The main principles of watershed management based on resource conservation, resource generation and resource utilization are:

- Utilizing the land based on its capability
- Protecting fertile top soil
- Minimizing silting up of tanks, reservoirs and lower fertile lands
- Protecting vegetative cover throughout the year
- In situ conservation of rain water
- Safe diversion of gullies and construction of check dam for in creasing ground water recharge
- In creasing cropping intensity through inter and sequence cropping.
- Alternate land use systems for efficient use of marginal lands.
- Water harvesting for supplementals irrigation.
- Maximizing farm income through agricultural related activities such as dairy, poultry, sheep, and goat forming.
- Improving infrastructural facilities for storage, transport and agricultural marketing,
- Improving socio-economic status of farmers

Controlling pollution

In agricultural systems, common practices include the use of buffer strips, grassed waterways, the reestablishment of wetlands, and forms of sustainable agriculture practices such as conservation tillage, crop rotation and intercropping. After certain practices are installed, it is important to continually monitor these systems to ensure that they are working properly in terms of improving environmental quality.

In urban settings, managing areas to prevent soil loss and control stormwater flow are a few of the areas that receive attention. A few practices that are used to manage stormwater before it reaches a channel are retention ponds, filtering systems and wetlands. It is important that stormwater is given an opportunity to infiltrate so that the soil and vegetation can act as a "filter" before the water reaches nearby streams or lakes. In the case of soil erosion prevention, a few common practices include the use of silt fences, landscape fabric with grass seed and hydroseeding. The main objective in all cases is to slow water movement to prevent soil transport.

Governance

The 2nd World Water Forum held in The Hague in March 2000 raised some controversies that exposed the multilateral nature and imbalance the demand and supply management of freshwater. While donor organisations, private and government institutions backed by the World Bank, believe that freshwater should be governed as an economic good by appropriate pricing, NGOs however, held that freshwater resources should be seen as a social good. The concept of network governance where all stakeholders form partnerships and voluntarily share ideas towards forging a common vision can be used to resolve this clash of opinion in freshwater management. Also, the implementation of any common vision presents a new role for NGOs because of their unique capabilities in local community coordination, thus making them a valuable partner in network governance.

Watersheds replicate this multilateral terrain with private industries and local communities interconnected by a common watershed. Although these groups share a common ecological space that could transcend state borders, their interests, knowledge and use of resources within the watershed are mostly disproportionate and divergent, resulting to the activities of a specific group adversely impacting on other groups. Examples being the Minamata Bay poisoning that occurred from 1932 to 1968, killing over 1,784 individuals and the Wabigoon River incidence of 1962. Furthermore, while some knowledgeable groups are shifting from efficient water resource exploitation to efficient utilization, net gain for the watershed ecology could be lost when other groups seizes the opportunity to exploit more resources. This gap in cooperative communication among multilateral stakeholders within an interconnected watershed, even with the likely presence of the usually reactive and political boundary-constraint state regulations, makes it necessary for the institutionalisation of an ecological-scale cooperative network of stakeholders. This concept supports an integrated management style for interconnected natural resources; resonating strongly with the Integrated Water Resources Management system proposed by Global Water Partnership.

Moreover, the need to create partnerships between donor organisations, private and government institutions and community representatives like NGOs in watersheds is to enhance an "organisational society" among stakeholders.

This posits a type of public-private partnership, commonly referred to as Type II partnership, which essentially brings together stakeholders that share a common watershed under a voluntary, idea sharing and collectively agreed vision aimed at granting mutual benefits to all stakeholders. Also, it explicates the concept of network governance, which is "the only alternative for collective action", requiring government to rescale its role in decision making and collaborate with other stakeholders on a level playing field rather than in an administrative or hierarchical manner.

Several riparian states have adopted this concept in managing the increasingly scarce resources of watersheds. These include, the nine Rhine states, with a common vision of pollution control, the Lake Chad and river Nile Basins, whose common vision is to ensure environmental sustainability. As a partner in the commonly shared vision, NGOs has adopted a new role in operationalising the implementation of regional watershed management policies at the local level. For instance, essential local coordination and education are areas where the services of NGOs have been effective. This makes NGOs the "nuclei" for successful watershed management.

Watershed management overview

Every body of water (e.g., rivers, lakes, ponds, streams, and estuaries) has a watershed. The watershed is the area of land that drains or sheds water into a specific receiving waterbody, such as a lake or a river. As rainwater or melted snow runs downhill in the watershed, it collects and transports sediment and other materials and deposits them into the receiving waterbody.

What is Watershed Management?

Watershed management is a term used to describe the process of implementing land use practices and water management practices to protect and improve the quality of the water and other natural resources within a watershed by managing the use of those land and water resources in a comprehensive manner.

What is Watershed Management Planning?

Watershed management planning is a process that results in a plan or a blueprint of how to best protect and improve the water quality and other natural resources in a watershed. Very often, watershed boundaries extend over political boundaries into adjacent municipalities and/or states. That is why a comprehensive planning process that involves all affected municipalities located in the watershed is essential to successful watershed management.

Why is watershed management important?

Runoff from rainwater or snowmelt can contribute significant amounts of pollution into the lake or river. Watershed management helps to control pollution of the water and other natural resources in the watershed by identifying the different kinds of pollution present in the watershed and how those pollutants are transported, and recommending ways to reduce or eliminate those pollution sources.

All activities that occur within a watershed will somehow affect that watershed's natural resources and water quality. New land development, runoff from already-developed areas, agricultural activities, and household activities such as gardening/lawn care, septic system use/maintenance, water diversion and car maintenance all can affect the quality of the resources within a watershed. Watershed management planning comprehensively identifies those activities that affect the health of the watershed and makes recommendations to properly address them so that adverse impacts from pollution are reduced.

Watershed management is also important because the planning process results in a partnership among all affected parties in the watershed. That partnership is essential to the successful management of the land and water resources in the watershed since all partners have a stake in the health of the watershed. It is also an efficient way to prioritize the implementation of watershed management plans in times when resources may be limited.

Because watershed boundaries do not coincide with political boundaries, the actions of adjacent municipalities upstream can have as much of an impact on the downstream municipality's land and water resources as those actions carried out locally. Impacts from upstream sources can sometimes undermine the efforts of downstream municipalities to control pollution. Comprehensive planning for the resources within the entire watershed, with participation and commitment from all municipalities in the watershed, is critical to protecting the health of the watershed's resources.

What are some key steps in watershed management?

Familiarize Yourself with Your Watershed

Comprehensive watershed plans should first identify the characteristics of the watershed and inventory the watershed's natural resources. It is important to establish a baseline of the overall nature and quality of the watershed in order to plan properly for the improvement of the resources in the watershed and to actually measure those improvements.

The first steps in watershed management planning are to:

- Delineate and map the watershed's boundaries and the smaller drainage basins within the watershed;
- and map the resources in the watershed;

- Inventory and map the natural and manmade drainage systems in the watershed;
- Inventory and map land use and land cover;
- Inventory and map soils;
- Identify areas of erosion, including stream banks and construction sites;
- Identify the quality of water resources in the watershed as a baseline; and
- Inventory and map pollution sources, both point sources (such as industrial discharge pipes) and nonpoint sources (such as municipal stormwater systems, failing septic systems, illicit discharges).

Much of this information may already be compiled and available through the DEP, the Natural Resources Conservation Service of the U. S. Department of Agriculture, and municipal offices such as planning and zoning, inland wetlands, and public works. Additional information specific to the watershed can be gathered during volunteer stream walks which allow for on the ground study of the general conditions of the receiving waters and the adjacent watershed areas.

Build Local Partnerships

Watershed planning should also identify and include the partners, or "stakeholders, " in the watershed. Development of local partnerships can also lead to greater awareness and support from the general public. Once individuals become aware of and interested in their watershed, they often become more involved in decision-making as well as hands-on protection and restoration efforts. Through such involvement, watershed management builds a sense of community, helps reduce conflicts, increases commitment to the actions necessary to meet environmental goals, and ultimately, improves the likelihood of success for the watershed management plan.

Local partnerships can include:

- Residents;
- Landowners;
- Federal, state, and municipal government officials;
- Watershed associations and other environmental and civic groups;
- Local business and industry leaders;
- Agricultural users;
- Developers;
- Teachers; and
- Recreational users.

Determine Priorities for Action

Watershed management planning should also determine what the opportunities are to reduce pollution or address other pressing environmental issues, prioritize those opportunities, and identify a time frame for accomplishing pollution reduction and resource and habitat improvements. Those issues that pose the greatest risk to human health or particular resources, or to desired uses of resources (i. e., swimming beaches), might be given highest priority for control and reduction. Watershed plans should establish clear goals, visions, and actions to be taken.

Examples of opportunities to reduce pollution and address other wide-ranging environmental issues include:

- Infrastructure improvements. More frequent maintenance of municipal stormwater systems or improving or replacing inadequate stormwater treatment systems, identifying and eliminating illicit (i. e., non-stormwater) connections to municipal stormwater systems;
- Reducing paved areas and other impervious cover, especially adjacent to waterbodies and wetlands. Zoning and subdivision regulations can be revised to address issues such as reducing lot coverage/impervious cover, reducing roadway widths, encouraging cluster and low impact development, limiting land disturbance such as grading and clearing, and increasing development setbacks from resources;
- Identifying appropriate areas for open space acquisition, greenways planning, and the establishment of vegetated buffers along waterbodies and wetland areas;
- Establishing sewer avoidance areas to limit development;
- Increasing inspections and maintenance of existing septic system and encouraging repairs to failing systems;
- Identifying other appropriate housekeeping practices for homeowners and landowners (encouraging the use of vegetated buffers adjacent to waterbodies and wetlands, reducing lawn areas and the amount of fertilizers and chemicals applied to them, recommending washing cars over lawns instead of driveways so rinse water can drain into the lawn and not run-off into storm drains, etc.);
- Identifying resource and wildlife habitat restoration priorities;
- Increasing and promoting public access and greenways and identifying areas where it is appropriate to do so; and

- Identifying and evaluating opportunities for nonstructural flood protection efforts;
- Improving waste management, pollution prevention, and recycling efforts at municipal facilities and businesses within the watershed.

Conduct Educational Programs

The degree of public education and participation in the planning process can greatly influence the success of watershed management. There are many ways to involve and educate the public in watershed management. The formation of citizen review groups and advisory committees can gain public support from the watershed and are an essential component to a successful, community-based, and locally led effort. These community-based groups and committees can also provide the means to keep the project going once the plan has been finalized to make sure that recommended actions are taken. It might also be helpful to identify a watershed coordinator to help in this effort.

- Outreach and education efforts can include:
- Periodic informational meetings;
- Stream walk assessments;
- Organized storm drain stenciling projects;
- Watershed clean-up days and riparian planting/habitat restoration days;
- Coordination with school systems within the watershed;
- Information kiosks and websites;
- Videos; and
- Newsletters and other printed materials to provide status and progress reports.

Ensure Implementation and Follow-up

It is important to establish a schedule with milestones and some sort of committee to ensure that projects proceed in a timely manner. A monitoring program should also be established to measure success through data gathering. It is also important to identify ways in which landowners can be assisted with undertaking necessary improvements, such as low interest loans or technical outreach information. Finally, it is important to ensure that the recommendations contained in the watershed plan, especially design standards, are integrated into municipal land use regulations (zoning, subdivision, inland wetlands).

Where can a municipality get additional information?

If you are interested in watershed planning, please contact the Department of Energy and Environmental Protection's Watershed Management and Coordination Program at 860-424-3020 or Office of Long Island Sound Programs at 860-424-3034.

Drainage basin

A drainage basin or catchment basin is an extent or an area of land where all surface water from rain, melting snow, or ice converges to a single point at a lower elevation, usually the exit of the basin, where the waters join another body of water, such as a river, lake, reservoir, estuary, wetland, sea, or ocean. Thus if a tributary stream joins a brook that in turn joins a small river which is a tributary of a larger river, there is a series of successively larger (and lower elevation) drainage basins. For instance, the Missouri and Ohio rivers are within their own drainage basins and also within the drainage basin of the Mississippi River.

Other terms used to describe drainage basins are catchment, catchment area, drainage area, river basin and water basin. In North America, the term watershed is commonly used to mean a drainage basin, though in other English-speaking countries, it is used only in its original sense, to mean a drainage divide, the former meaning an area, the latter the high elevation perimeter of that area. Drainage basins drain into other drainage basins in a hierarchical pattern, with smaller sub-drainage basins combining into larger drainage basins.

In closed ("endorheic") drainage basins the water converges to a single point inside the basin, known as a sink, which may be a permanent lake, a dry lake, or a point where surface water is lost underground. The drainage basin includes all the streams and rivers that convey the water towards the sink, as well as the land surfaces from which water drains into those channels. The drainage basin acts as a funnel by collecting all the water within the area covered by the basin and channelling it to a single point. Each drainage basin is separated topographically from adjacent basins by a perimeter, the drainage divide, making up a succession of higher geographical features (such as a ridge, hill or mountains) forming a barrier.

Drainage basins are similar but not identical to hydrologic units, which are drainage areas delineated so as to nest into a multi-level hierarchical drainage system. Hydrologic units are defined to allow multiple inlets, outlets, or sinks. In a strict sense, all drainage basins are hydrologic units but not all hydrologic units are drainage basins.

Major drainage basins of the world

The following is a list of the major ocean basins:

- About 48.7% of the world's land drains to the Atlantic Ocean. In North America, surface water drains to the Atlantic via the Saint Lawrence River and Great Lakes basins, the Eastern Seaboard of the United States, the Canadian Maritimes, and most of Newfoundland and Labrador. Nearly all of South America east of the Andes also drains to the Atlantic, as does most of Western and Central Europe and the greatest portion of western Sub-Saharan Africa, as well as Western Sahara and part of Morocco. The two major mediterranean seas of the world also flow to the Atlantic:

- The Caribbean Sea and Gulf of Mexico basin includes most of the U. S. interior between the Appalachian and Rocky Mountains, a small part of the Canadian provinces of Alberta and Saskatchewan, eastern Central America, the islands of the Caribbean and the Gulf, and a small part of northern South America.
- The Mediterranean Sea basin includes much of North Africa, east-central Africa (through the Nile River), Southern, Central, and Eastern Europe, Turkey, and the coastal areas of Israel, Lebanon, and Syria.
- The Arctic Ocean drains most of Western and Northern Canada east of the Continental Divide, northern Alaska and parts of North Dakota, South Dakota, Minnesota, and Montana in the United States, the north shore of the Scandinavian peninsula in Europe, and much of central and northern Russia, and parts of Kazakhstan and Mongolia in Asia, which totals to about 17% of the world's land.
- Just over 13% of the land in the world drains to the Pacific Ocean. Its basin includes much of China, eastern and southeastern Russia, Japan, the Korean Peninsula, most of Indochina, Indonesia and Malaysia, the Philippines, all of the Pacific Islands, the northeast coast of Australia, and Canada and the United States west of the Continental Divide (including most of Alaska), as well as western Central America and South America west of the Andes.
- The Indian Ocean's drainage basin also comprises about 13% of Earth's land. It drains the eastern coast of Africa, the coasts of the Red Sea and the Persian Gulf, the Indian subcontinent, Burma, and most of Australia.
- The Southern Ocean drains Antarctica. Antarctica comprises approximately eight percent of the Earth's land.

LARGEST RIVER BASINS

The five largest river basins (by area), from largest to smallest, are the basins of the Amazon, the Río de la Plata, the Congo, the Nile, and the Mississippi. The three rivers that drain the most water, from most to least, are the Amazon, Ganga, and Congo rivers.

Endorheic drainage basins

Endorheic drainage basins are inland basins that do not drain to an ocean. Around 18% of all land drains to endorheic lakes or seas or sinks. The largest of these consists of much of the interior of Asia, which drains into the Caspian Sea, the Aral Sea, and numerous smaller lakes. Other endorheic regions include the Great Basin in the United States, much of the Sahara Desert, the drainage basin of the Okavango River (Kalahari Basin), highlands near the African Great Lakes, the interiors of Australia and the Arabian Peninsula, and parts in Mexico and the Andes. Some of these, such as the Great Basin, are not single drainage basins but collections of separate, adjacent closed basins.

In endorheic bodies of standing water where evaporation is the primary means of water loss, the water is typically more saline than the oceans. An extreme example of this is the Dead Sea.

Importance of drainage basins

Geopolitical boundaries

Drainage basins have been historically important for determining territorial boundaries, particularly in regions where trade by water has been important. For example, the English crown gave the Hudson's Bay Company a monopoly on the fur trade in the entire Hudson Bay basin, an area called Rupert's Land. Bioregional political organization today includes agreements of states (e.g., international treaties and, within the U.S.A., interstate compacts) or other political entities in a particular drainage basin to manage the body or bodies of water into which it drains. Examples of such interstate compacts are the Great Lakes Commission and the Tahoe Regional Planning Agency.

Hydrology

In hydrology, the drainage basin is a logical unit of focus for studying the movement of water within the hydrological cycle, because the majority of water that discharges from the basin outlet originated as precipitation falling on the basin. A portion of the water that enters the groundwater system beneath the drainage basin may flow towards the outlet of another drainage basin because groundwater flow directions do not always match those of their overlying drainage network. Measurement of the discharge of water from a basin may be made by a stream gauge located at the basin's outlet.

Rain gauge data is used to measure total precipitation over a drainage basin, and there are different ways to interpret that data. If the gauges are many and evenly distributed over an area of uniform precipitation, using the arithmetic mean method will give good results. In the Thiessen polygon method, the drainage basin is divided into polygons with the rain gauge in the middle of each polygon assumed to be representative for the rainfall on the area of land included in its polygon. These polygons are made by drawing lines between gauges, then making perpendicular bisectors of those lines form the polygons. The isohyetal method involves contours of equal precipitation are drawn over the gauges on a map. Calculating the area between these curves and adding up the volume of water is time consuming. Isochrone maps can be used to show the time taken for runoff water within a drainage basin to reach a lake, reservoir or outlet, assuming constant and uniform effective rainfall.

Geomorphology

Drainage basins are the principal hydrologic unit considered in fluvial geomorphology. A drainage basin is the source for water and sediment that moves from higher elevation through the river system to lower elevations as they reshape the channel forms.

Ecology

Drainage basins are important in ecology. As water flows over the ground and along rivers it can pick up nutrients, sediment, and pollutants. With the water, they are transported towards the outlet of the basin, and can affect the ecological processes along the way as well as in the receiving water source. Modern use of artificial fertilizers, containing nitrogen, phosphorus, and potassium, has affected the mouths of drainage basins. The minerals are carried by the drainage basin to the mouth, and may accumulate there, disturbing the natural mineral balance. This can cause eutrophication where plant growth is accelerated by the additional material.

RESOURCE MANAGEMENT

Because drainage basins are coherent entities in a hydro-logical sense, it has become common to manage water resources on the basis of individual basins. In the U. S. state of Minnesota, governmental entities that perform this function are called "watershed districts". In New Zealand, they are called catchment boards. Comparable community groups based in Ontario, Canada, are called conservation authorities. In North America, this function is referred to as "watershed management". In Brazil, the National Policy of Water Resources, regulated by Act 9. 433 of 1997, establishes the drainage basin as the territorial division of Brazilian water management.

When a river basin crosses at least one political border, either a border within a nation or an international boundary, it is identified as a transboundary river. Management of such basins becomes the responsibility of the countries sharing it. Nile Basin Initiative, OMVS for Senegal River, Mekong River Commission are a few examples of arrangements involving management of shared river basins.

Management of shared drainage basins is also seen as a way to build lasting peaceful relationships among countries.

Catchment factors

The catchment is the most significant factor determining the amount or likelihood of flooding. Catchment factors are: topography, shape, size, soil type, and land use (paved or roofed areas). Catchment topography and shape determine the time taken for rain to reach the river, while catchment size, soil type, and development determine the amount of water to reach the river.

Topography

Generally, topography plays a big part in how fast runoff will reach a river. Rain that falls in steep mountainous areas will reach the primary river in the drainage basin faster than flat or lightly sloping areas (e.g., > 1% gradient).

Shape

Shape will contribute to the speed with which the runoff reaches a river. A long thin catchment will take longer to drain than a circular catchment.

Size

Size will help determine the amount of water reaching the river, as the larger the catchment the greater the potential for flooding. It also determined on the basis of length and width of the drainage basin.

Soil type

Soil type will help determine how much water reaches the river. Certain soil types such as sandy soils are very free-draining, and rainfall on sandy soil is likely to be absorbed by the ground. However, soils containing clay can be almost impermeable and therefore rainfall on clay soils will run off and contribute to flood volumes. After prolonged rainfall even free-draining soils can become saturated, meaning that any further rainfall will reach the river rather than being absorbed by the ground. If the surface is impermeable the precipitation will create surface run-off which will lead to higher risk of flooding; if the ground is permeable, the precipitation will infiltrate the soil.

Land use

Land use can contribute to the volume of water reaching the river, in a similar way to clay soils. For example, rainfall on roofs, pavements, and roads will be collected by rivers with almost no absorption into the groundwater.

8

WATER CONSERVATION PRACTICES

Low-flow plumbing fixtures and retrofit programs are permanent, one-time conservation measures that can be implemented with little or no additional cost over the lifetime of the fixtures. In some cases, these fixtures can even save the residents money over the long term. The most commonly recommended low-flow plumbing fixtures are pressure reduction devices, faucet aerators, toilet displacement devices, low-flush toilets, low-flow showerheads, and plumbing modifications for gray water reuse.

Pressure Reduction. Homeowners can reduce the water pressure in a home by installing pressure reducing valves. A reduction in water pressure can save water in other ways: it can reduce the likelihood of leaking water pipes, leaking water heaters, and dripping faucets.

Faucet Aerators. Faucet aerators, which break the flowing water into fine droplets and entrain air while maintaining wetting effectiveness, are inexpensive devices that can be installed in sinks to reduce the volume of water used. Aerators are easily installed and can reduce the volume of water use at a faucet by as much as 60% while still maintaining a strong flow. More efficient kitchen and bathroom faucets that use only 7.5 l/min, in contrast to standard faucets, which use 12 to 20 l/min, are also available.

Toilet Displacement Devices. Non-toxic bricks or plastic containers (e.g., milk jugs filled with water or pebbles) can be placed in a toilet tank to reduce the amount of water used per flush. By placing between one and three such containers in the tank, more than 41 of water can be saved per flush. A toilet dam, which holds back a reservoir of water when the toilet is flushed, can also be used instead of the displacement device to save water.

Low-Flush Toilets. Conventional toilets use 15 to 201 of water per flush, but low-flush toilets use only 61 of water or less. Since low-flush toilets use less water, they also reduce the volume of wastewater produced. A schematic of a low-flush toilet is shown in Figure 43. Even in existing residences, replacement of conventional toilets with low-flush toilets is a practical and economical water-saving alternative.

Low-Flow Showerheads. Showers account for about 20% of the total indoor water use in an household. By replacing the standard 18 l/min showerheads with 10 l/min showerheads, which cost less than $5 each, a family of four can save approximately 80,000 l/year. Properly designed low-flow showerheads, currently available, are able to provide the quality of water delivery found in higher volume models.

Gray Water Use. Domestic wastewater composed of washwater from kitchen sinks and tubs, clothes washers, and laundry tubs is called gray water. Gray water can be used by homeowners for home gardening, lawn maintenance, landscaping, and other uses that do not require potable water. The level of contamination of gray waters is minimal; however, the plumbing modifications needed to make use of this water should not allow its contamination by wastes from the toilets, which have the potential to spread disease, cause undesirable odors, and result in aesthetic degradation of homestead yards and gardens.

1. The 6 liter flush design of this gravity toilet has a different flush mechanism.
2. Steep bowl sides and a narrow trapway to allow the siphoned water to gain velocity for more effective removal of waste.
3. This is where the water pushes waste into the trapway.
4. Stored water flows into the bowl.

Source: USEPA, Cleaner Water Through Conservation, Washington B.C., 1995 (Report No. EPA-841/B-95-002).

Landscaping Water Conservation Practices

Drought-Tolerant Plants. Water conservation in landscaping can be accomplished through the use of plants that need little water, thereby saving not only water but labor and fertilizer as well. Careful landscape design can significantly reduce water use; it can also take advantage of native plants which have evolved water-saving or water-tolerant characteristics ideally suited for the local climatic conditions. Use of native plants can also help to minimize the spread of exotic plant species that disrupt local ecosystems. In addition to the selection of the plant species to be used in landscaping, practices such as the use of low precipitation rate sprinklers that have better distribution uniformity, bubbler/soaker systems, and/or drip or point irrigation systems can also conserve water used for landscaping purposes.

Xeriscaping. Xeriscaping is an innovative approach to landscaping that promotes water conservation and pollution prevention. Traditional landscapes might incorporate one or two principles of water conservation, but xeriscaping uses planning and design, soil analysis, selection of suitable plants, practical turf areas, efficient irrigation, use of mulches, and appropriate maintenance to create an appropriate landscape for a given climatic condition. Xeriscaping is most successfully practiced in arid and semi-arid areas, where it has proved useful for minimizing irrigation and external maintenance needs while presenting an attractive appearance.

Agricultural Water Conservation Practices

Water saving irrigation practices fall into three categories: field practices, management strategies, and system modifications. Examples of these practices include, respectively, the chisel plow aeration of extremely compacted soils, furrow diking to prevent uncontrolled runoff, and leveling of the land surface to distribute water more evenly. A number of these practices have been previously detailed in chapters 2 and 3.

Irrigation Scheduling. Improved irrigation scheduling can reduce the amount of water required to irrigate a crop effectively by reducing evaporative losses, supplying water when most needed by the irrigated plants, and applying the water in a manner best suited to the plants being irrigated. A careful choice of the rate and timing of irrigation can help farmers to maintain yields with less water. In making scheduling decisions, irrigators should consider:

- The uncertainty of rainfall and the timing of crop water demands.
- The limited water storage capacity of many irrigated soils.
- The finite pumping capacity of most irrigation systems.
- The price of water and changes in water prices as additional operators increase water demand.

Irrigation Management. Management strategies involve monitoring soil and water conditions and collecting information on water use and efficiency. The methods include measuring rainfall, determining soil moisture levels, monitoring pumping plant efficiency, and scheduling irrigation. Typical system modifications include the addition of drop tubes to a center pivot irrigation system, retrofitting wells with smaller pumps, installing a surge or demand irrigation system, and/or constructing a tailwater or return flow recovery system.

Industrial and Commercial Users Water Conservation Practices

Water recycling is the reuse of water for the same application for which it was originally used. Recycled water might require treatment before it can be reused.

Cooling water recirculation and washwater recycling are the most widely used water recycling practices. The following guidelines should be used when considering water reuse and recycling in industrial and commercial applications:

- Identification of water reuse opportunities: Are there areas within the factory or in the production process that currently use water only once that would be amenable to reuse?
- Determination of the minimum water quantity needed for the given use: Are there areas within the factory or in the production process where more water is being supplied than is needed to accomplish the purpose?
- Identification of wastewater sources that satisfy the water quality requirements: Does the process require potable water or water of a lesser standard? Can the same result be achieved with lower-quality water?
- Determination of how the water can be transported to the new use: What modifications, if any, in the process or factory may be needed to permit recovery and recirculation/recycling of the water currently sent to waste? What additional treatment may be necessary to reuse this water? What is the relative cost of the required modifications versus the cost of the raw water over the life of the modifications?

Cooling Water Recirculation. Recycling water within a recirculating cooling system can greatly reduce water consumption by using the same water to perform several cooling operations. The water savings are generally sufficiently substantial to result in an overall cost saving to industry. Such savings can be even greater if the waste heat is used as a heat source elsewhere in the manufacturing process. Three cooling water conservation approaches are typically used to reduce water consumption: evaporative cooling, ozonation, and heat exchange.

Washwater Recycling. Another common use of water by industry is in the use of fresh or deionized water for removing contaminants from products and equipment. Deionized water can generally be recycled after its first use, although the reclamation treatment cost of recycling this water may be as great as or greater than the cost of purchasing raw water from a producer and treating it. The same processes required to produce deionized water from municipal water can be used to produce deionized water from used washwater. It is also possible to blend used washwater with raw water, which also would result in an overall water saving. The reuse of once-used deionized water for a different application within the same factory should also be considered as a water conservation option. For example, used washwater may be perfectly acceptable for washing vehicles or the factory premises.

Water Conservation Practices for Water Utilities

Common practices used by water supply utilities include metering, leak detection, repairing water lines, well capping, retrofitting programs, pricing, wastewater reuse, and developing public education programs and drought management plans.

Metering. The measurement of water use with a meter provides essential data for charging fees based on actual customer use. Submetering may also be used in multiple-unit operations such as apartment buildings, condominia, and mobile homes to indicate water use by individual units within a complex. In such cases, the entire complex of units might be metered by the main supplier, while the individual units might be monitored by either the owner or the water utility.

Leak Detection. It has been estimated that in many distribution systems up to half of the water supplied by the water treatment plant is lost to leakage; even more may be lost due to unauthorized abstraction. One way to detect leaks and identify unauthorized connections is to use listening equipment to survey the distribution system, identify leak sounds, and pinpoint the locations of hidden underground leaks. Metering can also be used to help detect leaks in a system. It is not unusual for unaccounted water losses to drop by up to 36% after the introduction of metering and leak detection programs.

Water Distribution Network Rehabilitation. A water utility can improve the management and rehabilitation of its water distribution network by a well-planned preventive maintenance program based on a sound knowledge of the distribution network. This knowledge is often embodied in a distribution system database that includes the following data:

- An inventory of the characteristics of the system components, including information on their location, size, and age and the construction material(s) used in the network.
- A record of regular inspections of the network, including an evaluation of the condition of mains and degree of corrosion (if any).
- An inventory of soil conditions and types, including the chemical characteristics of the soils.
- A record of the quality of the product water in the system.
- A record of any high or low pressure problems in the network.
- Operating records, such as of pump and valve operations, failures, or leaks, and of maintenance and rehabilitation costs.
- A file of customer complaints.
- Metering data.

Through the monitoring of such records, advance warning of possible problems can be achieved. For example, excessive water use, or numerous complaints or demands for spare parts, could be early warning signs of an impending breakdown in the system. This system should also include a regular program of preventive maintenance to minimize the possibility of system failures.

Well Capping. Well capping is the sealing of abandoned wells. In the case of artesian wells, rusted casings can spill water in a constant flow into drainage ditches, resulting in evaporative loss or runoff losses. In non-artesian wells, uncapped abandoned wells form points of entry for contaminants into the groundwater system.

Pricing. Placing an economic value on freshwater is the principal means of achieving water conservation. Pricing provides a financial incentive to conserve water. Rate structures may be variable and/or graduated, with prices fixed on the basis of class of service (residential versus industrial or agricultural, for example) and quantity used (for example, the unit price for quantities below 400 l/day might be significantly lower than for quantities which exceed that amount for a single-family residence). Pricing has the advantage of minimizing the costs of overt regulation, restrictions, and policing, while providing a high degree of freedom of choice for individual water customers.

Retrofit Programs. Retrofitting involves the replacement of existing plumbing fixtures with equipment that uses less water. The most successful water-saving fixtures are those which operate in the same manner as the fixtures being replaced; for example, toilet tank inserts, faucet aerators, and low flow showerheads do not significantly change the operation of the systems into or onto which they are placed, but they do result in substantial water savings.

Water Audit Programs. Various types of audits can be undertaken. For example, residential water audits may involve sending trained water auditors into participating households, free of charge, to encourage water conservation efforts, or providing them with record sheets to note down their water use for external analysis. Water audits may also be undertaken in commercial and industrial facilities, and may be combined with an assessment of the potential for implementing water reuse and recycling programs. A pre-implementation and post-implementation water audit in factories adopting a reuse and recycling program would be a valuable means of demonstrating and quantifying the water savings achieved.

Public Information and Education. Public information and education programs can be undertaken to inform the public about the basics of water use and conservation. Programs should be developed for specific applications and may be targeted at specific user groups or age groups; for example, at housekeepers, to encourage domestic water conservation, or at schoolchildren, to provide information on the wider implications of water conservation for future consumption, the environment and other uses. Basic information should include the following:

- How water is delivered and how wastewater is disposed of.
- The costs of water and water supply services.
- Why water conservation is important.

The programs should provide guidance on how the user groups and individuals can participate in conservation efforts. It should be noted that there is a large body of public information and education materials available, particularly in the United States, which may be obtained from a variety of public agencies and NGOs at little or no cost and form the basis of a local public awareness initiative.

DROUGHT MANAGEMENT

Given the vagaries of the modem climate, in this period of climate change, it seems that droughts may be more severe or extensive than is the past. Many water conservation projects constructed to alleviate drought-induced water shortages are themselves victims of drought. Whether this may simply reflect changes in land use within a watershed that allow less water to infiltrate into the groundwater system, or results from population growth, which places greater demands on finite water resources, is not clear and rarely proved. In any case, many communities are currently experiencing a need to have drought management plans in place to ensure the greatest possible availability of freshwater during periods of below average rainfall.

Drought Management Planning. When rainfall is less than usual, there is less water to maintain normal soil moisture levels, stream flows, and reservoir levels and to recharge groundwater. Because of these varied sources and the multiple demands placed upon freshwater resources, a drought management plan should address a range of issues, from political and technical matters to public involvement. Some of the components of a typical drought management plan include the following:

- Identification of the available water resources.
- Tabulation of the multiple sectoral demands for freshwater.
- Description of possible shortfalls between supply and demand.
- Definition of the management measures required for various eventualities, and an agreed allocation schedule in the event that water rationing becomes necessary.
- Provision for user and public involvement in the drought management program.
- Promulgation of legislation, agreements, rules, and procedures to ensure a timely and equitable response to the onset of drought conditions.

- Issuance of a drought management event plan and public information materials to make it known.

Extent of Use

Water conservation measures have been practiced primarily in the United States, although some Latin American countries have implemented specific measures. For example, in Brazil, the pharmaceutical, food processing, and dairy industries were required to pay effluent charges that contributed to reductions in water use and wastewater production of between 42% and 62%. In Mexico, increased water prices contributed to an increase in wastewater reuse and the recycling of cooling water.

Chile is the only country in the region with a comprehensive water law that has encouraged the development of water markets. The 1981 National Water Law established secure, tradable, and transferable water use rights for both surface and groundwaters. As a result, during periods of low rainfall, farmers shift from the production of water-intensive crops, such as corn and oilseeds, to higher-valued and less water-intensive crops, such as fruits and vegetables.

Water recycling is used at a Container Corporation of America Mill in Santa Clara, California (U.S.A), that manufactures paperboard from the recycled fibers of newspapers, corrugated cardboard clippings, and ledger paper. Historically, water has been used in this process for a variety of purposes. In recent years, however, the mill has begun recycling water used in its rinsing processes after clarification. The mill has also installed a closed loop cooling tower, which has resulted in an additional reduction in water use. These water conservation and use efficiency practices have resulted in an estimated saving of approximately 2.8 million l/day, compared to its 1980 water use rates. These water reductions amount to approximately 900 million l/year and saved the company approximately $348 200 per year.

OPERATION AND MAINTENANCE

Given the variety of measures that might be undertaken to address conservation needs within a specific geographic area, of which a number are mechanical but many may be technological or informational, it is difficult to identify specific operational requirements. However, some of the more obvious requirements include the following: low-flow water conservation devices require periodic maintenance and repair; leak detection equipment and meters require periodic testing and repair; drought and water conservation management strategies, such as pricing and user charges, require monitoring and enforcement; and well-capping programs require monitoring and trained personnel in order to be effective. Maintenance requirements range from regular inspections of mechanical devices to the review of legislation and conservation plans to ensure their continued relevance.

Level of Involvement

The installation and maintenance of low-flow household and irrigation devices may require governmental incentives in order to be accepted. In some cases, employees of the water utility may install and maintain these systems at little or no charge in order to effect the desired water savings. Alternatively, government regulations may be necessary to provide incentives for the implementation of industrial and agricultural water conservation measures. Government action is required in the promulgation of plumbing codes for new construction that will contribute to the adoption of residential water conservation measures. Government or utility involvement is also needed for leakage detection and the repair of distribution systems. Metering, in addition to requiring technical personnel and equipment to be effective, generally requires governmental action to implement and government authority to establish or regulate water tariffs. However, community participation and voluntary conservation are a key element if this technology is to be effective.

Effectiveness of the Technology

Water conservation measures are highly effective. However, this technology may not be too popular with consumers, who may be asked to pay a higher price for the water they consume, and can be, politically, very unpalatable. Nevertheless, studies carried out in Seattle, Washington, U.S.A., reported the following results from water conservation measures:

- According to detailed data on the performance of low-flow water devices in 308 single-family residences, indoor per capita water use dropped 6.4% after low-flow showerheads were installed.
- Easily installed aerators reduced water use at a faucet by as much as 60% while still maintaining a strong flow.
- A reduction in water pressure from 100 pounds per square inch (psi) to 50 psi at an outlet resulted in a water flow reduction of about one-third of the pre-existing use.
- Gray water reuse saved a volume of water equivalent to that needed to supply more than 7 000 residences and businesses.
- Outdoor water use was reduced by restricting watering times to the early morning or late evening; watering on cooler days, when possible, also reduced outdoor water use. All these measures contributed to reduced evaporative losses.
- As many as 600 l of water were saved when washing a car by turning the hose off between rinses; additional benefits and water savings were

achieved by washing the car on the lawn, which both watered the lawn and reduced runoff.

- Sweeping sidewalks and driveways, instead of hosing them down, saved about 200 l of water every 5 minutes.

In other studies, such as an industrial water conservation project in California, the conversion of an industrial process from a single-pass freshwater cooling system to a closed-loop cooling system, with circulating chilled water, has saved an estimated 20 000 to 28 000 l/day, while cities in the hemisphere that have large, old, deteriorating systems, leak detection programs have been especially efficient in minimizing water losses.

Suitability

Water conservation measures are suitable and recommended for all public water supply systems, industries with high water use, agricultural enterprises, and individual residential users in Latin American countries and the Caribbean islands.

ADVANTAGES

Residential water users:

- Low-flow devices result in water use savings of 20% to 40%.
- Pressure reductions save up to 33% of the water normally consumed.
- Conservation-based landscape irrigation practices also produce significant water use savings.

Industrial/commercial users:

- Deionized water can be recycled after its first use at little or no additional cost, using the same equipment used to produce the deionized water from the municipal supply.
- Water recycling greatly reduces water use.
- Proper scheduling of landscape irrigation optimizes water use by minimizing evaporative losses.

Agricultural users

- Water savings can be achieved through a combination of field practices, monitoring, and system modifications.
- Wastewater reuse can produce significant water savings. Water supply plants:
- Widespread leakages and illegal connections may account for 30% to 50% of the water loss in a distribution system.

- Metering allows for greater accountability and assists in the development of a pricing structure that is fair and appropriate to the individual water supply system and that provides incentives for conservation.
- Equipment repairs to water mains and valves, and capping unused wells, can reduce unnecessary water loss, and prevent contamination of both piped water and groundwater.
- Retrofit programs can produce long-term savings of water and money.

DISADVANTAGES

Residential users:

- The initial cost of low-flow devices can be high.
- Changes or modifications in water use habits are not readily accepted.

Agricultural users

- Low-volume irrigation systems may be costly in some cases.
- The use of wastewater for irrigation may pose potential health risks.

Industrial/commercial users

- Modifications to manufacturing processes may be required in some cases, incurring an initial capital charge to the user.
- Changes in the piping system within a plant can be costly.

Water supply plants

" Implementation of leak detection, control and metering is costly.

" Meters and leak detection devices require regular maintenance.

Cultural Acceptability

Most conservation measures have been applied in response to government regulations or conservation programs. As was noted above, public acceptance is limited despite the economic benefits.

Further Development of the Technology

Improved equipment for use in leak detection and metering is required. Such devices need to be more robust and less costly. Meters should be able to withstand tampering. It would also be desirable for low-flow plumbing devices to be more cost effective so as to be more attractive to consumers. Implementation of educational programming on the necessity and the economic and environmental benefits of water conservation is also likely to lower consumer resistance to water conservation technologies.

How to Conserve Water and Use It Effectively

Water users can be divided into two basic groups: system users (such as residential users, industries, and farmers) and system operators (such as municipalities, state and local governments, and privately owned suppliers). These users can choose from among many different water use efficiency practices, which fall into two categories:

1. Engineering practices: practices based on modifications in plumbing, fixtures, or water supply operating procedures
2. Behavioral practices: practices based on changing water use habits

This chapter explores a number of water use efficiency practices. The practices have been evaluated by many researchers, and there is a growing body of literature that presents the results of many studies related to water use efficiency.

Practices for Residential Users

The following sections present examples of conservation and water use efficiency practices that can benefit residential users. Both engineering and behavioral practices are described.

Engineering Practices

Plumbing

An engineering practice for individual residential water users is the installation of indoor plumbing fixtures that save water or the replacement of existing plumbing equipment with equipment that uses less water. Low-flow plumbing fixtures and retrofit programs are permanent, one-time conservation measures that can be implemented automatically with little or no additional cost over their life times (Jensen, 1991). In some cases, they can even save the resident money over the long term.

The City of Corpus Christi, for example, has estimated that an average three-member household can reduce its water use by 54,000 gallons annually and can lower water bills by about $60 per year if water-efficient plumbing fixtures are used (Jensen, 1991). Further support for this conclusion is provided below.

Low-Flush Toilets. Residential demands account for about three-fourths of the total urban water demand. Indoor use accounts for roughly 60 percent of all residential use, and of this, toilets (at 3.5 gallons per flush) use nearly 40 percent. Toilets, showers, and faucets combined represent two-thirds of all indoor water use. More than 4.8 billion gallons of water is flushed down toilets each day in the United States. The average American uses about 9,000 gallons of water to flush 230 gallons of waste down the toilet per year (Jensen, 1991). In new construction and building rehabilitation or remodeling there is a great potential to reduce water consumption by installing low-flush toilets.

Conventional toilets use 3.5 to 5 gallons or more of water per flush, but low-flush toilets (see figure above) use only 1.6 gallons of water or less. Since low-flush toilets use less water, they also reduce the volume of wastewater produced (Pearson, 1993).

Effective January 1, 1994, the Energy Policy Act of 1992 (Public Law 102-486) requires that all new toilets produced for home use must operate on 1.6 gallons per flush or less (Shepard, 1993). Toilets that operate on 3.5 gallons per flush will continue to be manufactured, but their use will be allowed for only certain commercial applications through January l, 1997 (NAPHCC, 1992).

Even in existing residences, replacement of conventional toilets with low-flush toilets is a practical and economical alternative. The effectiveness of low-flush toilets has been demonstrated in a study in the City of San Pablo, California. In a 30-year-old apartment building, conventional toilets that used about 4.5 gallons per flush were replaced with low-flush toilets that use approximately 1.6 gallons per flush. The change resulted in a decrease in water consumption from approximately 225 gallons per day per average household of 3° persons to 148 gallons per day per household a savings of 34 percent! Although the total cost for replacement of the conventional toilets with low-flush toilets was about $250 per unit (including installation), the water conservation fixtures saved an average of $46 per year from each unit's water bill. Therefore, the cost for the replacement of the conventional toilet with a low-flush toilet could be recovered in 5.4 years.

Toilet Displacement Devices. Plastic containers (such as plastic milk jugs) can be filled with water or pebbles and placed in a toilet tank to reduce the amount of water used per flush. By placing one to three such containers in the tank (making sure that they do not interfere with the flushing mechanisms or the flow of water), more than l gallon of water can be saved per flush. A toilet dam, which holds back a reservoir of water when the toilet is flushed, can also be used instead of a plastic container to save water. Toilet dams result in a savings of 1 to 2 gallons of water per flush (USEPA, l991b).

Low-Flow Showerheads. Showers account for about 20 percent of total indoor water use. By replacing standard 4.5-gallon-per-minute showerheads with 2.5-gallon-per-minute heads, which cost less than $5 each, a family of four can save approximately 20,000 gallons of water per year (Jensen, 1991). Although individual preferences determine optimal shower flow rates, properly designed low-flow showerheads are available to provide the quality of service found in higher-volume models.

Whitcomb (1990) developed a model to estimate water use savings resulting from the installation of low-flow showerheads in residential housing. Detailed data from 308 single-family residences involved in a pilot program in Seattle, Washington, were analyzed. The estimated indoor water use per person dropped 6.4 percent after low-flow showerheads were installed (Whitcomb, 1990).

Faucet Aerators. Faucet aerators, which break the flowing water into fine droplets and entrain air while maintaining wetting effectiveness, are inexpensive devices that can be installed in sinks to reduce water use. Aerators can be easily installed and can reduce the water use at a faucet by as much as 60 percent while still maintaining a strong flow. More efficient kitchen and bathroom faucets that use only 2 gallons of water per minute--unlike standard faucets, which use 3 to 5 gallons per minute--are also available (Jensen, 1991).

Pressure Reduction. Because flow rate is related to pressure, the maximum water flow from a fixture operating on a fixed setting can be reduced if the water pressure is reduced. For example, a reduction in pressure from 100 pounds per square inch to 50 psi at an outlet can result in a water flow reduction of about one-third (Brown and Caldwell, 1984).

Homeowners can reduce the water pressure in a home by installing pressure-reducing valves. The use of such valves might be one way to decrease water consumption in homes that are served by municipal water systems. For homes served by wells, reducing the system pressure can save both water and energy. Many water use fixtures in a home, however, such as washing machines and toilets, operate on a controlled amount of water, so a reduction in water pressure would have little effect on water use at those locations.

A reduction in water pressure can save water in other ways: it can reduce the likelihood of leaking water pipes, leaking water heaters, and dripping faucets. It can also help reduce dishwasher and washing machine noise and breakdowns in a plumbing system.

A study in Denver, Colorado, illustrates the effect of water pressure on water savings. Water use in homes was compared among different water pressure zones throughout the city. Elevation of a home with respect to the elevation of a pumping station and the proximity of the home to the pumping station determine the pressure of water delivered to each home. Homes with high water pressure were compared to homes with low water pressure. An annual water savings of about 6 percent was shown for homes that received water service at lower pressures when compared to homes that received water services at higher pressures.

Gray Water Use. Domestic wastewater composed of wash water from kitchen sinks and tubs, clothes washers, and laundry tubs is called gray water (USEPA, 1989). Gray water can be used by homeowners for home gardening, lawn maintenance, landscaping, and other innovative uses. The City of St. Petersburg, Florida, has implemented an urban dual distribution system for reclaimed water for nonpotable uses. This system provides reclaimed water for more than 7,000 residential homes and businesses (USEPA, 1992).

Landscaping

Lawn and landscape maintenance often requires large amounts of water, particularly in areas with low rainfall. Outdoor residential water use varies greatly depending on geographic location and season. On an annual average basis, outdoor water use in the arid West and Southwest is much greater than that in the East or Midwest. Nationally, lawn care accounts for about 32 percent of the total residential outdoor use. Other outdoor uses include washing automobiles, maintaining swimming pools, and cleaning sidewalks and driveways.

Landscape Irrigation. One method of water conservation in landscaping uses plants that need little water, thereby saving not only water but labor and fertilizer as well (Grisham and Fleming, 1989). A similar method is grouping plants with similar water needs. Scheduling lawn irrigation for specific early morning or evening hours can reduce water wasted due to evaporation during daylight hours. Another water use efficiency practice that can be applied to residential landscape irrigation is the use of cycle irrigation methods to improve penetration and reduce runoff. Cycle irrigation provides the right amount of water at the right time and place, for optimal growth. Other practices include the use of low-precipitation-rate sprinklers that have better distribution uniformity, bubbler/soaker systems, or drip irrigation systems (RMI, 1991).

Xeriscape Landscapes. Careful design of landscapes could significantly reduce water usage nationwide. Xeriscape landscaping is an innovative, comprehensive approach to landscaping for water conservation and pollution prevention. Traditional landscapes might incorporate one or two principles of water conservation, but xeriscape landscaping uses all of the following: planning and design, soil analysis, selection of suitable plants, practical turf areas, efficient irrigation, use of mulches, and appropriate maintenance (Welsh et al., 1993).

Benefits of xeriscape landscaping include reduced water use, decreased energy use (less pumping and treatment required), reduced heating and cooling costs because of carefully placed trees, decreased storm water and irrigation runoff, fewer yard wastes, increased habitat for plants and animals, and lower labor and maintenance costs (USEPA, 1993).

More than 40 states have initiated xeriscape projects. Some communities use contests and demonstration gardens to promote public awareness. El Paso Water Utilities and the Council of El Paso Garden Clubs sponsor an annual "Accent

Sun Country" contest. The contest spotlights homes that have water-conserving landscapes consisting of plants and grasses that require only a minimum of supplemental water and yet beautify the homes. The winning entries are publicized, and cash prizes are awarded. People are invited to tour the grounds to get ideas on how they, too, can save water, time, and money while maintaining an attractive landscape (RMI, 1991). The offices of the Southwest Florida Water Management District in Tampa and Brooksville offer free xeriscape tours every month. The tours begin with a slide show on the principles of xeriscape and continue with a walking tour of water-saving landscaping (Xeriscape tours, 1993).

Behavioral Practices

Behavioral practices involve modifying water use habits to achieve more efficient use of water, thus reducing overall water consumption by an industrial/commercial facility. Changes in behavior can save water without modifying the existing equipment at a facility.

Monitoring the amount of water used by an industrial/commercial facility can provide baseline information on quantities of overall company water use, the seasonal and hourly patterns of water use, and the quantities and quality of water use in individual processes. Baseline information on water use can be used to set company goals and to develop specific water use efficiency measures. Monitoring can make employees more aware of water use rates and makes it easier to measure the results of conservation efforts. The use of meters on individual pieces of water-using equipment can provide direct information on the efficiency of water use. Records of meter readings can be used to identify changes in water use rates and possible problems in a system (Brown and Caldwell, 1990).

Many of the practices described in the section for residential users can also be applied by commercial users. These include low-flow fixtures, water-efficient landscaping, and water reuse and recycling (e.g., using recycled wash water for pre-rinse).

PRACTICES FOR AGRICULTURAL USERS

Engineering Practices

Irrigation

Water-saving irrigation practices fall into three categories: field practices, management strategies, and system modifications. Field practices are techniques that keep water in the field, distribute water more efficiently across the field, or encourage the retention of soil moisture. Examples of these practices include the chiseling of extremely compacted soils, furrow diking to prevent runoff, and leveling of the land to distribute water more evenly. Typically, field practices are not very costly.

Management strategies involve monitoring soil and water conditions and collecting information on water use and efficiency. The information helps in making decisions about scheduling applications or improving the efficiency of the irrigation system. The methods include measuring rainfall, determining soil moisture, checking pumping plant efficiency, and scheduling irrigation.

System modifications require making changes to an existing irrigation system or replacing an existing system with a new one. Because system modifications require the purchase of equipment, they are usually more expensive than field practices and management strategies. Typical system modifications include adding drop tubes to a center pivot system, retrofitting a well with a smaller pump, installing surge irrigation, or constructing a tailwater recovery system (Kromm and White, 1990).

Water Reuse and Recycling

Agricultural irrigation represents approximately 40 percent of the total water demand nationwide. Given that high demand, significant water conservation benefits could result from irrigating with reused or recycled water.

Water reuse is the use of wastewater or reclaimed water from one application for another application. Reused water must be used for a beneficial purpose and in accordance with applicable rules (USEPA, 1991a). Water recycling is the reuse of water for the same application for which it was originally intended.

Factors that should be considered in an agricultural water reuse program include:

- The identification of water reuse opportunities
- Determination of the minimum water quality needed for the given use
- Identification of wastewater sources that satisfy the water quality requirements
- Determination of how the water can be transported to the new use (Brown and Caldwell, 1990)

Water reuse for irrigation is already in widespread use in rural areas and is also applicable in areas where agricultural sites are near urban areas and can easily be integrated with urban reuse applications (USEPA, 1992).

Behavioral Practices

Behavioral practices involve changing water use habits to achieve more efficient use of water. Behavioral practices for agricultural water users can be applied to irrigation application rates and timing. Changes in water use behavior can be implemented without modifying existing equipment.

For example, better irrigation scheduling can result in a reduction in the amount of water that is required to irrigate a crop effectively. The careful choice of irrigation application rates and timing can help farmers to maintain yields with less water. In making scheduling decisions, irrigators should consider:

- The uncertainty of rainfall and crop water demand
- The limited water storage capacity of many irrigated soils
- The limited pumping capacity of irrigation systems
- Rising pumping costs as a result of higher energy prices

Local NRCS-Conservation Districts and Cooperative Extension Service offices can play an important role in promoting better irrigation scheduling. Accurate information on crop water use requires information on solar radiation and other weather variables that can be collected by local weather stations. An additional method that can be used to improve irrigation scheduling and might result in high returns is the use of equipment such as resistance blocks, tensiometers, and neutron probes to monitor soil moisture conditions to help in determining when water should be applied (Bosch and Ross, 1990).

PRACTICES FOR SYSTEM OPERATORS

Engineering Practices

Metering

Metering. The measurement of water use with a meter provides essential data for charging fees based on actual customer use. Billing customers based on their actual water use has been found to contribute directly to water conservation. Meters also aid in detecting leaks throughout a water system. In 1977, for example, Boston, Massachusetts, could not account for the use of 50 percent of the water in its municipal water system. After installing meters, the city identified leaks and undertook a vigorous leak detection program (Grisham and Fleming, 1989). Unaccounted-for water dropped to 36 percent after metering and leak detection programs were started.

Submetering. Submetering is used in units such as apartments, condominiums, and trailer homes to indicate water use by those individual units; the entire complex of units is metered by the main supplier. Submetering of water use in apartment or business complexes makes it possible to bill tenants for the water that they actually use rather than for a percentage of the total water use for the complex. Submetering makes water users more aware of how much water they use and its cost, and tenants who conserve water can benefit from lower water use costs. Submetering is reported to reduce water usage by 20 to 40 percent (Rathnau, 1991).

Leak Detection

One way to detect leaks is to use listening equipment to survey the distribution system, identify leak sounds, and pinpoint the exact locations of hidden underground leaks. As mentioned in the previous section, metering can also be used to help detect leaks in a system.

An effective way to conserve water is to detect and repair leaks in municipal water systems. Repairing leaks controls the loss of water that water agencies have paid to obtain, treat, and pressurize. The early detection of leaks also reduces the chances that leaks will cause major property damage. When water leaks from a system before it reaches the consumer, water agencies lose revenue and incur unnecessary costs. Such costs should provide an incentive for system operators to implement a leak detection program.

Programs for finding and repairing leaks in water mains and laterals (conduits) might be cost-effective in spite of their high initial costs. Leak detection programs have been especially important in cities that have large, old, deteriorating systems (RMI, 1991).

Water Main Rehabilitation

A water utility can improve the management and rehabilitation of a water distribution network by using a distribution system database. Using the database can help to lower maintenance costs and can result in more efficient use of the water resource. The database can help the utility manager to set priorities and efficiently allocate rehabilitation funds (Habibian, 1992). A comprehensive database should include information on the following:

- The characteristics of the system's components, such as size, age, and material
- The condition of mains, such as corrosion
- Soil conditions or type
- Failure and leak records
- Water quality
- High/low pressure problems
- Operating records, such as pump and valve operations
- Customer complaints
- Meter data
- Operating and rehabilitation costs

Water Reuse

Another practice that should be considered by water system operators who operate publicly owned treatment works is the reuse of treated wastewater. As discussed earlier, water reuse is the use of wastewater or reclaimed water from one application for another application. Some potential applications for water reuse include landscape irrigation, agricultural irrigation, aesthetic uses such as fountains, industrial uses, and fire protection (USEPA, l991a). These factors should be considered in a water reuse program:

- The identification of water reuse opportunities
- The determination of the minimum water quality needed for the given use
- The identification of wastewater sources that satisfy the water quality requirements
- The determination of how the water can be transported to the new use (Brown and Caldwell, 1990)

Well Capping

Well capping is the capping of abandoned artesian wells whose rusted casings spill water in a constant flow into drainage ditches. In Seminole County, Florida, state hydrologists estimate that 1,500 abandoned artesian wells are discharging 54 Mgal/d. To put that in perspective, utilities in Seminole County pump less than 40 Mgal/d. The cost to plug such wells is about $750 (1990 dollars) per well. The state legislature has required that all such wells be capped beginning in 1993 (Florida Commission, 1990).

Planning and Management Practices

In addition to engineering practices, system operators can use several other practices to conserve water or improve water use efficiency.

Pricing

Information and education promoting conservation do not appear to be effective by themselves in achieving a conservation goal without at the same time imposing significant price increases to provide a financial incentive to conserve water (Martin and Kulakowski, 1991). Customers use less water when they have to pay more for it and use more when they know they can afford it. However, most people consider water to be a "free good" and are not willing to pay higher prices that reflect the true costs associated with the water delivered to their homes. Rate structures have the advantage of avoiding the costs of overt regulation, restrictions, and policing while retaining a greater degree of individual freedom of choice for water customers.

Overall charges for water service increased at an average compound rate of 7 percent per year during the 1980s nearly double the rate of inflation (Russet and Woodcock, 1992). There is concern over "price gouging" due to increased water rates (Collinge, 1992). Some pricing has been objected to on the grounds that it can lead to a substantial excess of revenues over costs an excess that might be inequitable and, in some states, unconstitutional (Collinge, 1992).

Water utility managers must establish and design water rates that meet revenue requirements and are fair and equitable to all customer classes in the water system. This task involves the following procedures:

- Determination of the water utility's total annual revenue requirements for the period for which the rates are to be in effect
- Determination of service costs by allocation of the total annual revenue requirements to the basic water system cost components and distribution of these costs to the various customer classes in accordance with their service requirements
- Design of water rates to recover the cost of service from each class of customer (Mui et al., 1991)

Several price rate structuring alternatives are available for water system operators.

Increasing Block Rate, or Tiered, Pricing. Increasing block rate, or tiered, pricing reduces water use by increasing per-unit charges for water as the amount used increases. For example, the first volume of water (block) used is charged a base rate, the second block is charged the base rate plus a surcharge, and the third block is charged the base rate plus a higher surcharge. It is necessary to increase real prices significantly to overcome the effects of conservation (Martin and Kulakowski, 1991).

For example, as the cost of water increased in Tucson, Arizona, residents used 33 percent less water between 1974 and 1980. A 10 percent increase in water rates provided about 3 percent more revenue while triggering a 7 percent reduction in use (Billings and Day, 1989). Using seasonal increasing block rate pricing during summer and winter months, to encourage year-round conservation, resulted in estimated water savings for the single-family residential class in Tucson of an average 2.23 Mgal/d during 1983-1986 (Cuthbert, 1989).

Decreasing Block Rate Pricing. Decreasing block rate prices reflect per-unit costs of production and delivery that go down as customers consume more water.

The monthly water use records of 101 customers were measured in a study of municipal water use in the city of Denton, Texas. Summer water use records from 1976 to 1980 during a decreasing block rate period were compared to summer use records from 1981 to 1985 during an increasing block rate period. It was found that the decreasing block rate scenario encouraged greater water use, whereas the increasing block rate scenario resulted in a reaction to the price increase and a corresponding decrease in water use (Nieswiadomy and Molina, 1989).

Time-of-Day Pricing. Time-of-day pricing charges users relatively higher prices during a utility's peak use periods. Because customers are sensitive to price increases, these charges curtail demand. Time-of-day pricing can cut generating capacity and reduce reliance on expensive secondary fuel sources (Sexton et al., 1989).

Water Surcharges. A water surcharge imposes a higher rate on excessive water use. The customer pays more money per gallon for water use that is considered higher-than-average.

Surcharges include unit surcharges, winter/ summer ratios, and alternative seasonal rates. The unit surcharge method establishes a threshold level for excess consumption based on average daily per capita or per-household consumption. A surcharge is imposed for all water use above that threshold level. For the winter/summer ratio, metered water use during the winter period is compared to consumption during the corresponding summer period, and a higher rate or surcharge is imposed for water consumption above the average winter use. Typically, an increase in usage of 14-20 percent occurs during the summer. Under an alternative seasonal rate structure, all water used during the summer or peak season is billed at a higher rate than that used during the other seasons. The increased rate is applied to all customers at all water-use levels (Schlette and Kemp, 1991).

Retrofit Programs

Retrofit programs are another tool system operators can use to promote water use efficiency practices. Retrofitting involves the replacement of existing plumbing equipment with equipment that uses less water. The most successful water-saving fixtures are those which operate in the same manner as the fixtures they are replacing--for example, toilet tank inserts, shower flow restrictors, and low-flow showerheads. (For more information, refer to the practices for residential users.) As discussed previously, retrofit programs are permanent, one-time conservation measures that can be implemented with little or no additional cost over their lifetimes (Jensen, 1991).

A retrofit program can involve the use of education programs to let users know which fixtures are best, where to get them, and how to install them. System operators can also purchase water-efficient fixtures and resell them at cost to the users, but the most successful retrofit programs have been those in which the system operator purchases, distributes, and installs the fixtures (AWWA, n.d.).

Retrofit programs have been shown to be cost-effective and useful in conserving water in many cases. An apartment building in New England with 151 units was retrofitted with low-flow showerheads and faucet aerators at a cost of $1,074. As a result of the retrofit 1,725,000 gallons of water, $8,590 for energy, and $980 for water were saved in 1 year (AWWA, n.d.). In another retrofit program, the Lower

Colorado River Authority installed low-flow showerheads and toilet dams in an apartment complex and public housing program in Marble Falls, Texas. Indoor per capita water use was reduced by 21 percent (from 81 to 64 gal/cap/day) in the apartment complex and was reduced 11 percent (from 102 to 91 gal/cap/day) in the public housing program (Jensen, 1991).

Current use of low-flow toilets throughout Texas could reduce the need to build new water and wastewater treatment plants by 15 percent, resulting in a savings of as much as $3.4 billion during the next 50 years. Residential water and sewer bills could also be reduced by as much as $200 million over the long term. The Texas Water Development Board estimates that the use of water-efficient plumbing fixtures should save a typical four-member household 55,800 gallons of water and $627 in reduced water and energy costs per year. The Board estimates that the use of low-flow fixtures might reduce water use statewide by 805 Mgal/d by the year 2040 (Jensen, 1991).

Residential Water Audit Programs

Residential water audit programs involve sending trained water auditors to participating family homes, free of charge, to encourage water conservation efforts. Auditors visit participating homes to identify water conservation opportunities, such as repairing leaks and low-flow plumbing, and to recommend changes in water use practices to reduce home water use. The audit programs try to stretch existing water supplies by getting water users to use water more efficiently (Whitcomb, 1990). The largest percentage of indoor use comes from bathing and toilet flushing. Therefore, the bathroom is an ideal place for water system operators to focus water conservation efforts (Grisham and Fleming, 1989).

Public Education

Public education programs can be used to inform the public about the basics of water use efficiency:

- How water is delivered to them
- The costs of water service
- Why water conservation is important
- How they can participate in conservation efforts

Public education is an essential component of a successful water conservation program. A number of tools can be used to educate the public [BROKEN]: bill inserts, feature articles and announcements in the news media, workshops, booklets, posters and bumper stickers, and the distribution of water-saving devices. Public school education is also an important means for instilling water conservation awareness (Grisham and Fleming, 1989).

Another way to provide public information and education, as well as to collect real-world data on water conservation and use efficiency, is through the use of demonstration projects. In Tucson, Arizona, the Casa del Agua, a single-family home, has been used to demonstrate and study water conservation and reuse techniques and technologies. In 1985, the University of Arizona designed and retrofitted the Casa del Agua with water-conserving fixtures, a rainwater harvesting system, gray water reuse and storage systems, and drought-tolerant plants. Measurements of water use and water quality at the Casa del Agua have provided a useful collection of data for evaluating the possible benefits of conservation techniques and technologies in a residential home (Karpiscak et al., 1991).

A study of water demand in the United States using American Water Works Association (AWWA) data indicated that water users are more sensitive to a change in price in the South and the West than in the other regions of the country. Public education appears to have reduced water usage in the West. A heightened awareness of water's scarcity might make educational programs more effective in the West than in the rest of the country (Nieswiadomy, 1992).

9

RAINWATER STORAGE AND TREATMENT

Storage is usually the most expensive component of a rainwater system and often determines the type of filtration and pumping system. We provide a wide range of storage solutions from a 75 gallon consumer tank to custom-designed modular systems capable of storing millions of gallons to run large commercial or multi-family structures or irrigation systems.

COMPARING RAINWATER STORAGE OPTIONS

Storage tanks, usually the most expensive component of the rainwater harvesting system, come in a wide variety of sizes and types. When deciding on the type of tank to use, the main factors to consider include where you live and your budget. When choosing the size of storage tank or cistern, consider several variables: rainwater supply (local precipitation), demand, projected length of dry spells without rain, catchment surface area, aesthetics, personal preference, and of course, your budget.

A myriad of variations on storage tanks and cisterns have been used over the centuries and in different geographical regions: earthenware cisterns in India, large pottery containers in Africa, above-ground vinyl-lined swimming pools in Hawaii, concrete or brick cisterns in the central United States, and, in Texas and Colorado, galvanized steel tanks and site-built stone-and-mortar cisterns.

Tanks can be above or below ground. Factors such as soil, outside temperature ranges, and cost should be used to determine whether a tank is placed above or below ground. Some tanks are suited for above-ground placement (i.e. vinyl-lined swimming pools), where others can be used both above and below ground (i.e. polyethylene). Some types of tanks are built to be buried (i.e. polyethylene tanks designed for burial).

Consequently, understanding all the information about the options available is critical to making a good decision about the type of tank to purchase, since it should prove to be something you live with for a long time. Below is a general overview of the various tank types to choose from and some of their characteristics.

Fiberglass

Fiberglass tanks are light-weight, reasonably priced, and long lasting. They are built in standard capacities from small 50 gallon barrels to much larger 15,000 gallon tanks and are available in both vertical cylinder and low-horizontal cylinder configurations. Smaller fiberglass tanks (i.e. under 1,000 gallons) are expensive for their size, so polyethylene might be preferred. Tanks for potable use need to have a USDA-approved food-grade resin lining and the tank should be opaque to inhibit algae growth. The durability of fiberglass tanks has been thoroughly tested and proven to last for years and can be easily repaired. The fittings on fiberglass tanks are an integral part of the tank, eliminating one common potential problem - leaking fittings.

Polyethylene

Polyethylene tanks are probably the most common type of tank being sold today and are readily available at most farm, ranch supply, and large landscape retailers. They vary greatly in size, shape, and color, and can be used above or below ground.

Most of the tanks stocked by farm and garden supply houses are usually for above-ground installations. For buried installation, specially designed and reinforced tanks are necessary to withstand soil expansion and contraction. Polyethylene tanks are comparatively inexpensive, lightweight, and long-lasting and are available in capacities from small 50 gallon barrels to large 10,000 gallon tanks. They are lighter in weight than other types of tanks, including fiberglass, and consequently, are cheaper and easier to transport.

Polyethylene tanks tend not retain paint well, so use pre-painted (i.e. pigmented) tanks manufactured with opaque plastic. Black and dark colored tanks will absorb heat and thus, should be shaded or buried. The fittings of these tanks are aftermarket modifications and are easy to plumb. However, the fittings are not always tight, and should be checked for leakage occasionally.

In-ground polyethylene

In-ground polyethylene tanks are more costly for two reasons: the cost of excavation and the cost of a more heavily reinforced tank.

The latter is required if the tank is to be buried more than two feet deep. Burying a tank in soil with high clay content is not recommended because of the expansion and contraction cycles of clay.

For below ground installation, the walls of poly tanks must be manufactured thicker, and sometimes, an interior bracing structure must be added.

Swimming Pools

Above ground swimming pools are commonly used in Hawaii for rainwater catchment; and although used throughout the islands, they are not a good rainwater catchment solution. A few of the issues related to these pools as a storage device are: 1) The pool liner typically only goes up to the lip of the pool and slips into the pool. 2) Liners are not food grade quality and may in fact be made to retard bacterial growth in the pool and can be toxic. 3) Large diameters of these pools are not easily covered and therefore it is difficult to keep out debris and animals.

However, swimming pools are very inexpensive, readily available and an easy to install solution. If this is the only option available then take the following precautions: 1) Make sure the liner is approved by the Food and Drug Administration (i.e. not the standard liner but one made for storing and holding water). 2) Plan to have a support system (i.e. a post in the middle of the pool) to hold up the cover so it does not sag into the water. 3) Make sure the cover can be firmly attached to the top of the pool to keep out unwanted guests and debris.

Wood

For personal aesthetic appeal, a wood tank are sometimes a desirable choice. Wood tanks, similar to water towers at railroad depots, were historically made of redwood. However, modern wood tanks are usually of pine, cedar, or cypress wrapped with steel tension cables. Wood tanks are lined with plastic to increase longevity. For potable use, a food-grade liner should be used.

Redwood, as a tank material, has a great reputation for being durable and attractive. It contains no resins and has high levels of tannin, a natural preservative resistant to insects and decay. It is also a good insulator, keeping the water cooler in the summer and protects it from freezing the winter.

These tanks are available in capacities from small 700 gallon tanks to very large 37,000 gallon tanks, and are usually site-built by skilled technicians. As with metal tanks, they can be dismantled and moved to a different location, if required. These tanks are for above-ground use and not for use in dry, hot climates.

Redwood is very attractive, but it is expensive and not readily available. Pine is commonly used and although it does not have some of the characteristics of redwood, it is readily available and less expensive.

Metal

As with wood, galvanized sheet metal tanks can also be an attractive option. They are available in sizes that range from small tanks of 150 gallons to medium-sized 2,500 gallon tanks, and are lightweight and easy to relocate if required.

Most tanks are corrugated galvanized steel dipped in hot zinc to improve corrosion resistance. These tanks should be lined with a food-grade liner, usually polyethylene or PVC, or coated on the inside with epoxy paint. The paint or liner will extend the life of the metal and, if being used for potable water, must be FDA and NSF approved for potability.

These tanks are for above-ground use. Old or recycled tanks may contain lead and should be avoided. Brass and bronze fittings should not be connected directly to the tank as they will cause corrosion. Additionally, care should be taken when cleaning these tanks, as a film develops naturally on the inside of the tank, which coats the tank and inhibits corrosion.

Concrete

The most versatile of tanks, concrete tanks can either be poured in place or prefabricated. They can be constructed above or below ground. They can be owner or contractor-built. Poured-in-place tanks can be very attractive and easily integrated into new construction. For example, the tank can be placed under a patio, or a basement. Concrete tanks, once poured, are considered permanent.

One unique advantage of poured concrete is that the concrete will over time decrease the corrosiveness of rainwater by leaching into the water. This advantage of concrete tanks results in a desirable taste imparted to the water by calcium in the concrete being dissolved in locations where there is slightly acidic rainwater. For potable systems, it is essential that the interior of the tank be plastered with a high-quality material approved for potable use.

Underground concrete tanks are prone to cracking and leaking, especially when in clay soil. Leaks can be easily repaired, although the tank may need to be drained to make the repair. If building the tank yourself, it is recommended to involve the expertise of a structural engineer to determine the size and spacing of reinforcing steel to match the structural loads of a poured-in-place concrete. A product that repairs leaks in concrete tanks, Xypex(TM), is now available and approved for potable use.

Ferrocement

Ferrocement is the term used to describe a steel and mortar composite material. These tanks can be above or below ground and can be done by contractors or homeowners. They were developed in third-world countries to be relatively low-cost and durable. They are listed separately from concrete, not just because of the materials used to construct them, but also because they have differing problems and advantages.

These tanks are typically built with concrete, but have multiple layers of wire mesh - typically chicken wire-wrapped around a light framework of rebar, embedded in the concrete. Walls can be as thin as 1" and still be strong. Consequently, it can cost less to build than a concrete-only tank. If buying a ferrocement tank, make sure it does not contain any toxic compounds in the concrete and that the wires are not visible on the inside of the tank.

Ferrocement, like concrete, will need maintenance and repair as cracks appear. It is important to ensure that the ferrocement mix does not contain any toxic components. Some sources recommend painting above-ground tanks white to reflect the sun's rays, reduce evaporation, and keep the water cool.

STONE AND MASON

Hand made stone or mason tanks are not as common as they once were. Increasing labor costs, decreasing costs and increasing availability of other types of tanks has limited their use to areas where labor is very, very cheap (i.e. third-world countries) or where budget is not an issue.

The mass of the stone gives these tanks two distinct advantages: it keeps the water cool in hot climates and they can be very attractive. As with ferrocement tanks, care should be taken to make sure the mix does not contain toxic materials.

These tanks are custom-built, so they can be as large as designed. Most tanks are designed to be circular, since this shape is more structurally sound. These tanks, if properly constructed and maintained, will last for decades.

Plastered Tire Cistern

Another type of hand-made tanks are plastered tire cisterns. They were pioneered with the earthships in the Taos, New Mexico area and some have been in use for decades. They are simply a circle of buried tires with a wire mesh inside covered with plaster. Just like stone or cement cisterns, they will need periodic maintenance to repair cracks on the inside. These tanks are meant to be buried and can be very economical for large tank sizes (i.e., 10,000+ gallons/37,900 liters), especially if owner-built. In earthships, tanks are typically built as an integral part of the home and can provide cooling in hot climates. In cooler climates, the tops and sides should be insulated to prevent cooling. Care should be taken in building a plastered tie cistern to ensure the wire mesh and tires are thoroughly covered with plaster.

As with stone and mason cisterns, these tanks are custom-built, so they can be as large as designed. They arc typically designed to be circular, but since they are buried, they can be almost any shape. These tanks, if properly constructed and maintained, will last for decades.

General Guidelines

General guidelines for all tanks include:

- Before you start, conserve, conserve, conserve. Cutting your water usage will reduce the size of tank you need and save you money.
- Remember - water is very heavy. (i.e., 500 gallons weighs over 2 tons!)
- Make sure the tank is easy to access and maintain.
- Tank should be opaque or darker, either upon purchase or painted later, to inhibit algae growth.
- For potable systems, storage tanks must never have been used to store toxic materials.
- Tanks must be covered and vents screened to discourage mosquito breeding.
- Tanks used for potable systems must be accessible for cleaning.
- Install first-flush and screening devices prior to water reaching the tanks to keep it as fresh and clean as possible.
- Keep tops of tanks free of debris to make it harder for animals to reach the top of the tank.
- Buried tanks should be located in well-drained soil and location.
- Water weighs about 8 pounds per gallon so plan your pad, if any, before installing your tank.
- Plan where storage tank overflow should be piped or directed to. Keep it away from underneath your holding tank to prevent pad erosion and to keep animals away.

Collecting and Using Rainwater at Home

Rainwater harvesting collecting and storing rain for later use is an ancient practice. It is still used in many rural places throughout the world, and today it is making a comeback in urban centres as an additional source of water.

At its simplest, rainwater harvesting consists of a rain barrel placed under the downspout of your home to collect rainwater for garden irrigation.

Larger, more sophisticated, systems can be incorporated into your home's plumbing system to provide water for a variety of household needs, from toilet flushing to laundry, and even drinking water (in these instances the rainwater will be treated prior to use). All rainwater harvesting systems, simple or complex, have the same basic components:

- A catchment area to capture the rainfall this is typically the roof of the house.
- A conveyance system to move the water from the roof to a storage area eavestroughs and downspouts, and maybe piping.
- A storage system to hold the rainwater for future use a barrel, a cistern or a tank.
- A distribution system to get the water from storage to where it is being used this can range from a watering can to full integration with the existing plumbing system in the house.

Rainwater harvesting is advantageous for a number of reasons:

- It can offset the amount of municipally treated water you use in your home helping to reduce your water bill
- Larger cisterns can help divert and retain runoff from your property reducing the impact on local stormwater infrastructure and combined sewer systems
- Stored rainwater can be used for landscape irrigation even during watering bans and helps to replenish groundwater supplies with water that would normally flow to stormwater systems
- For those on well water, rainwater harvesting offers a way to reduce well pump energy consumption and reduce the amount of water the well has to deliver an important consideration if your well is affected by dry spells or water table fluctuations.

Getting Started

When thinking about harvesting rainwater for your home, there are a number of questions to consider.

1. What would you like to use rainwater for?

Are you simply looking to harvest rainwater for outdoor watering purpose, or are you interested in a system that provides water for indoor uses as well (e.g., toilets, laundry)? Or are you wondering about using rainwater on a seasonal basis for a variety of purposes at the cottage? Each scenario involves different set-up, equipment, maintenance and costs.

2. Which rainwater uses are approved in your area?

The use of rainwater and other non-potable water supplies is regulated differently in each province and territory. The 2010 National Plumbing Code of Canada permits the use of non-potable water (including rainwater) for toilet and urinal flushing as well as outdoor irrigation. However, provincial, territorial and local jurisdictions may have different positions and requirements so it is prudent to check with local building authorities to see what's allowed in your specific area and what's not - before constructing your system.

3. How much rainwater do you need?

Depending on what you wish to use your rainwater for, the amount you need will vary depending on:

Household water consumption habits e.g., how often do you water outdoors and how much water is used, how many toilet flushes are there in a given period, how many loads of laundry?

The number of people in your household.

The age and efficiency of your plumbing fixtures (e.g., toilets/washing machines). If they are not water efficient, you may wish to replace them with some that are to make the most of both the rainwater you harvest and the municipal water you purchase.

CMHC's Collecting and Using Rainwater At Home, A Guide for Homeowners includes a worksheet on residential water consumption per household activity based on average consumption rates. This will help you calculate how much rainwater is needed to meet various requirements in your household on a daily, weekly, monthly and yearly basis depending on what is allowed by the building authority in your area.

4. How much rainwater can you collect?

The amount of rain you can harvest is calculated based on the square footage of your catchment area and the climatic rainfall conditions in your area. The general rule of thumb is that about 80% of the annual rain that falls on a catchment area can be collected; the rest is lost due to overflows, leaks, absorption by the roofing materials, and so on. Refer to the guide (see above) on how to determine how much rainwater you may be able to collect over the course of the year.

5. How much rainwater can you store?

Rainwater storage tanks are available in many sizes, from less than one hundred litres suitable for supplementary outdoor plant watering to thousands of litres for larger residential systems. The Rainwater Guide offers help with determining optimal tank size for a specific volume of water needed based on your roof area and local climatic conditions.

Designing Your System

This section provides an overview of the key components and features of a rainwater harvesting system.

ROOF CATCHMENT AREA

Certain roofing materials can impact rainwater quantity and quality. For instance, porous materials such as asbestos shingles, clay or concrete tiles, or wood shingles will soak up some of the rainwater, which reduces the amount harvested during a rainfall. Other non-absorptive roofing materials, such as steel, shed water readily for collection. Certain roofing materials may contain potential contaminants (heavy metals, and pesticides) that can be absorbed by the rainwater as it flows over the roof surface, making it unsuitable as drinking water, or for use on edible plants.

Eavestroughs and Downspouts

To get the full benefit of the catchment area, your home needs to be equipped with properly sized and sloped eavestroughs (gutters) and an adequate number of downspouts. Otherwise water may overflow during heavy rainstorms and get absorbed in the ground or run off into the storm sewers. It may be necessary to have an in-ground collection system that gathers all the water drained by downspouts located at different positions around the house to take full advantage of the rainwater that falls on the roof.

Rainwater Conveyance Network Piping

The conveyance network includes the plumbing used to take the rainwater to the storage tank and to distribute it from the tank to the various end-use points. Pipe materials, sizes and slopes are regulated by local and provincial codes. Above-ground piping should be UV-resistant and structurally strong enough to keep it from deteriorating or being damaged.

Rainwater Storage Tanks

Tanks can be located above or below ground (see figure 3), outside or inside the home, garage or shed, and are available in many different sizes and materials.

Polyethylene or polypropylene tanks are the most common. They are durable, lightweight and available in a range of sizes, shapes and colours, and typically suitable for both above- and below-ground installation (see figure 4). Concrete tanks are generally only used in below-ground systems. They can be cast in place as part of a building's foundation or installed separately elsewhere on your property.

Other types of materials include fiberglass (above or below ground), galvanized steel (low cost, above ground only), stainless steel (more costly, above ground only), and wood (aesthetically pleasing, above ground only).

Protective Measures

A well-designed system includes an overflow pipe to protect against damage caused by the tank overflowing during periods of heavy rains or low usage. All overflow water should be discharged away from foundations and other structures.

Tanks, conveyance networks and distribution systems must be protected against freezing. All below-ground components should be installed below the frost line. Seasonal systems should be completely drained at the end of the season.

In plumbing systems where there is a possibility that the rainwater conveyance piping could become interconnected with the household potable water piping, the use of backflow prevention devices ensures that non-potable water cannot flow backwards into the potable water system and contaminate it this is required by the 2010 National Plumbing Code of Canada.

Rainwater Treatment and Filtration Components

Rainwater can become contaminated from leaf litter and debris on your roof as well as from dust and local air pollution. Care should be taken to remove contaminants before they collect in your storage tank. Pre-filtration devices can be incorporated in the conveyance system to remove contaminants that may collect on the roof, before the water reaches the storage tank. Options include gutter guards, downspout filters and "first-flush" devices to divert the first several litres of rainwater from the roof away from the conveyance system at the start of each rainfall.

Post-storage Treatment

Depending on the intended use of your rainwater, you may wish to consider some form of filtration and/or disinfection of the water that comes from the storage tank to reduce potential health risks, particularly if you wish to use the harvester rainwater for drinking. Check with your local health and building departments.

Preparing for Installation

Proper installation of a rainwater harvesting system is essential. While a simple rain barrel can be placed almost anywhere with very few restrictions (apart from ensuring it does not spill water against the wall or foundation of the house), a large system must be installed safely, in accordance with local and provincial codes and standards. Whether you choose an above-ground, indoor or underground tank, there are a number of issues to consider.

Site Selection

Proper site selection will help prevent the rainwater system from causing damage to, being damaged by, or interfering with existing property features. Factors include location of utility services, wells and septic tanks; topography, depth to bedrock and soil classification; property easements and rights-of-ways, and visibility and aesthetics for you and your neighbours.

Tank Placement

Water is heavy 1,000 litres (enough to fill a 1 metre by 1 metre by 1 metre cube) weighs 1,000 kg so the tank will need adequate structural support. Depending on local regulations, above-ground tanks may be placed on a flat, level concrete or compacted earthen or gravel base. In some municipalities, the base must be designed by a structural engineer. Whether indoors or outdoors, the tank also needs to be secured to prevent it from falling over. Anchoring devices or ballasts are required for below-ground tanks to ensure that buoyancy doesn't push them to the surface.

Health and Safety

The health and safety implications of your rainwater harvesting system need to be addressed. All plumbing and electrical work must be undertaken by certified plumbers and electricians, in accordance with local codes and regulations.

Further, the 2010 National Plumbing Code of Canada requires that plumbing that provides non-potable water within a residence must be labeled accordingly. This will prevent plumbers or Do-It-Yourselfers from accidentally connecting non-potable supplies to the potable supply and thereby possibly contaminating the drinking water and placing the household at risk. Check your local regulations for precautions that must be taken in your area.

Maintenance

A rainwater harvesting system needs ongoing monitoring and regular maintenance to protect against blockage, contamination and unwanted leaks, and to ensure optimal performance and water quality.

Most of the maintenance routines are quick and easy to perform, and require mainly a visual inspection of the system components every 3,6 and 12 months, with replacements and repairs as needed. Some municipalities may require an annual mandatory inspection by a licensed plumber. It is also recommended that the storage tank be drained every 3 to 5 years to check for deterioration and clean out accumulated debris. CMHC's Collecting and Using Rainwater At Home, A Guide for Homeowners, contains a detailed inspection and maintenance schedule as well as troubleshooting guidance.

Costing Your System

Because each system is unique, its cost is unique as well. A rainwater harvesting system can cost anywhere from a few hundred to thousands of dollars, depending on its size and level of sophistication. Costs may also vary by region and by availability of trades and rainwater harvesting experts.

Get Estimates

Talk with several suppliers and/or rainwater harvesting experts and get quotes, so you can compare and can choose based on the best value for your investment. Use the costing worksheet in CMHC's Collecting and Using Rainwater At Home, A Guide for Homeowners to make sure you are including everything needed for your system and to keep track of quotes.

After-Storage Treatment

Treatment of the water after the storage and before use is critical for both health of the users and maintenance of the system. The level of treatment will depend of the intended use of the water. For example, water used for landscape drip irrigation would not need the same level of treatment as water used for potable indoor purposes. In any case, we recommend a multi-barrier approach where more than one method of treatment are used to maximize effectiveness. For example, a common treatment-train for potable water in rural Texas is a set of cartridge filters followed by a ultraviolet light. Below are the different types of water treatment available.

Filtration

Filtration is similar to screening but on a smaller scale. There are various levels of filtration, and they are measured on a "micron" level. This measures the diameter of a particle that would be blocked by the filter. For example, a 5-micron filter would block particles in the water that are 5 microns or larger. (a micron is one-millionth of a meter). Filters can remove microorganisms, sediment, metals, and other organic matter. If larger sized filters are used, small microorganisms, such as bacteria can pass through, so a disinfection method is needed (explained next). It is important that they are checked and changed on a regular basis so that they maintain their effectiveness.

The goal of disinfection is to destroy the microbiological organisms that have the potential to cause illness or harm. Some of the smaller microorganisms, such as bacteria, may pass through large cartridge filters so they must be targeted with disinfection. Three disinfection methods common to RWH systems are chlorination, ultraviolet light (UV), and ozonation.

Chlorination uses the chemical of chlorine in either dry, liquid, or gas to kill microorganisms. It is very effective with viruses and bacteria. An advantage to chlorine is that is leaves a residual so that your entire distribution system remains disinfected. However, using chlorine can be dangerous, so it's important to follow manufacturer's guidelines. Only use chlorine that is approved for potable water use (standard ANSI/NSF 60). For more information see this guidebook by the TCEQ.

UV light is a common disinfection method used by Texas residents. The device works by sending water through a tube that contains the UV bulb (see picture on left). The light disrupts the DNA of microorganisms so that they can't reproduce. There are a Class A and Class B UV light. The Class A light is rated to destroy pathogens and is required to have an alarm to alert users if the device is not working. A Class B light cannot be used as a sole method of disinfection but helps to polish water that has already been treated. An important point to keep in mind is that UV lights are most effective when the water is clear. Any sediment in the water can block the pathogens from the light. This is why have the filtration BEFORE the UV light is so critical. The UV light is harmful to look at directly, so use approved safety glasses when maintaining the device.

Ozonation disinfects by introducing ozone gas to the water. It is usually done at the point where water is used in the distribution system or in the storage tank. It is a colorless gas that disinfects, oxidizes, deodorizes, and decolorizes. Ozone gas is toxic and installation and maintenance of this type of system must be done by a licensed professional.

Adsorption uses activated carbon (charcoal filters) to bind to harmful contaminants, such as VOCs. These are the types of filters commonly used on faucets. They can also be used to remove chlorine from water used for disinfection.

When I was growing up, I remember drinking out of a rain barrel with a ladle. My great aunt would yell out the door, "Remember not to drink off the top!" That was then and now is now.

A lot has changed in 4 decades. There are a lot more pollutants, and we are more aware of the risks. We now know that E. coli and other harmful bacteria can be passed along in untreated contaminated water. A report by Peter H. Gleick estimates that if no action is taken to address unmet basic human needs for water, as many as 135 million people will die from water-related diseases by 2020.

Rainwater harvesting is viewed by many, including the EPA, as a partial solution to the problems posed by water scarcity: droughts and desertification, erosion from runoff, over-reliance on depleted aquifers, and the costs of new irrigation, diversion, and water treatment facilities.

Harvested rainwater in the U.S. is used mostly for irrigation; however, there is a growing interest in using rainwater for drinking and other indoor uses. Over 50% of household water is used indoors; bringing rain indoors could save the expense and environmental costs of treating and transporting water.

Can rainwater be made safe to drink? Yes. How safe? As safe as your well or tap water. How do you make it safe for indoor use? By filtering and purifying it.

Contaminants in water may include algae, air pollution, bird excrement, and leaves, sand, and dust. Local wells have dealt with these problems for decades. Installation of filtration and purification equipment can remove these contaminants at home as well.

First, take measures to keep foreign matter out of the incoming rainwater. First flush devices, gutter screens and other screening mechanisms keep the rainwater as clean as possible before it enters the conveyance system. Using screens and filters will greatly reduce maintenance and lengthen the life of the pump and filtration/purification system.

Even the best screening systems will allow unwanted particulates into the cistern. To keep sediment where it belongs, at the bottom of your tank, screen incoming rainwater, give the remaining sediment time to settle, avoid disturbing it, and don't pull water from the bottom of the tank. Use a floating filter, which extracts water from the middle of the tank, leaving sediment undisturbed.

Next is filtration, which removes debris from the water. Disinfection or purification follows, which kills contaminants and removes harmful substances that may be present.

To determine what type of system you need, test the rainwater at a reliable laboratory. Without testing, you could spend a lot of money on equipment that will not give you safe water.

Filtration is included in every system, even simple irrigation systems. Examples of filtration systems include: screen filters, paper filters, and carbon or charcoal filters.

Almost all systems use multiple filters. For example, after gutter screens and/or a first flush device, a system often includes two in-line filters of increasing fineness, a carbon filter and a UV light. Each of these are described below to assist you in evaluating what might be the right alternative for your planned water use and required water quality.

In starting to evaluate filter options, it is imperative to know exactly what the filter system you select will actually remove from the water. National Sanitation Foundation/American National Standards Institutions (NSF/ANSI) standards are the best, most stringent in the industry. Almost all water-filtration products are certified under NSF Standard 61 for Drinking Water System Components (see Related Topics). But the critical standards for contaminant removal are Standard 42, "Drinking Water Treatment Units – Aesthetic Effects," and Standard 53, "Drinking Water Treatment Units – Health Effects."

Standard 42 covers specific aesthetic contaminants (chlorine taste and odor, and visible particulates). Standard 53 covers health-related contaminants, such as Cryptosporidium, Giardia, lead, and volatile organic chemicals that may be present in drinking water. Systems that meet both of these standards are available, but expensive. Fortunately, the NSF website (see Related Topics) provides an easy way to search for units made by a specific manufacturer or that remove a specific contaminant.

Filters and Disinfection

Filters are measured in microns. One micron is about 1/25,000th of an inch. For comparison, sand is about 100 – 1,000 microns, a human hair is about 100 microns, a particle of dust is about 1 micron and a virus can be smaller than .01 micron.

The first filters in a system are cartridge filters. They range widely in what they are capable of removing and are used in a series (e.g., a 20 micron followed immediately by a 5 micron filter).

Filters are rated by the smallest size of particle they are capable of filtering. The smaller the micron size the better the filter. However, the finer the filter, the higher its cost and the slower its process. Filters have to be changed regularly, as an old, used filter is an excellent environment for microorganisms and potentially harmful pathogens.

For wells and rainwater systems a larger (e.g., a 50 micron) filter or equivalent screen (e.g., 300 mesh) should be used first to eliminate sand and large particles. This screen should be easily accessible and cleaned quarterly. Next is a 20 or 10 micron filter, followed immediately by a 10 or 5 micron filter. These are cleaned less frequently, but at least annually.

Filters will not eliminate all substances in the water. To create drinking quality water, filtration is always followed by disinfection. The EPA requires surface and ground water to be disinfected before it is consumed. Consequently, public water systems add disinfectants to destroy microorganisms that can cause disease in people and animals.

This is also necessary for rainwater, as the natural environment contains many microorganisms. Most are not harmful to us. Some, however, such as Giardia lamblia, can be deadly. These need to be eliminated from water before it is consumed.

Kinds of disinfection include chlorinization, ozonization, ultraviolet (UV) light, and membrane filtration. In evaluating disinfection methods, be aware that some actually create unhealthy byproducts that need to be treated.

The effectiveness of disinfection is judged by looking for an indicator organism that, if present, indicates other more harmful pathogens may be present. In getting a water test, this indicator organism is Total Coliform Bacteria that, if present, indicates other pathogens may be present as well.

Chlorine has been used as a disinfectant in public water systems for most of the past century. The introduction of chlorine to disinfect water has virtually eliminated waterborne diseases such as cholera, typhoid, dysentery and hepatitis, saving thousands of lives. However, it is often maligned due to suspected side effects.

For disinfection purposes, 2.3 fluid ounces of household bleach must be added per 1,000 gallons of water. Chlorine dosage rate will vary depending on quantity of water to be treated, pH and temperature.

A major downside of chlorine is that it is very reactive and easily combines with naturally occurring organic material to create harmful trihalomethanes (THMs) like chloroform. Chloroform is formed when chlorine reacts with either humic and/or fulvic acids, which are commonly found in water.

Because chlorine is reactive, it quickly dissipates. Keeping the dosage rate correct is critical when using this method of disinfection. THMs should be tested for in the water source if you are going to use Chlorine.

To reduce the possibility of harmful byproducts with the use of Chlorine, do the following:

Remove the byproducts after they have been created. This is costly, typically meaning other purification systems must be employed (e.g., Reverse Osmosis or other purifcation systems) or

The concentration of particulates/organics in the water before it is treated. This is accomplished by using filters to remove these substances from the water prior to chlorine treatment.

The Chlorine smell and taste can be removed with an activated carbon filter, often referred to as a charcoal filter. Granulated activated carbon filters are sometimes made from coconut shells and can be considered a "green" solution. Carbon block filters are compressed activated carbon, fused with a binding substance into a solid block.

ULTRAVIOLET LIGHT

An alternative for disinfecting water is Ultraviolet (UV) light. UV lights have been used for nearly a century in Europe and are now common in the US. With UV lights, the water must always pass through a filtration system first. If no filter is used, pathogens and bacteria will cast shadows in the flowing water, thereby allowing live organisms to pass through unharmed.

UV light works by penetrating an organism's cell walls and disrupting the cell's genetic makeup, making it impossible to reproduce and rendering it harmless. Often it is claimed that it kills the microorganism, but it doesn't - it just makes them unable to reproduce and thus harmless. UV lights do not change the chemical composition of the water and leave behind no by-products.

For UV to be effective the right light dose must be used to a specific unit of water and the water must be clear of suspended solids and other particulates. Most UV units are usually insensitive to temperature and pH differences in the water, but manufacturers' fine print should be read and followed.

There are several issues with UV lights should be taken into consideration:

Replace the bulb at the manufacturer's specified intervals – generally after 9,000 hours, or about every 12 months;

UV light is not visible to the human eye, so it may appear to be lit and in fact is not working;

The glass enclosure around the light needs to be cleaned occasionally for the UV light to be effective;

If no backup light is installed the water needs to be shut off upstream of the bulb prior to replacing the unit. Generally it is prudent to disinfect the water downstream after the system has been shut down for any reason.

UV systemCorrect UV treatment is effective in reducing harmful pathogens from the water. It is generally recommended that home units include alarms to notify the user when a bulb needs to be serviced or the unit is not working. Purchase a unit that has an automatic bulb cleaner, to reduce maintenance requirements. Two units should be installed, so when one unit needs servicing the second unit can be turned on so there is no disruption in disinfecting the drinking water.

UV light manufacturers rate their systems to a given dosage at a given flow rate (e.g., 10 gallons per minute). When installing a UV light, make sure the flow rate of the UV unit is matched to your flow rate of water (i.e., the pump flow rate). If the pump rate is greater than that of the UV light, install a pressure regulator or flow restrictor.

To properly treat the rainwater, it must contain particulates no larger than 50 microns and contain no tannins, sulfur or sulfur-related bacteria, have less than 0.3 parts per million of iron, and less than 0.005 parts per million of manganese. Knowing whether these are in the water and need to be treated is a great reason to test your water before installing a system. If any of the above is present in the water, the filters must deal with these elements before the water is treated by a UV light. Most of these will not be present in rainwater, but could result from local air pollution or contamination of the conveyance system. Don't assume anything until your water has been lab tested.

The UV light unit is typically installed after all filtration and the resulting water is clean, bug-free and ready to use. Entry-level units will handle about 10 gallons per minute. The price of the unit will increase as options and flow rates increase.

Membrane Filtration

Membrane filtration is another alternative. Membrane filtration involves pushing water through a layer of material. Pressure-driven membrane technologies include microfiltration, ultrafiltration, nanofiltration and reverse osmosis. It is one of the few technologies capable of removing pharmaceuticals, and creates no byproducts.

Membrane technologies are more costly than other alternatives, but prices are rapidly declining. Most water purification experts expect membrane technology to become the prevalent technology in smaller systems over time as their price drops.

Choosing the right membrane technology is not straightforward, as the technology is changing and there are no real standards. Make sure you know what you need and match it to the type of system you are evaluating. Again, it is critical to test your water to know what you need before evaluating options.

Microfiltration (MF) is a membrane separation process using a pore size of .03 to 10 microns. Although this does not sound like a big range, when it comes to water purification, it is. The smaller the pore size, the more the system will remove. Microfiltration membranes are good for the removal of sand, silt, clay, algae, cysts and some bacteria.

Ultrafiltration (UF) is a membrane separation process using a pore size of approximately .002 to .1 microns. UF will remove all materials removed by an MF system, plus some viruses.

Nanofiltration membranes (NF) have an approximate pore size of only .001 microns. These small pore sizes require much more power to push water through the membrane and generate more waste than either MF or UF filtration systems. These systems eliminate virtually all cysts, bacteria, viruses, and other materials, including minerals. Consequently, the resulting water has a low pH that can be corrosive and needs to be remineralized, commonly using limestone, to raise the pH. Due to the greater power requirements, NF has yet to become mainstream.

Reverse Osmosis (RO) is the most widely used membrane technology today. Under counter RO systemThese systems remove particles as fine as .001 microns, are compact, simple to operate and have been in use for over a decade. RO systems remove radium, natural organics, pesticides, cysts, bacteria and viruses. To ensure contaminant reduction, seek out units certified by NSF for contaminant reduction and not just safety. RO systems produce waste water that needs to be processed; however, the newer units are becoming "greener," producing less, but still significant, waste. These units vary greatly in their efficiency, so make sure to ask about waste and efficiency when shopping for an RO system.

RO waste water contains a high concentration of the contaminants removed from the water, so dealing with this waste must be planned for when installing an RO system. Options for dealing with this water include plumbing through a greywater system to the irrigation system or directly to the septic system.

RO systems come in small under-the-counter units or whole-house systems. Prices will vary greatly for these units and only NSF-certified units should be considered. Under-the-counter units generally include a sediment filter, a carbon filter, the RO membrane and another carbon filter, and will generally cost under $1,000. A whole-house unit contains all the same components, but is capable of handling much larger water flow rates, and generally includes a calcite or equivalent filter to reduce the pH of the water, and a large storage tank (e.g., 20 – 50 gallons). The cost of a whole-house unit can run upwards of $8,000, depending on size of the house and family.

Regardless of system size, maintenance needs to be performed regularly. The most frequent maintenance is changing cartridges. Filters are used to protect the RO membrane from particle fouling. As these filters trap particles from the water supply, a reduction in pressure occurs. Many RO units include a low-pressure switch that prevents the RO from running if the pressure drops too low. Check the allowable pressure drop across the cartridge and compare this to the incoming feed pressure. If it is lower than manufacturer recommendations, the filters need to be replaced.

Distillation

The last commonly available purification technology is distillation. Distillation separates the water from the impurities through heating and then collecting the condensation. It is very energy intensive and loses about 5-10% of the water due to evaporation. Distillation removes almost all substances from the water with the exception of volatile organic chemicals (VOCs) that evaporate easily. To this end, some distillation systems are also equipped with carbon filters to remove the VOCs.

Distillation works slowly to reduce energy requirements and, like RO systems, will store the purified water in a tank for later use. In addition to using a lot of electricity to operate, distillation systems generate heat.

Distillation units producing 5 -12 gallons of water a day will typically cost about $1,500 - $2,000. Cost will increase as capacity increases and as options are added. High-end automatic home units with larger storage capacity may cost upwards of $4,000. New solar distillers give you the option of reducing the electrical requirements.

Standard Practice for Household Use

A common practice in off the grid homes is to filter all the incoming rainwater and then store it in a small pressure tank. From the pressure tank the outgoing water is split into two separate paths - one path for potable and the other for non-potable water. A purification process is added to produce potable water. The major advantage of this approach is that it requires a much smaller unit and costs less, since it treats less water than a whole-house unit. But the disadvantage is that it requires a dual plumbing system – one to supply filtered but non-potable water to the toilets, clothes washer, irrigation faucets, etc., and one to supply potable water to the faucets.

An apparently low-cost, entry-level system is a countertop or pitcher type unit for potable water. However, when measured on gallons of water processed between changing filters, these units tend to be much more expensive in the long run. For example, a typical faucet unit available at most large hardware stores needs its filter changed every 100 gallons. For a family, this would be more than once a month and each filter costs about $30. This could cost nearly $500 a year, just for filters!

Before investing in filtration or purification equipment, invest in removing particulates before they enter into the system by installing gutter screens, leaf screens and roof washers. Removing materials before they enter the system is far easier and less expensive than dealing with them afterwards.

There is no perfect solution for disinfecting water, as all solutions have some environmental cost. Some require substantial energy, some create harmful by-products and some waste water. To save money, test your water (have you heard that before?) and get the right unit to solve your specific problem. Generally, the smaller the capacity the less expensive the unit will be overall, so get only what you need.

Lastly, remember that as the owner of a water system, it is your responsibility to maintain it. When you pay for utility-purified and -delivered water, maintenance is included in your bill. But when you own your water system, it is your responsibility to maintain it on a regular basis.

Rainwater can be safely used outdoors and indoors if the correct steps are taken to handle, store and clean it. Although not yet common in the US, indoor use of rainwater is practiced worldwide. As population growth continues, water rates increase and the desire to be "more green" and self-reliant increases, rainwater use will become more common here in the United States.

The Adoption of Water Conservation Application Techniques

A number of farm application irrigation technologies that could contribute to the conservation of water in both surface and piped systems have been developed. These techniques are generally adopted for the cultivation of high value crops. About 20 million hectares worldwide are now irrigated by sprinkler irrigation. Micro-irrigation as a technology has matured into reliable water and fertigation management system for crop production over the past decades and its usage continues to increase rapidly. A total of 3.2 million hectares are being irrigated by micro-irrigation techniques throughout the world, which represent one percent of the total areas irrigated in the world. A growth of more than six times over the last 20 years has been achieved, which is quite a success. Micro-irrigation is an important tool in the drive towards food security. Micro-irrigation systems have great potential to create significant opportunities for smallholder agriculture.

During the 1980s micro-irrigation started penetrating developing countries, mainly India. A few years later, fearing a water and food shortage, the Chinese Government declared drip irrigation technology to be amongst the top priorities for development in the last decade. However, Vidal et al., (2001) have shown that micro-irrigation "is not a miracle technology: excellent as well as poor results were obtained from these technologies, and their adoption really depends on farmers capacity to finance and operate them, as well as on the type of crop production".

According to the ICID database on modern irrigation techniques, it appears that the use of pressurized water application techniques is widely different between countries. The percentage of modern technique decreases with the total area irrigated in developed countries from a low of 21 percent in the USA and 16 percent in Italy to 90 percent or over in Northern European countries (France and Germany). Sprinkler has a much greater application compared to that of the drip systems with the exception of Cyprus, Israel and Jordan.

In South Asia and Africa low cost bucket-and-drip sets are becoming increasingly popular with farmers. In areas where shallow groundwater is plentiful, thousands of poor farmers in Bangladesh have used low-cost treadle pumps to supply water for crops for their own food security and additional income.

Advanced surface irrigation techniques such as precise land levelling, gated pipes or cablegation have also been developed to improve the typical poorly distributed application of water under conventional methods by furrow, border or basin irrigation. Experience shows that these techniques are too sophisticated to be adopted by most smallholders in developing countries. Adaptive research in the Middle East and North Africa has shown the difficulties in introducing surface water conservation techniques. Many attempts to introduce gated pipes in Morocco for example have failed for a number of reasons, mainly the poor maintenance of the quaternary earth canals. Considerable efforts for training of farmers in water conservation in surface systems are needed.

Farmers have adopted pressurised systems much faster than surface application techniques. For example, about 65 percent of the farmers in the Jordan valley have shifted from surface to drip irrigation over a 10-year period. Farmers built on-farm storage reservoirs to provide the flexibility required for drip irrigation.

Although there are no statistical data available on it, it is likely that most of the areas irrigated through water conservation techniques are served from wells or piped systems. There is still a huge potential for development of water conservation techniques in the areas presently irrigated from groundwater. Application of these techniques will continue to expand with the market for high value crops. However, the adoption of these techniques is limited to the cultivation of certain crops because of technical and financial considerations.

Canal Irrigation Systems: Water Efficiency and Water Productivity

Much more challenging than that of groundwater is the improvement of performance of canal systems, which will remain the dominant water transfer technique in developing countries in the future. The standards used for design of irrigation canals in many countries have not kept up with the development of new technologies. Most traditional delivery systems have no or little flexibility built into them. Their control strategies and control structures are inadequate for good water management practices. Water deliveries at farm level are not compatible with the key requirements of modern agriculture. Thoughts on how to assess the performance of canal projects are evolving.

The engineering concept of efficiency at different levels of a canal system is important for planning, designing and even operating a canal irrigation system. However, water resources experts have criticised this concept which could be misleading. These experts argue that seepage losses contribute to the recharge of aquifers and losses from a project can be recovered downstream if the water quality has not deteriorated beyond certain limits. These arguments are valid when considering the development of water resources at basin level or even at broader levels when aquifers extend beyond basin boundaries. Advocates of water conservation measures, such as canal lining, then respond that water quality may decline in the recycling process, lands adjacent to canals may be affected by waterlogging and salinization, and an important volume of water can be lost from wasteland. Furthermore, farmers have to support the cost of groundwater pumping to which opponents can argue that farmers use groundwater more efficiently than canal water which is too often unreliable. The application of water conservation measures in irrigation can have negative or positive impact on third parties. The theoretical debate is endless.

Efficiency alone is not a sufficient indicator to define the performance of an irrigation system. Indicators of water productivity expressed in terms of production or value of production per unit of water are very important in increasingly water-scarce situation. A canal irrigation system may have high conveyance efficiency with a minimum of seepage and operational losses. However, if water delivery is too rigid or unreliable, there will be considerable waste further down at the farm level.

A water productivity indicator provides a global indication of the effectiveness of water conservation measures and of the quality of service provided to the users, as well as the farm use of water and other inputs.

25 ways to Conserve Water in the Home and Yard

Water conservation has become an essential practice in all regions, even in areas where water seems abundant.

In addition to saving money on your utility bill, water conservation helps prevent water pollution in nearby lakes, rivers and local watersheds.

Conserving water can also extend the life of your septic system by reducing soil saturation, and reducing any pollution due to leaks. Overloading municipal sewer systems can also cause untreated sewage to flow to lakes and rivers. The smaller the amount of water flowing through these systems, the lower the likelihood of pollution. In some communities, costly sewage system expansion has been avoided by communitywide household water conservation.

Water conservation in the home...

1. Check faucets and pipes for leaks: A small drip from a worn faucet washer can waste 20 gallons of water per day. Larger leaks can waste hundreds of gallons.
2. Don't use the toilet as an ashtray or wastebasket: Every time you flush a cigarette butt, facial tissue or other small bit of trash, five to seven gallons of water is wasted.
3. Check your toilets for leaks: Put a little food coloring in your toilet tank. If, without flushing, the color begins to appear in the bowl within 30 minutes, you have a leak that should be repaired immediately. Most replacement parts are inexpensive and easy to install.
4. Use your water meter to check for hidden water leaks: Read the house water meter before and after a two-hour period when no water is being used. If the meter does not read exactly the same, there is a leak.
5. Install water-saving shower heads and low-flow faucet aerators: Inexpensive water-saving low-flow shower heads or restrictors are easy for the homeowner to install. Also, long, hot showers can use five to ten gallons every unneeded minute. Limit your showers to the time it takes to soap up, wash down and rinse off. "Low-flow" means it uses less than 2.5 gallons per minute.

You can easily install a ShowerStart showerhead, or add a ShowerStart converter to existing showerheads, which automatically pauses a running shower once it gets warm. Also, all household faucets should be fit with aerators. This single best home water conservation method is also the cheapest!

6. Put plastic bottles or float booster in your toilet tank: To cut down on water waste, put an inch or two of sand or pebbles inside each of two plastic bottles to weigh them down. Fill the bottles with water, screw the lids on, and put them in your toilet tank, safely away from the operating mechanisms. Or, buy an inexpensive tank bank or float booster. This may save ten or more gallons of water per day. Be sure at least 3 gallons of water remain in the tank so it will flush properly. If there is not enough water to get a proper flush, users will hold the lever down too long or do multiple flushes to get rid of waste. Two flushings at 1.4 gallons is worse than a single 2.0 gallon flush. A better suggestion would be to buy an adjustable toilet flapperthat allow for adjustment of their per flush use. Then the user can adjust the flush rate to the minimum per flush setting that achieves a single good flush each time. For new installations, consider buying "low flush" toilets, which use 1 to 2 gallons per flush instead of the usual 3 to 5 gallons. Replacing an 18 liter per flush toilet with an ultra-low volume (ULV) 6 liter flush model represents a 70% savings in water flushed and will cut indoor water use by about 30%.
7. Insulate your water pipes: It's easy and inexpensive to insulate your water pipes with pre-slit foam pipe insulation. You'll get hot water faster plus avoid wasting water while it heats up.
8. Take shorter showers: One way to cut down on water use is to turn off the shower after soaping up, then turn it back on to rinse. A four-minute shower uses approximately 20 to 40 gallons of water.
9. Turn off the water after you wet your toothbrush: There is no need to keep the water running while brushing your teeth. Just wet your brush and fill a glass for mouth rinsing.
10. Rinse your razor in the sink: Fill the sink with a few inches of warm water. This will rinse your razor just as well as running water, with far less waste of water.
11. Use your dishwasher and clothes washer for only full loads: Automatic dishwashers and clothes washers should be fully loaded for optimum water conservation. Most makers of dishwashing soap recomend not pre-rinsing dishes which is a big water savings.

    With clothes washers, avoid the permanent press cycle, which uses an added 20 liters (5 gallons) for the extra rinse. For partial loads, adjust water levels to match the size of the load. Replace old clothes washers. New Energy Star rated washers use 35 - 50% less water and 50% less energy per load. If you're in the market for a new clothes washer, consider buying a water-saving frontload washer.

12. Minimize use of kitchen sink garbage disposal units: In-sink 'garburators' require lots of water to operate properly, and also add considerably to the volume of solids in a septic tank which can lead to maintenance problems. Start a compost pile as an alternate method of disposing food waste.
13. When washing dishes by hand, don't leave the water running for rinsing: If your have a double-basin, fill one with soapy water and one with rinse water. If you have a single-basin sink, gather washed dishes in a dish rack and rinse them with a spray device or a panful of hot water. Dual-swivel aerators are available to make this easier. If using a dishwasher, there is usually no need to pre-rinse the dishes.
14. Don't let the faucet run while you clean vegetables: Just rinse them in a stoppered sink or a pan of clean water. Use a dual-setting aerator.
15. Keep a bottle of drinking water in the fridge: Running tap water to cool it off for drinking water is wasteful. Store drinking water in the fridge in a safe drinking bottle. If you are filling water bottles to bring along on outdoor hikes, consider buying a LifeStraw personal water filter which enables users to drink water safely from rivers or lakes or any available body of water.

    Water conservation in the yard and garden...

16. Plant drought-resistant lawns, shrubs and plants: If you are planting a new lawn, or overseeding an existing lawn, use drought-resistant grasses such as the new "Eco-Lawn".

    Many beautiful shrubs and plants thrive with far less watering than other species. Replace herbaceous perennial borders with native plants. Native plants will use less water and be more resistant to local plant diseases. Consider applying the principles of xeriscape for a low-maintenance, drought resistant yard.

    Plant slopes with plants that will retain water and help reduce runoff.

    Group plants according to their watering needs.

17. Put a layer of mulch around trees and plants: Mulch will slow evaporation of moisture while discouraging weed growth. Adding 2 - 4 inches of organic material such as compost or bark mulch will increase the ability of the soil to retain moisture. Press the mulch down around the dripline of each plant to form a slight depression which will prevent or minimize water runoff.
18. Don't water the gutter: Position your sprinklers so water lands on the lawn or garden, not on paved areas. Also, avoid watering on windy days.
19. Water your lawn only when it needs it: A good way to see if your lawn needs watering is to step on the grass. If it springs back up when you move, it doesn't need water. If it stays flat, the lawn is ready for watering. Letting the grass grow taller (to 3") will also promote water retention in the soil.

Most lawns only need about 1” of water each week. During dry spells, you can stop watering altogether and the lawn will go brown and dormant. Once cooler weather arrives, the morning dew and rainfall will bring the lawn back to its usual vigor. This may result in a brown summer lawn, but it saves a lot of water.

20. Deep-soak your lawn: When watering the lawn, do it long enough for the moisture to soak down to the roots where it will do the most good. A light sprinkling can evaporate quickly and tends to encourage shallow root systems. Put an empty tuna can on your lawn - when it’s full, you’ve watered about the right amount. Visit our natural lawn care page for more information.
21. Water during the early parts of the day; avoid watering when it’s windy: Early morning is generally better than dusk since it helps prevent the growth of fungus. Early watering, and late watering, also reduce water loss to evaporation. Watering early in the day is also the best defence against slugs and other garden pests. Try not to water when it’s windy - wind can blow sprinklers off target and speed evaporation.
22. Add organic matter and use efficient watering systems for shrubs, flower beds and lawns: Adding organic material to your soil will help increase its absorption and water retention. Areas which are already planted can be ‘top dressed’ with compost or organic matter.

You can greatly reduce the amount of water used for shrubs, beds and lawns by:

- The strategic placement of soaker hoses
- Installing a rain barrel water catchment system
- Installing a simple drip-irrigation system

Avoid over-watering plants and shrubs, as this can actually diminish plant health and cause yellowing of the leaves.

When hand watering, use a variable spray nozzle for targeted watering.

23. Don’t run the hose while washing your car: Clean the car using a pail of soapy water. Use the hose only for rinsing - this simple practice can save as much as 150 gallons when washing a car. Use a spray nozzle when rinsing for more efficient use of water. Better yet, use a waterless car washing system; there are several brands, such as EcoTouch, which are now on the market.
24. Use a broom, not a hose, to clean driveways and sidewalks.
25. Check for leaks in pipes, hoses, faucets and couplings: Leaks outside the house may not seem as bad since they’re not as visible. But they can be just

as wasteful as leaks indoors. Check frequently to keep them drip-free. Use hose washers at spigots and hose connections to eliminate leaks.

Water conservation comes naturally when everyone in the family is aware of its importance, and parents take the time to teach children some of the simple water-saving methods around the home which can make a big difference.

Water Conservation Summary

In 1990, 30 states in the US reported 'water-stress' conditions. In 2000, the number of states reporting water-stress rose to 40. In 2009, the number rose to 45. There is a worsening trend in water supply nationwide. Taking measures at home to conserve water not only saves you money, it also is of benefit to the greater community.

Saving water at home does not require any significant cost outlay. Although there are water-saving appliances and water conservation systems such as rain barrels, drip irrigation and on-demand water heaters which are more expensive, the bulk of water saving methods can be achieved at little cost. For example, 75% of water used indoors is in the bathroom, and 25% of this is for the toilet. The average toilet uses 4 gallons per flush (gpf). You can invest in a ULF (ultra-low flush) toilet which will use only 2 gpf. But you can also install a simple tank bank, costing about $2, which will save .8 gpf. This saves 40% of what you would save with the ULF toilet. Using simple methods like tank banks, low-flow showerheads and faucet aerators you can retrofit your home for under$50.

By using water-saving features you can reduce your in-home water use by 35%. This means the average household, which uses 130,000 gallons per year, coulod save 44,00 gallons of water per year. On a daily basis, the average household, using 350 gallons per day, could save 125 gallons of water per day. The average individual, currently using 70 gallons per day, could save 25 gallons of water per day.

When buying low-flow aerators, be sure to read the label for the actual 'gpm' (gallons per minute) rating. Often, the big box retailers promote "low-flow" which are rated at 2.5 gpm, which is at the top of the low-flow spectrum. This may be needed for the kitchen sink, but we find that a 1.5 gpm aerator works fine for the bathroom sink and most water outlets, delivering the same spray force in a comfortable, soft stream. Finally, it should be noted that installing low-flow aerators, showerheads, tank banks and other water-saving devices usually is a very simple operation which can be done by the homeowner and does not even require the use of tools. Water conservation at home is one of the easiest measures to put in place, and saving water should become part of everday family practice.

Water Conservation Solutions / Water is Life

Because of these circumstances, resources of consumable water are very small. The closest estimate to date is that there is about 1 per cent of water on the earth that is suitable for drinking. This is serious, because in many ways water represents life. Albert Szent-Gyorgyi, the Hungarian biochemist and Nobel Prize winner for medicine says, "Water is life's matter and matrix, mother and medium. There is no life without water."

Consumable and fresh water is the need of every living being on earth. Humans, animals, other creatures, and agriculture all need water to survive and thrive. Without water our planet would become a vacant and barren wasteland.

And if you are thinking that water scarcity is just an issue in third world countries…think again. "Twenty-one percent of irrigation in the United States is achieved by pumping groundwater at rates that exceed the water's ability to recharge," wrote water expert Tony Clarke of the Polaris Institute.

Having water available in its clean and purest form is a blessing, and should be treated with the utmost respect. It is our obligation to save and conserve water as much as we can. In this way we will be helping ourselves and all of humanity.

Below are simple water conservation solutions. And while you practice these water conservation methods give a silent expression of thanks for having the clean water that so many others do not.

Water Conservation Solutions / Simple Things You Can Do

- Take shorter showers.
- Close the tap at once after using the water.
- While brushing, only open the tap when you need water.
- Do not over-water plants.
- Wash your car less frequently and use smaller amounts of water, or use no water when you wash your car.
- Fix leaky pipes immediately upon discovery.
- Slow the flow of water to consume less water, using less energy.
- Operate the washing machine only when you have a full load of laundry.
- Don't throw towels in the laundry after only one use.
- Run your dishwasher only when you have a full load.

These are some ideas to conserve water. Now, let's look at what solutions are available on the market that could help you use less water.

Water Conservation Solutions / Products Available that Help the Cause

- Use the water efficient taps. These taps turn off automatically when not in use.
- Buy appliances that are marked with water efficiency labels. These energy efficient machines use about half the amount of the water as standard appliances.

- Use water efficient showerheads.
- Install a dual flush toilet.
- Install aerators, flow regulators, flow restrictors, and pressure-limiting valves to your taps. These measures will minimize the flow of water considerably.
- Purchase washing machines with front-load washing. These machines use much smaller amounts of water compared to the other machines.
- Keep your appliances at the recommended level to ensure greater efficiency.
- Don't use water to wash your car. Yes, you read that correctly. Find out about this innovative new product at Freedom Waterless and see how it works.

Consider each of these points as a thank you to the planet. When individuals begin to take part in creating a healthier environment, the law of cause and effect will start to spread in a life-enhancing way.

10

RECYCLING WATER AND RAIN WATER HARVESTING

Water is a common substance that is essential to all known forms of life. Water covers 70 percent of the planet's surface. On the planet Earth water is mostly found in oceans and large bodies of water. 1.6 percent of the Earth's water is found underground in aquifers. A tiny percentage of the world's water is found in the bodies of plants and animals and in manufactured goods.

Humans and other animals need access to a supply of fresh and clean drinking water for their survival.

The availability of safe drinking water in almost all areas of the world has increased to a large extent over the last few decades.

Despite this fact some observers have interpolated that by 2025 a large part of the world's population will be facing shortages of water.

One household in the USA for example uses on average 127,000 gallons of water per year. But if greater use was made of low-flow fixtures on their taps, washing machine, shower head and toilet, one USA household could save 40,000 gallons of water per year and 30% on their water bill. If every household used low-flow fixtures, we could save 5.4 billion gallons/day which is worth $11.3 million dollars!

The United Nations calls this situation a water crisis. The role of water in the world's economy is enormous because it is used as a dilutant for a wide variety of chemical substances in industrial and medical applications.

Agriculture is the biggest consumer of freshwater using over 70 percent of freshwater supplies.

Recycling water is the use of treated and processed wastewater for useful purposes like agricultural irrigation, industrial processes, toilet flushing, and groundwater replenishment which is also called a groundwater recharge. Water recycling is also known as water reusing or water reclamation. These terms are often used interchangeably.

How Can Water Recycling Benefit Us?

Water recycling provides enormous environmental benefits. It also provides an additional source of water for various purposes. This a list of some benefits that water recycling can present.

- Water recycling decreases the extraction of water form sources that may be dwindling and may stop being viable as habitats for valuable and endangered wildlife.
- Recycling wastewater can decrease the discharge of effluents that may damage and pollute the ecosystems of the sensitive bodies of water.
- Recycled water can be used to create new wetlands or to enhance and improve the quality of existing ones.
- Water recycling can reduce and prevent pollution by leaving damaging pollutants at the treatment plant.

Recycling Water / What it Does For the Future

Recycling water has been established as an effective and safe way to provide non-potable water for a wide range of industrial, home, and agricultural use. The non-potable utilization of recycled water can only continue to grow. In numerous parts of the United States the uses of recycled water are increasing to accommodate the environmental situation and growing demands for water.

Technological advances in wastewater treatment and health studies in the use of indirect potable water have indicated the possible planned indirect potable reuse might soon become feasible. As the need for more water grows and environmental pressures increase, the necessity for water recycling can only increase as well. Water recycling will need to have a greater role in providing water for all the needs of society. We must work together to make water recycling and water conservation effective enough to help us sustain our valuable and vital water resources.

Water Conservation Facts and Tips

- Less than 2% of the Earth's water supply is fresh water.
- Of all the earth's water, 97% is salt water found in oceans and seas.

- Only 1% of the earth's water is available for drinking water. Two percent is frozen.
- The human body is about 75% water.
- A person can survive about a month without food, but only 5 to 7 days without water.
- Every day in the United States, we drink about 110 million gallons of water.
- Landscaping accounts for about half the water Californians use at home. Showers account for another 18 percent, while toilets use about 20 percent.
- Showering and bathing are the largest indoor uses (27%) of water domestically.
- The average American uses 140-170 gallons of water per day.
- If every household in America had a faucet that dripped once each second, 928 million gallons of water a day would leak away.
- There are 7.48 gallons in a cubic foot of water. Therefore, 2000 cubic feet of water is 14,960 gallons.
- An acre foot of water is about 326,000 gallons. One-half acre foot is enough to meet the needs of a typical family for a year. There are 7.48 gallons in a cubic foot of water.
- It takes 3.3 acre feet of water to grow enough food for an average family for a year.
- A leaky faucet can waste 100 gallons a day.
- One flush of the toilet uses 6 ½ gallons of water.
- An average bath requires 37 gallons of water.
- An average family of four uses 881 gallons of water per week just by flushing the toilet.
- The average 5-minute shower takes 15-25 gallons of water--around 40 gallons are used in 10 minutes.
- Take short showers instead of baths.
- A full bathtub requires about 36 gallons of water.
- You use about 5 gallons of water if you leave the water running while brushing your teeth.
- If you water your grass and trees more heavily, but less often, this saves water and builds stronger roots.
- Each person needs to drink about 2 ½quarts (80 ounces) of water every day.

- Water your lawn only when it needs it. If you step on the grass and it springs back up when you move, it doesn't need water. If it stays flat, it does need water.
- Run your dishwasher and washing machine only when they are full.
- When washing a car, use soap and water from a bucket. Use a hose with a shut-off nozzle for rinsing.
- Never put water down the drain when there may be another use for it such as watering a plant or garden, or cleaning.
- Avoid flushing the toilet unnecessarily. Dispose of tissues, insects and other such waste in the trash rather than the toilet.
- When washing dishes by hand, fill one sink or basin with soapy water. Quickly rinse under a slow-moving stream from the faucet.
- An automatic dishwasher uses 9 to 12 gallons of water while hand washing dishes can use up to 20 gallons.
- Store drinking water in the refrigerator rather than letting the tap run every time you want a cool glass of water.
- Water lawns during the early morning hours, or evening when temperatures and wind speed are the lowest.
- This reduces losses from evaporation.
- Do not hose down your driveway or sidewalk.
- Use a broom to clean leaves and other debris from these areas.
- Using a hose to clean a driveway wastes hundreds of gallons of water.
- Don't leave the water running when brushing your teeth or shaving. Get in the habit of turning off the water when it's not being used.
- Use of bowl of water to clean fruits & vegetables rather than running water over them. You can reuse this for your house plants.
- Public water suppliers process 38 billion gallons of water per day for domestic and public use.
- Approximately 1 million miles of pipelines and aqueducts carry water in the U.S. & Canada. That's enough pipe to circle the earth 40 times.
- About 800,000 water wells are drilled each year in the United States for domestic, farming, commercial, and water testing purposes.
- More than 13 million households get their water from their own private wells and are responsible for treating and pumping the water themselves.
- Industries released 197 million pounds of toxic chemicals into waterways in 1990.

- You can refill an 8-oz glass of water approximately 15,000 times for the same cost as a six-pack of soda pop.
- A dairy cow must drink four gallons of water to produce one gallon of milk.
- 300 million gallons of water are needed to produce a single day's supply of U.S. newsprint.
- One inch of rainfall drops 7,000 gallons or nearly 30 tons of water on a 60' by 180' piece of land.

Facts on Saving Water

According to the United States Environmental Protection Agency (EPA), a family of four uses 400 gallons of water every day. Despite the majority of the Earth being covered in water, only 1 percent of it is useable - the rest is salty ocean water or permanently frozen. Water is a limited resource, and with the population continuing to grow, it is essential to practice water conservation. Here are some water conservation facts and ways to reduce the amount of water you use or waste.

WATER CONSERVATION FACTS

Many of your everyday activities use water - from bathing and cleaning the house to drinking and cooking. On a broader scale, water is necessary for agriculture, power plants and industrial or mining. The American lifestyle requires a great deal of this precious resource. Here are some surprising water usage statistics from the EPA.

- The average American family spends $500 per year on its water and sewer bill.
- Treating and delivering water takes a lot of energy - letting your faucet run for 5 minutes takes as much energy as letting a 60-watt lightbulb run for 14 hours.
- More than 30 percent of household water is for outdoor use, such as watering lawns.
- Turning off the tap water while you brush your teeth can save 8 gallons of water a day - that equals about 200 gallons of water saved per month.
- A faucet that leaks 1 drip per second wastes 3,000 gallons of water each year.
- Leaky toilets can waste up to 200 gallons per day, the equivalent of flushing the toilet 50 times for no reason.
- Older toilets use 60 percent more water per flush as compared to newer high-efficiency model (look for the WaterSense label when buying a new toilet).

- The average bath requires 70 gallons of water while a shower uses 10 to 25 gallons.
- A running hose outdoors - for washing cars or watering plants - can waste 6 gallons of water per minute when left unattended.
- US power plants use 136 billion gallons of water each day.
- US agriculture uses 142 billion gallons of water per day for irrigation.
- US industrial facilities require more than 20 billion gallons of water every day.
- Lower water levels in fresh water reservoirs can contribute to concentrated levels of both natural and human-based pollutants in the water supply.

Water Saving Tips

Environmental advocate Lesley Chilcott, co-founder of Unscrew America and documentary filmmaker for former vice-president Al Gore's An Inconvenient Truth, recommends the following water-saving tips.

1. Shut the Water off

Instead of letting the water run while you brush your teeth, shave, peel vegetables or scrub dishes, turn it off until you need it.

2. Shower Efficiently

Don't run the shower for five minutes to warm it up while you are doing something else - adjust the temperature when you get in. And keep your shower time to five minutes. Turn the shower off if you need to shave or deep condition your hair.

3. Install Energy-efficient Appliances

Use an EnergyStar washing machine and dishwasher and only run full loads, or at the very least use the lowest water level setting you need for smaller loads.

4. Update your Toilet

Install a low-flow or high-efficiency toilet to save nearly 2 gallons of water per flush.

5. Change your Showerhead

In addition to shorter showers, install a low-flow showerhead, which will also reduce water waste. While you're at it, repair your leaky faucets.

6. Plant Sense

Change your landscaping and house plants to low-water, native or drought-tolerant plants. Then do your watering in the morning or evening when there is less chance of evaporation. Additionally, use your broom instead of the hose to clear debris from your driveway and side-walks.

7. Become a Flexitarian

The food production water usage for plant-based foods is significantly less than that of meat-based foods. A plant-based diet is healthier, too. (Click for more information on The Flexitarian Diet.)

8. Install a Water Filtration System

Instead of buying bottle after bottle of drinking water, install a water filter and use refillable beverage containers.

9. Use a Bucket

Wash your car with a bucket of soapy water instead of letting the hose run.

10. Set up a Rain Water Barrel

Collect rain water and use it later to water the garden. Another option is to install a gray water system.

11. Use Biodegradable Cleaning Products

Take advantage of the growing selection of eco-friendly household cleaning products, dish washing soaps, and laundry detergents. Biodegradable products won't pollute the water.

Rain Water Harvesting

The term rainwater harvesting is being frequently used these days, however, the concept of water harvesting is not new for India. Water harvesting techniques had been evolved and developed centuries ago.

Ground water resource gets naturally recharged through percolation. But due to indiscriminate development and rapid urbainzation, exposed surface for soil has been reduced drastically with resultant reduction in percolation of rainwater, thereby depleting ground water resource. Rainwater harvesting is the process of augmenting the natural filtration of rainwater in to the underground formation by some artificial methods. "Conscious collection and storage of rainwater to cater to demands of water, for drinking, domestic purpose & irrigation is termed as Rainwater Harvesting."

Rainwater harvesting is the accumulation and storage of rainwater for reuse before it reaches the aquifer. Uses include water for garden, water for livestock, water for irrigation, cleaning of bathrooms as in first flush, etc. In many places the water collected is just redirected to a deep pit with percolation. The harvested water can be used for drinking water as well if the storage is a tank that can be accessed and cleaned when needed. Rainwater harvesting provides an independent water supply during regional water restrictions, and in developed countries is often used to supplement the mains supply. Rainwater harvesting systems are appealing as they are easy to understand, install and operate. They are effective in 'green droughts' as water is captured from rainfall where runoff is insufficient to flow into dam storages. The quality of captured rainwater is usually sufficient for most household needs, reducing the need for detergents because rainwater is soft. Financial benefits to the users include that rain is 'renewable' at acceptable volumes despite climate change forecasts, and rainwater harvesting systems generally have low running costs, providing water at the point of consumption (Ferguson 2012).

Benefits of widespread rainwater harvesting to the regional reticulated supply system may include reduced treatment, pumping, operation and augmentation costs, reducing peak storm water runoff and storm water processing costs, as well as reduced greenhouse gas emissions due to reduced dependence on pumping and potential augmentation through sources such as desalination (Coombes 2007, White, 2009).

The principle of collecting and using precipitation from a catchments surface. An old technology is gaining popularity in a new way. Rain water harvesting is enjoying a renaissance of sorts in the world, but it traces its history to biblical times. Extensive rain water harvesting apparatus existed 4000 years ago in the Palestine and Greece. In ancient Rome, residences were built with individual cisterns and paved courtyards to capture rain water to augment water from city's aqueducts. As early as the third millennium BC, farming communities in Baluchistan and Kutch impounded rain water and used it for irrigation dams.

System of collection rainwater and conserving for future needs has traditionally been practiced in India. The traditional systems were time-tested wisdom of not only appropriate technology of Rainwater Harvesting, but also water management systems, where conservation of water was the prime concern. Traditional water harvesting systems were Bawaries, step wells, jhiries, lakes, tanks etc. These were the water storage bodies to domestic and irrigation demands. People were themselves responsible for maintenance to water sources and optimal use of water that could fulfill their needs.

ARTIFICIAL RECHARGE TO GROUND WATER

Artificial recharge to ground water is a process by which the ground water reservoir is augmented at a rate exceeding that obtaining under natural conditions or replenishment. Any man-made scheme or facility that adds water to an aquifer may be considered to be an artificial recharge system.

Why Harvest Rainwater ?

Rain water harvesting is essential because :

Surface water is inadequate to meet our demand and we have to depend on ground water.

Due to rapid urbanization, infiltration of rain water into the sub-soil has decreased drastically and recharging of ground water has diminished.

As you read this guide, seriously consider conserving water by harvesting and managing this natural resource by artificially recharging the system. The examples covering several dozen installations successfully operating in India constructed and maintained by CGWB, provide an excellent snapshot of current systems.

This is perhaps one of the most frequently asked question, as to why one should harvest rainwater. There are many reasons but following are some of the important ones.

- To arrest ground water decline and augment ground water table
- To beneficial water quality in aquifers
- To conserve surface water runoff during monsoon
- To reduce soil erosion
- To inculcate a culture of water conservation

How to harvest rainwater:

Broadly there are two ways of harvesting rainwater:

(i) Surface runoff harvesting

(ii) Roof top rainwater harvesting

Surface runoff harvesting

In urban area rainwater flows away as surface runoff. This runoff could be caught and used for recharging aquifers by adopting appropriate methods.

Roof top rainwater harvesting (RTRWH)

It is a system of catching rainwater where it falls. In rooftop harvesting, the roof becomes the catchments, and the rainwater is collected from the roof of the house/ building. It can either be stored in a tank or diverted to artificial recharge system. This method is less expensive and very effective and if implemented properly helps in augmenting the ground water level of the area.

Rainwater harvesting is a technology used for collecting and storing rainwater from rooftops, the land surface or rock catchments using simple techniques such as jars and pots as well as more complex techniques such as underground check dams. The techniques usually found in Asia and Africa arise from practices employed by ancient civilizations within these regions and still serve as a major source of drinking water supply in rural areas. Commonly used systems are constructed of three principal components; namely, the catchment area, the collection device, and the conveyance system.

Quality

The concentration of contaminants is reduced significantly by diverting the initial flow of runoff water to waste. Improved water quality can also be obtained by using a floating draw-off mechanism (rather than from the base of the tank) and by using a series of tanks, with draw from the last in series. The stored rainwater may need to be analyzed properly before use in a way appropriate to its safety.

System Setup

Rainwater harvesting systems can be installed with minimal skills. The system should be sized to meet the water demand throughout the dry season since it must be big enough to support daily water consumption. Specifically, the rainfall capturing area such as a building roof must be large enough to maintain adequate flow. Likewise, the water storage tank should be large enough to contain the captured water.

Vendors

There are three main types of companies operating in the rainwater harvesting industry: makers of water storage, makers of accessories, and integrators. Water storage companies make tanks, barrels, and underground cisterns. Accessories are added to facilitate or improve the water capturing process. Integrators are regional practitioners which install systems.

HISTORY

Early Era

Around the third century BCE, farming communities in Baluchistan (in present-day Pakistan, Afghanistan and Iran), and Kutch (in present-day India) used rainwater harvesting for irrigation.

In ancient Tamil Nadu (India), rainwater harvesting were done by Chola kings. Rainwater from the Brihadeeswarar temple was collected in Sivaganga tank. During the later Chola period, the V?r?nam tank was built (1011 to 1037 CE) in Cuddalore district of Tamil Nadu to store water for drinking and irrigation purposes. V?r?nam is a 16-kilometre (9.9 mi) long tank with a storage capacity of 1,465,000,000 cubic feet (41,500,000 m).

Present day

- Currently in China and Brazil, rooftop rainwater harvesting is being practiced for providing drinking water, domestic water, water for livestock, water for small irrigation and a way to replenish ground water levels. Gansu province in China and semi-arid north east Brazil have the largest rooftop rainwater harvesting projects ongoing.

- In Bermuda, the law requires all new construction to include rainwater harvesting adequate for the residents.
- The U.S. Virgin Islands have a similar law.
- In Senegal and Guinea-Bissau, the houses of the Diola-people are frequently equipped with homebrew rainwater harvesters made from local, organic materials.
- In the Irrawaddy Delta of Myanmar, the groundwater is saline and communities rely on mud-lined rainwater ponds to meet their drinking water needs throughout the dry season. Some of these ponds are centuries old and are treated with great reverence and respect.
- Until 2009 in Colorado, water rights laws almost completely restricted rainwater harvesting; a property owner who captured rainwater was deemed to be stealing it from those who have rights to take water from the watershed. Now, residential well owners that meet certain criteria may obtain a permit to install a rooftop precipitation collection system (SB 09-080). Up to 10 large scale pilot studies may also be permitted (HB 09-1129). The main factor in persuading the Colorado Legislature to change the law was a 2007 study that found that in an average year, 97% of the precipitation that fell in Douglas County, in the southern suburbs ofDenver, never reached a stream-it was used by plants or evaporated on the ground. In Colorado you cannot even drill a water well unless you have at least 35 acres. In New Mexico, rainwater catchment is mandatory for new dwellings in Santa Fe.
- In Beijing, some housing societies are now adding rain water in their main water sources after proper treatment.
- In Ireland, Professor Micheal McGinley established a project to design a rain water harvesting prototype in the Biosystems design Challenge Module at University College Dublin.
- In Australia rainwater harvesting is typically used as a supplement to the reticulated mains supply, and it is mandated in many building codes. In south east Queensland, households that harvested rainwater doubled each year from 2005 to 2008, reaching 40% penetration at that time (White, 2009 (PhD)).

India

- In India, rain water harvesting was first introduced by Andhra Pradesh ex-Chief Minister N. Chandrababu Naidu. He made a rule that every house which is going to built in cities of that state must have a percolation pit/ rainwater harvesting system. This rule increased the ground water level in good phase. After his term as Chief Minister, the next leaders neglected this system.

- In the state of Tamil Nadu, rainwater harvesting was made compulsory for every building to avoid ground water depletion. It proved excellent results within five years, and every other state took it as role model. Since its implementation, Chennai saw a 50 percent rise in water level in five years and the water quality significantly improved.
- In Rajasthan, rainwater harvesting has traditionally been practiced by the people of the Thar Desert. There are many ancient water harvesting systems in Rajasthan, which have now been revived
- Kerala:
- At present, in Pune (in Maharashtra), rainwater harvesting is compulsory for any new society to be registered.
- An attempt has been made at Dept. of Chemical Engineering, IISc, Bangalore to harvest rainwater using upper surface of a solar still, which was used for water distillation

RAIN WATER HARVESTING TECHNIQUES

There are two main techniques of rain water harvestings.

- Storage of rainwater on surface for future use.
- Recharge to ground water.

The storage of rain water on surface is a traditional techniques and structures used were underground tanks, ponds, check dams, weirs etc. Recharge to ground water is a new concept of rain water harvesting and the structures generally used are :-

Pits :- Recharge pits are constructed for recharging the shallow aquifer. These are constructed 1 to 2 m, wide and to 3 m. deep which are back filled with boulders, gravels, coarse sand.

Trenches:- These are constructed when the permeable stram is available at shallow depth. Trench may be 0.5 to 1 m. wide, 1 to 1.5m. deep and 10 to 20 m. long depending up availability of water. These are back filled with filter. materials.

Dug wells:- Existing dug wells may be utilised as recharge structure and water should pass through filter media before putting into dug well.

Hand pumps :- The existing hand pumps may be used for recharging the shallow/deep aquifers, if the availability of water is limited. Water should pass through filter media before diverting it into hand pumps.

Recharge wells :- Recharge wells of 100 to 300 mm. diameter are generally constructed for recharging the deeper aquifers and water is passed through filter media to avoid choking of recharge wells.

Recharge Shafts :- For recharging the shallow aquifer which are located below clayey surface, recharge shafts of 0.5 to 3 m. diameter and 10 to 15 m. deep are constructed and back filled with boulders, gravels & coarse sand.

Lateral shafts with bore wells :- For recharging the upper as well as deeper aquifers lateral shafts of 1.5 to 2 m. wide & 10 to 30 m. long depending upon availability of water with one or two bore wells are constructed. The lateral shafts is back filled with boulders, gravels & coarse sand.

Spreading techniques :- When permeable strata starts from top then this technique is used. Spread the water in streams/Nalas by making check dams, nala bunds, cement plugs, gabion structures or a percolation pond may be constructed.

Diversion of Run off into Existing Surface Water Bodies

Construction activity in and around the city is resulting in the drying up of water bodies and reclamation of these tanks for conversion into plots for houses.

Free flow of storm run off into these tanks and water bodies must be ensured. The storm run off may be diverted into the nearest tanks or depression, which will create additional recharge.

Urbanisation effects on Groundwater Hydrology :

- Increase in water demand
- More dependence on ground water use
- Over exploitation of ground water
- Increase in run-off, decline in well yields and fall in water levels
- Reduction in open soil surface area
- Reduction in infiltration and deterioration in water quality

Methods of artificial recharge in urban areas :

- Water spreading
- Recharge through pits, trenches, wells, shafts
- Rooftop collection of rainwater
- Roadtop collection of rainwater
- Induced recharge from surface water bodies.

Computation of artificial recharge from Roof top rainwater collection :

- Factors taken for computation .
- Roof top area 100 sq.m. for individual house and 500 sq.m. for multi-storied building.

- Average annual monsoon rainfall - 780 mm.
- Effective annual rainfall contributing to recharge 70% - 550 mm.

Benefits of Artificial Recharge in Urban Areas :

- Improvement in infiltration and reduction in run-off.
- Improvement in groundwater levels and yields.
- Reduces strain on Special Village Panchayats/ Municipal / Municipal Corporation water supply
- Improvement in groundwater quality
- Estimated quantity of additional recharge from 100 sq. m. roof top area is 55.000 liters.

Harvesting Rainwater Harnessing Life

A Noble Goal - A Common Responsibility

Ground water exploitation is inevitable is Urban areas. But the groundwater potential is getting reduced due to urbanisation resulting in over exploitation. Hence, a strategy to implement the groundwater recharge, in a major way need to be launched with concerted efforts by various Governmental and Non-Governmental Agencies and Public at large to build up the water table and make the groundwater resource, a reliable and sustainable source for supplementing water supply needs of the urban dwellers.

Recharge of groundwater through storm run-off and roof top water collection, diversion and collection of run-off into dry tanks, play grounds, parks and other vacant places are to be implemented by Special Village Panchayats/ Municipalities /Municipal Corporations and other Government Establishments with special efforts.

The Special Village Panchayats /Municipalities/Municipal Corporations will help the citizens and builders to adopt suitable recharge method in one's own house or building through demonstration and offering subsidies for materials and incentives, if possible.

Components of a Rainwater Harvesting System

A rainwater harvesting system comprises components of various stages - transporting rainwater through pipes or drains, filtration, and storage in tanks for reuse or recharge. The common components of a rainwater harvesting system involved in these stages are illustrated here.

Catchments

The catchment of a water harvesting system is the surface which directly receives the rainfall and provides water to the system. It can be a paved area like a terrace or courtyard of a building, or an unpaved area like a lawn or open ground. A roof made of reinforced cement concrete (RCC), galvanised iron or corrugated sheets can also be used for water harvesting.

Conveyance Systems

Conveyance systems are required to transfer the rainwater collected on the rooftops to the storage tanks. This is usually accomplished by making connections to one or more down-pipes connected to the rooftop gutters. When selecting a conveyance system, consideration should be given to the fact that, when it first starts to rain, dirt and debris from the rooftop and gutters will be washed into the down-pipe. Thus, the relatively clean water will only be available some time later in the storm. There are several possible choices to selectively collect clean water for the storage tanks. The most common is the down-pipe flap. With this flap it is possible to direct the first flush of water flow through the down-pipe, while later rainfall is diverted into a storage tank. When it starts to rain, the flap is left in the closed position, directing water to the down-pipe, and, later, opened when relatively clean water can be collected. A great disadvantage of using this type of conveyance control system is the necessity to observe the runoff quality and manually operate the flap. An alternative approach would be to automate the opening of the flap as described below.

A funnel-shaped insert is integrated into the down-pipe system. Because the upper edge of the funnel is not in direct contact with the sides of the down-pipe, and a small gap exists between the down-pipe walls and the funnel, water is free to flow both around the funnel and through the funnel. When it first starts to rain, the volume of water passing down the pipe is small, and the *dirty* water runs down the walls of the pipe, around the funnel and is discharged to the ground as is normally the case with rainwater guttering. However, as the rainfall continues, the volume of water increases and *clean* water fills the down-pipe. At this higher volume, the funnel collects the clean water and redirects it to a storage tank. The pipes used for the collection of rainwater, wherever possible, should be made of plastic, PVC or other inert substance, as the pH of rainwater can be low (acidic) and could cause corrosion, and mobilization of metals, in metal pipes.

In order to safely fill a rainwater storage tank, it is necessary to make sure that excess water can overflow, and that blockages in the pipes or dirt in the water do not cause damage or contamination of the water supply. The design of the funnel system, with the drain-pipe being larger than the rainwater tank feed-pipe, helps to ensure that the water supply is protected by allowing excess water to bypass the storage tank. A modification of this design is shown in Figure 5, which illustrates a simple overflow/bypass system. In this system, it also is possible to fill the tank from a municipal drinking water source, so that even during a prolonged drought the tank can be kept full. Care should be taken, however, to ensure that rainwater does not enter the drinking water distribution system.

Transportation

Rainwater from rooftop should be carried through down take water pipes or drains to storage/harvesting system. Water pipes should be UV resistant (ISI HDPE/PVC pipes) of required capacity. Water from sloping roofs could be caught through gutters and down take pipe. At terraces, mouth of the each drain should have wire mesh to restrict floating material.

First-flushing

A first flush device is a valve that ensures that runoff from the first spell of rain is flushed out and does not enter the system. This needs to be done since the first spell of rain carries a relatively larger amount of pollutants from the air and catchment surface.

First flush is a device used to flush off the water received in first shower. The first shower of rains needs to be flushed-off to avoid contaminating storable/ rechargeable water by the probable contaminants of the atmosphere and the catchment roof. It will also help in cleaning of silt and other material deposited on roof during dry seasons Provisions of first rain separator should be made at outlet of each drainpipe.

Filter

There is always some skepticism regarding Roof Top Rainwater Harvesting since doubts are raised that rainwater may contaminate groundwater. There is remote possibility of this fear coming true if proper filter mechanism is not adopted. Secondly all care must be taken to see that underground sewer drains are not punctured and no leakage is taking place in close vicinity. Filters are used fro treatment of water to effectively remove turbidity, colour and microorganisms. After first flushing of rainfall, water should pass through filters. There are different types of filters in practice, but basic function is to purify water.

Sand Gravel Filter

These are commonly used filters, constructed by brick masonry and filleted by pebbles, gravel, and sand as shown in the figure. Each layer should be separated by wire mesh.

Charcoal Filter

Charcoal filter can be made in-situ or in a drum. Pebbles, gravel, sand and charcoal as shown in the figure should fill the drum or chamber. Each layer should be separated by wire mesh. Thin layer of charcoal is used to absorb odor if any.

PVC- Pipe filter

This filter can be made by PVC pipe of 1 to 1.20 m length; Diameter of pipe depends on the area of roof. Six inches dia. pipe is enough for a 1500 Sq. Ft. roof and 8 inches dia. pipe should be used for roofs more then 1500 Sq. Ft. Pipe is divided into three compartments by wire mesh. Each component should be filled with gravel and sand alternatively as shown in the figure. A layer of charcoal could also be inserted between two layers. Both ends of filter should have reduce of required size to connect inlet and outlet. This filter could be placed horizontally or vertically in the system.

Sponge Filter

It is a simple filter made from PVC drum having a layer of sponge in the middle of drum. It is the easiest and cheapest form filter, suitable for residential units.

METHODS OF ROOF TOP RAINWATER HARVESTING

Storage of Direct use

In this method rain water collected from the roof of the building is diverted to a storage tank. The storage tank has to be designed according to the water requirements, rainfall and catchment availability. Each drainpipe should have mesh filter at mouth and first flush device followed by filtration system before connecting to the storage tank. It is advisable that each tank should have excess water over flow system.

Excess water could be diverted to recharge system. Water from storage tank can be used for secondary purposes such as washing and gardening etc. This is the most cost effective way of rainwater harvesting. The main advantage of collecting and using the rainwater during rainy season is not only to save water from conventional sources, but also to save energy incurred on transportation and distribution of water at the doorstep. This also conserve groundwater, if it is being extracted to meet the demand when rains are on.

Recharging ground water aquifers

Ground water aquifers can be recharged by various kinds of structures to ensure percolation of rainwater in the ground instead of draining away from the surface. Commonly used recharging methods are:

(a) Recharging of bore wells

(b) Recharging of dug wells.

(c) Recharge pits

(d) Recharge Trenches

(e) Soak ways or Recharge Shafts

(f) Percolation Tanks

Recharging of bore wells

Rainwater collected from rooftop of the building is diverted through drainpipes to settlement or filtration tank. After settlement filtered water is diverted to bore wells to recharge deep aquifers. Abandoned bore wells can also be used for recharge.

Optimum capacity of settlement tank/filtration tank can be designed on the basis of area of catchement, intensity of rainfall and recharge rate as discussed in design parameters. While recharging, entry of floating matter and silt should be restricted because it may clog the recharge structure. "first one or two shower should be flushed out through rain separator to avoid contamination. This is very important, and all care should be taken to ensure that this has been done."

Recharge Pits

Recharge pits are small pits of any shape rectangular, square or circular, contracted with brick or stone masonry wall with weep hole at regular intervals. to of pit can be covered with perforated covers. Bottom of pit should be filled with filter media.

The capacity of the pit can be designed on the basis of catchment area, rainfall intensity and recharge rate of soil. Usually the dimensions of the pit may be of 1 to 2 m width and 2 to 3 m deep depending on the depth of pervious strata. These pits are suitable for recharging of shallow aquifers, and small houses.

Soak away or Recharge Shafts

Soak away or recharge shafts are provided where upper layer of soil is alluvial or less pervious. These are bored hole of 30 cm dia. up to 10 to 15 m deep, depending on depth of pervious layer. Bore should be lined with slotted/perforated PVC/MS pipe to prevent collapse of the vertical sides. At the top of soak away required size sump is constructed to retain runoff before the filters through soak away. Sump should be filled with filter media.

Recharging of dug wells

Dug well can be used as recharge structure. Rainwater from the rooftop is diverted to dug wells after passing it through filtration bed. Cleaning and desalting of dug well should be done regularly to enhance the recharge rate. The filtration method suggested for bore well recharging could be used.

Recharge Trenches

Recharge trench in provided where upper impervious layer of soil is shallow. It is a trench excavated on the ground and refilled with porous media like pebbles, boulder or brickbats. it is usually made for harvesting the surface runoff. Bore wells can also be provided inside the trench as recharge shafts to enhance percolation. The length of the trench is decided as per the amount of runoff expected. This method is suitable for small houses, playgrounds, parks and roadside drains. The recharge trench can be of size 0.50 to 1.0 m wide and 1.0 to 1.5 m deep.

Percolation tanks

Percolation tanks are artificially created surface water bodies, submerging a land area with adequate permeability to facilitate sufficient percolation to recharge the ground water. These can be built in big campuses where land is available and topography is suitable.

Surface run-off and roof top water can be diverted to this tank. Water accumulating in the tank percolates in the solid to augment the ground water. The stored water can be used directly for gardening and raw use. Percolation tanks should be built in gardens, open spaces and roadside green belts of urban area.

Do's and Don'ts

Harvested rainwater is used for direct usage or for recharging aquifers. It is most important to ensure that the rainwater caught is free from pollutants. Following precautionary measures should be taken while harvesting rainwater:

- Roof or terraces uses for harvesting should be clean, free from dust, algal plants etc.
- Roof should not be painted since most paints contain toxic substances and may peel off.
- Do not store chemicals, rusting iron, manure or detergent on the roof.
- Nesting of birds on the roof should be prevented.
- Terraces should not be used for toilets either by human beings or by pets.
- Provide gratings at mouth of each drainpipe on terraces to trap leaves debris and floating materials.
- Provision of first rain separator should be made to flush off first rains.
- Do not use polluted water to recharge ground water.
- Ground water should only be recharged by rainwater.
- Before recharging, suitable arrangements of filtering should be provided.
- Filter media should be cleaned before every monsoon season.

11

CANAL BUILDING

Canals and navigations are human-made channels for water conveyance (supply), or to service water transport vehicles. In the vernacular, both are referred to as 'canals', and in most cases, the engineered works will have a series of dams and locks that create areas of low speed current flow. These areas are referred to as 'slack water levels', often just called 'levels'.

The main difference between them is that a navigation parallels a river and shares part of its waters, its drainage basin and leverages its resources by building dams and locks to increase and lengthen its long stretches of deep enough slack water levels while staying in its valley. In contrast, a canal cuts across a drainage divide atop a ridge (however humble), generally requiring an external water source above the highest elevation. Many canals have been built at elevations towering over valleys and others water ways crossing far below. Cities need a lot of fresh water and many canals with sources of water at a higher level can deliver water to a destination where there is a lack of water. The Roman Empire's Aqueducts were such water supply canals.

TYPES OF ARTIFICIAL WATERWAYS

A navigation is a series of channels that run roughly parallel to the valley and stream bed of an unimproved river. A navigation always shares the drainage basin of the river. A vessel uses the calm parts of the river itself as well as improvements, traversing the same changes in height.

A true canal is a channel that cuts across a drainage divide, making a navigable channel connecting two different drainage basins.

Most commercially important canals of the first half of the 19th century were a little of each, using rivers in long stretches, and divide crossing canals in others. This is true for many canals still in use.

Structures used in artificial waterways

Both navigations and canals use engineered structures to improve navigation:

weirs and dams to raise river water levels to usable depths;

looping descents to create a longer and gentler channel around a stretch of rapids or falls;

locks to allow ships and barges to ascend/descend.

Since they cut across drainage divides, canals are more difficult to construct and often need additional improvements, like viaducts and aqueducts to bridge waters over streams and roads, and ways to keep water in the channel.

Types of canals

There are two broad types of canal:

Waterways: canals and navigations used for carrying vessels transporting goods and people. These can be subdivided into two kinds:

Those connecting existing lakes, rivers, other canals or seas and oceans.

Those connected in a city network: such as the Canal Grande and others of Venice Italy; the gracht of Amsterdam, and the waterways of Bangkok.

Aqueducts: water supply canals that are used for the conveyance and delivery of potable water for human consumption, municipal uses, hydro power canals and agricultureirrigation.

Loading Anthracite on the Lehigh Canal to feed the early United States industries in the pioneer-era.

1. Design High Water Level (HWL) 2. Low water channel 3. Flood channel 4. Riverside slope 5. Riverside banquette 6. Levee crown 7. Landside slope 8. Landside banquette 9. Berm 10. Low water revetment 11. Riverside land 12. Levee 13. Protected lowland 14. River zone

Importance

Historically canals were of immense importance to commerce and the development, growth and vitality of a civilization. In 1855 the Lehigh Canal carried over 1.2 million tons of clean burning anthracite coal; by the 1930s the company which built and operated it over a century pulled the plug. The few canals still in operation in our modern age are a fraction of the numbers that once fueled and enabled economic growth, indeed were practically a prerequisite to further urbanization and industrialization—for the movement of bulk raw materials such as coal and ores are difficult and marginally affordable without water transport. Such raw materials fueled the industrial developments and new metallurgy resulting of the spiral of increasing mechanization during 17th–20th century, leading to new research disciplines, new industries and economies of scale, raising the standard of living for any industrialized society.

The surviving canals, including most ship canals, today primarily service mostly bulk cargo and large ship transportation industries, whereas the once critical smaller inland waterways conceived and engineered as boat and barge canals have largely been supplanted and filled in, abandoned and left to deteriorate, or kept in service and staffed by state employees, where dams and locks are maintained for flood control or pleasure boating. Their replacement was gradual, beginning first in the United States in the mid-1850s where canal shipping was first augmented by, then began being replaced by using much faster, less geographically constrained & limited, and generally cheaper to maintain railways.

By the early 1880s, canals which had little ability to economically compete with rail transport, were off the map. In the next couple of decades, coal was increasingly diminished as the heating fuel of choice by oil, and growth of coal shipments levelled off. Later, after World War I when motor-trucks came into their own, the last small U.S. barge canals saw a steady decline in cargo ton-miles alongside many railways, the flexibility and steep slope climbing capability of lorries taking over cargo hauling increasingly as road networks were improved, and which also had the freedom to make deliveries well away from rail lined road beds or ditches in the dirt which couldn't operate in the winter.

Construction

Canals are built in one of three ways, or a combination of the three, depending on available water and available path:

Human made streams

A canal can be created where no stream presently exists. Either the body of the canal is dug or the sides of the canal are created by making dykes or levees by piling dirt, stone, concrete or other building materials. The finished shape of the canal as seen in cross section is known as the canal prism. The water for the canal must be provided from an external source, like streams or reservoirs. Where the new waterway must change elevation engineering works like locks, lifts or elevators are constructed to raise and lower vessels. Examples include canals that connect valleys over a higher body of land, like Canal du Midi, Canal de Briare and the Panama Canal.

A canal can be constructed by dredging a channel in the bottom of an existing lake. When the channel is complete, the lake is drained and the channel becomes a new canal, serving both drainage of the surrounding polder and providing transport there. Examples include the Lage Vaart (nl). One can also build two parallel dikes in an existing lake, forming the new canal in between, and then drain the remaining parts of the lake. The eastern and central parts of the North Sea Canal were constructed in this way. In both cases pumping stations are required to keep the land surrounding the canal dry, either pumping water from the canal into surrounding waters, or pumping it from the land into the canal.

Canalization and navigations

A stream can be canalized to make its navigable path more predictable and easier to maneuver. Canalization modifies the stream to carry traffic more safely by controlling the flow of the stream by dredging, damming and modifying its path. This frequently includes the incorporation of locks and spillways, that make the river a navigation. Examples include the Lehigh Canal in Northeastern Pennsylvania's coal Region, Basse Saône, Canal de Mines de Fer de la Moselle, and Aisne River. Riparian zone restoration may be required.

Lateral canals

When a stream is too difficult to modify with canalization, a second stream can be created next to or at least near the existing stream. This is called a lateral canal, and may meander in a large horshoe bend or series of curves some distance from the source waters stream bed lengthening the effective length in order to lower the ratio of rise over run (slope or pitch). The existing stream usually acts as the water source and the landscape around its banks provide a path for the new body. Examples include the Chesapeake and Ohio Canal, Canal latéral à la Loire, Garonne Lateral Canal and Juliana Canal.

Smaller transportation canals can carry barges or narrowboats, while ship canals allow seagoing ships to travel to an inland port (e.g., Manchester Ship Canal), or from one sea or ocean to another (e.g., Caledonian Canal, Panama Canal).

Features

A canal boat traverses the longest and highest aqueduct in the UK, at Pontcysyllte in Denbighshire, Wales.

At their simplest, canals consist of a trench filled with water. Depending on the stratum the canal passes through, it may be necessary to linethe cut with some form of watertight material such as clay or concrete. When this is done with clay, it is known as puddling.

The Corinth Canal seen from the air.

Canals need to be level, and while small irregularities in the lie of the land can be dealt with through cuttings and embankments, for larger deviations other approaches have been adopted. The most common is the pound lock, which consists of a chamber within which the water level can be raised or lowered connecting either two pieces of canal at a different level or the canal with a river or the sea. When there is a hill to be climbed, flights of many locks in short succession may be used.

Prior to the development of the pound lock in 984 AD in China by Chhaio Wei-Yo and later in Europe in the 15th century, either flash locksconsisting of a single gate were used or ramps, sometimes equipped with rollers, were used to change the level. Flash locks were only practical where there was plenty of water available.

Locks use a lot of water, so builders have adopted other approaches for situations where little water is available. These include boat lifts, such as the Falkirk Wheel, which use a caisson of water in which boats float while being moved between two levels; and inclined planeswhere a caisson is hauled up a steep railway.

To cross a stream, road or valley (where the delay caused by a flight of locks at either side would be unacceptable) the valley can be spanned by a navigable aqueduct - a famous example in Wales is the Pontcysyllte Aqueduct (now a UNESCO World Heritage Site) across the valley of the River Dee.

Another option for dealing with hills is to tunnel through them. An example of this approach is the Harecastle Tunnel on the Trent and Mersey Canal. Tunnels are only practical for smaller canals.

Some canals attempted to keep changes in level down to a minimum. These canals known as contour canals would take longer, winding routes, along which the land was a uniform altitude. Other, generally later, canals took more direct routes requiring the use of various methods to deal with the change in level.

Canals have various features to tackle the problem of water supply. In cases, like the Suez Canal, the canal is simply open to the sea. Where the canal is not at sea level, a number of approaches have been adopted. Taking water from existing rivers or springs was an option in some cases, sometimes supplemented by other methods to deal with seasonal variations in flow. Where such sources were unavailable, reservoirs—either separate from the canal or built into its course—and back pumping were used to provide the required water. In other cases, water pumped from mines was used to feed the canal. In certain cases, extensive "feeder canals" were built to bring water from sources located far from the canal.

Where large amounts of goods are loaded or unloaded such as at the end of a canal, a canal basin may be built. This would normally be a section of water wider than the general canal. In some cases, the canal basins contain wharfs and cranes to assist with movement of goods.

When a section of the canal needs to be sealed off so it can be drained for maintenance stop planks are frequently used. These consist of planks of wood placed across the canal to form a dam. They are generally placed in pre-existing grooves in the canal bank. On more modern canals, "guard locks" or gates were sometimes placed to allow a section of the canal to be quickly closed off, either for maintenance, or to prevent a major loss of water due to a canal breach.

History

The transport capacity of pack animals and carts is limited. A mule can carry an eighth-ton [250 pounds (113 kg)] maximum load over a journey measured in days and weeks, though much more for shorter distances and periods with appropriate rest. Besides, carts need roads. Transport over water is much more efficient and cost-effective for large cargoes. It goes back to the earliest days of recorded history.

Ancient canals

The oldest known canals were irrigation canals, built in Mesopotamia circa 4000 BC, in what is now Iraq and Iran. The Indus Valley Civilization, Ancient India, (circa 2600 BC) had sophisticated irrigation and storage systems developed, including the reservoirs built at Girnar in 3000 BC. In Egypt, canals date back at least to the time of Pepi I Meryre (reigned 2332–2283 BC), who ordered a canal built to bypass the cataract on the Nile near Aswan.

In ancient China, large canals for river transport were established as far back as the Warring States period (481–221 BC), the longest one of that period being the Hong Gou (Canal of the Wild Geese), which according to the ancient historian Sima Qian connected the old states of Song, Zhang, Chen, Cai, Cao, and Wei. By far the longest canal was the Grand Canal of China, still the longest canal in the world today, and the oldest extant one. It is 1,794 kilometres (1,115 mi) long and was built to carry the Emperor Yang Guang between Zhuodu (Beijing) and Yuhang (Hangzhou). The project began in 605 and was completed in 609, although much of the work combined older canals, the oldest section of the canal existing since at least 486 BC. Even in its narrowest urban sections it is rarely less than 30 metres (98 ft) wide.

Greek engineers were the first to use canal locks, by which they regulated the water flow in the Ancient Suez Canal as early as the 3rd century BC.

"There was little experience moving bulk loads by carts, while a packhorse would [i.e. 'could'] carry only an eighth of a ton. On a soft road a horse might be able to draw 5/8ths of a ton. But if the load were carried by a barge on a waterway, then up to 30 tons could be drawn by the same horse

—technology historian Ronald W. Clark referring to transport realities before the industrial revolution and the Canal age.."

Middle Ages

In the Middle Ages, water transport was several times cheaper and faster than transport overland. Overland transport by animal drawnconveyances was used around settled areas, but unimproved roads required pack animal trains, usually of mules to carry any degree of mass, and while a mule could carry an eighth ton, it also needed teamsters to tend it and one man could only tend perhaps five mules,meaning overland bulk transport was also expensive, as men expect compensation in the form of wages, room and board. This was because long-haul roads were unpaved, more often than not too narrow for carts, much less wagons, and in poor condition, wending their way through forests, marshy or muddy quagmires as often as unimproved but dry footing. In that era, as today, greater cargoes, especially bulk goods and raw materials, could be transported by ship far more economically than by land; in the pre-railroad days of the industrial revolution, water transport was the gold standard of fast transportation. The first artificial canal in Western Europe was the Fossa Carolinabuilt at the end of the 8th century under personal supervision of Charlemagne.

In Britain, the Glastonbury Canal is believed to be the first post-Roman canal and was built in the middle of the 10th century to link the River Brue at Northover with Glastonbury Abbey, a distance of about 1.75 kilometres (1,900 yd). Its initial purpose is believed to be the transport of building stone for the abbey, but later it was used for delivering produce, including grain, wine and fish, from the abbey's outlying properties. It remained in use until at least the 14th century, but possibly as late as the mid-16th century.

More lasting and of more economic impact were canals like the Naviglio Grande built between 1127 and 1257 to connect Milan with the Ticino River. The Naviglio Grande is the most important of the lombard "navigli" and the oldest functioning canal in Europe.

Later, canals were built in the Netherlands and Flanders to drain the polders and assist transportation of goods and people.

Canal building was revived in this age because of commercial expansion from the 12th century. River navigations were improved progressively by the use of single, or flash locks. Taking boats through these used large amounts of water leading to conflicts with watermill owners and to correct this, the pound or chamber lock first appeared, in the 10th century in China and in Europe in 1373 in Vreeswijk, Netherlands. Another important development was the mitre gate, which was, it is presumed, introduced in Italy by Bertola da Novate in the 16th century. This allowed wider gates and also removed the height restriction of guillotine locks.

To break out of the limitations caused by river valleys, the first summit level canals were developed with the Grand Canal of China in 581–617 AD whilst in Europe the first, also using single locks, was the Stecknitz Canal in Germany in 1398.

Early modern period

c. 1500–1800 The first canal to use pound locks was the Briare Canal connecting the Loire and Seine (1642), followed by the more ambitious Canal du Midi (1683) connecting the Atlantic to the Mediterranean. This included a staircase of 8 locks at Béziers, a 157 metres (515 ft) tunnel and three major aqueducts.

Canal building progressed steadily in Germany in the 17th and 18th centuries with three great rivers, the Elbe, Oder and Weser being linked by canals. In post-Roman Britain, the first early modern period canal built appears to have been the Exeter Canal, which was surveyed in 1563, and open in 1566.

The oldest canal, technically a mill race built for industrial purposes in North America is Mother Brook, also known as mill brook, in between the two Boston, Massachusetts neighbourhoods of Dedham and Hyde Park connecting the higher waters of the Charles River and the mouth of the Neponset River and the sea. It was constructed in 1639 to provide water power for mills.

In Russia, the Volga–Baltic Waterway, a nationwide canal system connecting the Baltic Sea and Caspian Sea via the Neva and Volgarivers, was opened in 1718.

Industrial Revolution

The modern canal system was mainly a product of the 18th century and early 19th century. It came into being because the Industrial Revolution (which began in Britain during the mid-18th century) demanded an economic and reliable way to transport goods and commodities in large quantities.

By the early 18th century, river navigations such as the Aire and Calder Navigation were becoming quite sophisticated, with pound locksand longer and longer "cuts" (some with intermediate locks) to avoid circuitous or difficult stretches of river. Eventually, the experience of building long multi-level cuts with their own locks gave rise to the idea of building a "pure" canal, a waterway designed on the basis of where goods needed to go, not where a river happened to be.

The claim for the first pure canal in Great Britain is debated between "Sankey" and "Bridgewater" supporters. The first true canal in what is now the United Kingdom was the Newry Canal in Northern Ireland constructed by Thomas Steers in 1741.

The Sankey Brook Navigation, which connected St Helens with the River Mersey, is often claimed as the first modern "purely artificial" canal because although originally a scheme to make the Sankey Brook navigable, it included an entirely new artificial channel that was effectively a canal along the Sankey Brook valley. However, "Bridgewater" supporters point out that the last quarter-mile of the navigation is indeed a canalized stretch of the Brook, and that it was the Bridgewater Canal (less obviously associated with an existing river) that captured the popular imagination and inspired further canals.

In the mid-eighteenth century the 3rd Duke of Bridgewater, who owned a number of coal mines in northern England, wanted a reliable way to transport his coal to the rapidly industrializing city of Manchester. He commissioned the engineer James Brindley to build a canal for that purpose. Brindley's design included an aqueduct carrying the canal over the River Irwell. This was an engineering wonder which immediately attracted tourists. The construction of this canal was funded entirely by the Duke and was called the Bridgewater Canal. It opened in 1761 and was the first major British canal.

The new canals proved highly successful. The boats on the canal were horse-drawn with a towpath alongside the canal for the horse to walk along. This horse-drawn system proved to be highly economical and became standard across the British canal network. Commercial horse-drawn canal boats could be seen on the UK's canals until as late as the 1950s, although by then diesel-powered boats, often towing a second unpowered boat, had become standard.

The canal boats could carry thirty tons at a time with only one horse pulling - more than ten times the amount of cargo per horse that was possible with a cart. Because of this huge increase in supply, the Bridgewater canal reduced the price of coal in Manchester by nearly two-thirds within just a year of its opening. The Bridgewater was also a huge financial success, with it earning what had been spent on its construction within just a few years.

This success proved the viability of canal transport, and soon industrialists in many other parts of the country wanted canals. After the Bridgewater canal, early canals were built by groups of private individuals with an interest in improving communications. In Staffordshire the famous potter Josiah Wedgwood saw an opportunity to bring bulky cargoes of clay to his factory doors and to transport his fragile finished goods to market in Manchester, Birmingham or further away, by water, minimizing breakages. Within just a few years of the Bridgewater's opening, an embryonic national canal network came into being, with the construction of canals such as the Oxford Canal and the Trent & Mersey Canal.

The new canal system was both cause and effect of the rapid industrialization of The Midlands and the north. The period between the 1770s and the 1830s is often referred to as the "Golden Age" of British canals.

For each canal, an Act of Parliament was necessary to authorize construction, and as people saw the high incomes achieved from canal tolls, canal proposals came to be put forward by investors interested in profiting from dividends, at least as much as by people whose businesses would profit from cheaper transport of raw materials and finished goods.

In a further development, there was often out-and-out speculation, where people would try to buy shares in a newly floated company simply to sell them on for an immediate profit, regardless of whether the canal was ever profitable, or even built. During this period of "canal mania", huge sums were invested in canal building, and although many schemes came to nothing, the canal system rapidly expanded to nearly 4,000 miles (over 6,400 kilometres) in length.

Many rival canal companies were formed and competition was rampant. Perhaps the best example was Worcester Bar in Birmingham, a point where the Worcester and Birmingham Canal and the Birmingham Canal Navigations Main Line were only seven feet apart. For many years, a dispute about tolls meant that goods travelling through Birmingham had to be portaged from boats in one canal to boats in the other.

Aqueduct over the Mohawk River at Rexford, New York, one of 32 navigable aqueducts on the Erie Canal.

Canal companies were initially chartered by individual states in the United States. These early canals were constructed, owned, and operated by private joint-stock companies. Three were completed when the War of 1812 broke out; these were the Santee Canal (opened 1800) in South Carolina, the Middlesex Canal (opened 1802) in Massachusetts and the Dismal Swamp Canal (opened 1805) in Virginia. The Erie Canal (opened 1825) was chartered and owned by the state of New York and financed by bonds bought by private investors. The Erie canal runs about 363 miles (584 km) from Albany, New York, on the Hudson River to Buffalo, New York, at Lake Erie. The Hudson River connects Albany to the Atlantic port of New York City and the Erie Canal completed a navigable water route from the Atlantic Ocean to the Great Lakes. The canal contains 36 locks and encompasses a total elevation differential of around 565 ft. (169 m). The Erie Canal with its easy connections to most of the U.S. mid-west and New York City soon quickly paid back all its invested capital (US$7 million) and started turning a profit. By cutting transportation costs in half or more it became a large profit centre for Albany and New York City as it allowed the cheap transportation of many of the agricultural products grown in the mid west of the United States to the rest of the world. From New York City these agricultural products could easily be shipped to other U.S. states or overseas. Assured of a market for their farm products the settlement of the U.S. mid-west was greatly accelerated by the Erie Canal. The profits generated by the Erie Canal project started a canal building boom in the United States that lasted until about 1850 when railroads started becoming seriously competitive in price and convenience. The Blackstone Canal (finished in 1828) in Massachusetts and Rhode Island fulfilled a similar role in the early industrial revolution between 1828 and 1848. The Blackstone Valley was a major contributor of the American Industrial Revolution where Samuel Slater built his first textile mill.

Power canals

Sluice in the canal of Gabèíkovo Dam (Slovakia) - the canal is conveying water to a hydroelectric power station.

A power canal refers to a canal used for hydraulic power generation, rather than for transport. Nowadays power canals are built almost exclusively as parts of hydroelectric power stations. Parts of the United States, particularly in the Northeast, had enough fast-flowing rivers that water power was the primary means of powering factories (usually textile mills) until after the American Civil War. For example, Lowell, Massachusetts, considered to be "The Cradle of the American Industrial Revolution," has 6 miles (9.7 km) of canals, built from around 1790 to 1850, that provided water power and a means of transportation for the city. The output of the system is estimated at 10,000 horsepower. Other cities with extensive power canal systems include Lawrence, Massachusetts, Holyoke, Massachusetts, Manchester, New Hampshire, and Augusta, Georgia. The most notable power canal was built in 1862 for the Niagara Falls Hydraulic Power and Manufacturing Company.

19th century

American canals circa 1825

Competition, from railways from the 1830s and roads in the 20th century, made the smaller canals obsolete for most commercial transport, and many of the British canals fell into decay. Only the Manchester Ship Canal and the Aire and Calder Canal bucked this trend. Yet in other countries canals grew in size as construction techniques improved. During the 19th century in the US, the length of canals grew from 100 miles (161 km) to over 4,000, with a complex network making the Great Lakes navigable, in conjunction with Canada, although some canals were later drained and used as railroad rights-of-way.

In the United States, navigable canals reached into isolated areas and brought them in touch with the world beyond. By 1825 the Erie Canal, 363 miles (584 km) long with 36 locks, opened up a connection from the populated Northeast to the Great Lakes. Settlers flooded into regions serviced by such canals, since access to markets was available. The Erie Canal (as well as other canals) was instrumental in lowering the differences in commodity prices between these various markets across America. The canals caused price convergence between different regions because of their reduction in transportation costs, which allowed Americans to ship and buy goods from farther distances much cheaper. Ohio built many miles of canal, Indiana had working canals for a few decades, and the Illinois and Michigan Canalconnected the Great Lakes to the Mississippi River system until replaced by a channelized river waterway.

A family rides a boat in one of the canals of Amsterdam

Three major canals with very different purposes were built in what is now Canada. The first Welland Canal, which opened in 1829 between Lake Ontario and Lake Erie, bypassing Niagara Falls and the Lachine Canal (1825), which allowed ships to skirt the nearly impassable rapids on the St. Lawrence River at Montreal, were built for commerce. The Rideau Canal, completed in 1832, connects Ottawa on the Ottawa River to Kingston, Ontario on Lake Ontario. The Rideau Canal was built as a result of the War of 1812 to provide military transportation between the British colonies of Upper Canada and Lower Canada as an alternative to part of the St. Lawrence River, which was susceptible to blockade by the United States.

A proposal for the Nicaragua Canal, from around 1870

In France, a steady linking of all the river systems — Rhine, Rhône, Saône and Seine — and the North Sea was boosted in 1879 by the establishment of the Freycinet gauge, which specified the minimum size of locks. Canal traffic doubled in the first decades of the 20th century.

Many notable sea canals were completed in this period, starting with the Suez Canal (1869) - which carries tonnage many times that of most other canals - and the Kiel Canal (1897), though the Panama Canal was not opened until 1914.

In the 19th century, a number of canals were built in Japan including the Biwako canal and the Tone canal. These canals were partially built with the help of engineers from the Netherlands and other countries.

MODERN USES

Canals can disrupt water circulation in marsh systems

Large-scale ship canals such as the Panama Canal and Suez Canal continue to operate for cargo transportation, as do European barge canals. Due to globalization, they are becoming increasingly important, resulting in expansion projects such as the Panama Canal expansion project. The expanded canal began commercial operation on 26 June 2016. The new set of locks allow transit of larger, Post-Panamax and New Panamax ships.

The narrow early industrial canals, however, have ceased to carry significant amounts of trade and many have been abandoned to navigation, but may still be used as a system for transportation of untreated water. In some cases railways have been built along the canal route, an example being the Croydon Canal.

A movement that began in Britain and France to use the early industrial canals for pleasure boats, such as hotel barges, has spurred rehabilitation of stretches of historic canals. In some cases, abandoned canals such as the Kennet and Avon Canal have been restored and are now used by pleasure boaters. In Britain, canalside housing has also proven popular in recent years.

The Seine–Nord Europe Canal is being developed into a major transportation waterway, linking France with Belgium, Germany, and the Netherlands.

Canals have found another use in the 21st century, as easements for the installation of fibre optic telecommunications network cabling, avoiding having them buried in roadways while facilitating access and reducing the hazard of being damaged from digging equipment.

Canals are still used to provide water for agriculture. An extensive canal system exists within the Imperial Valley in the Southern California desert to provide irrigation to agriculture within the area.

Cities on water

A canal (Gracht) in Amsterdam, Netherlands.

Griboyedov Canal in St. Petersburg, Russia.

Canals are so deeply identified with Venice that many canal cities have been nicknamed "the Venice of…". The city is built on marshy islands, with wooden piles supporting the buildings, so that the land is man-made rather than the waterways. The islands have a long history of settlement; by the 12th century, Venice was a powerful city state.

Amsterdam was built in a similar way, with buildings on wooden piles. It became a city around 1300. Many Amsterdam canals were built as part of fortifications. They became grachten when the city was enlarged and houses were built alongside the water.

Canal of La Peyrade in Sète, France

Other cities with extensive canal networks include: Alkmaar, Amersfoort, Bolsward, Brielle, Delft, Den Bosch, Dokkum, Dordrecht, Enkhuizen, Franeker, Gouda, Haarlem, Harlingen, Leeuwarden, Leiden, Sneek and Utrecht in the Netherlands; Brugge and Gent in Flanders, Belgium; Birminghamin England; Saint Petersburg in Russia; Aveiro in Portugal; Hamburg and Berlin in Germany; Fort Lauderdale and Cape Coral in Florida, United States and Lahore in Pakistan.

Liverpool Maritime Mercantile City is a UNESCO World Heritage Site near the centre of Liverpool, England, where a system of intertwining waterways and docks is now being developed for mainly residential and leisure use.

Canal Estates (commonly known as bayous) are a form of subdivision popular in cities like Miami, Florida, Texas City, Texas and the Gold Coast, Queensland; the Gold Coast has over 700 km of residential canals. Wetlands are difficult areas upon which to build housing estates, so dredging part of the wetland down to a navigable channel provides fill to build up another part of the wetland above the flood level for houses. Land is built up in a finger pattern that provides a suburban street layout of waterfront housing blocks.

Boats

Inland canals have often had boats specifically built for them. An example of this is the British narrowboat, which is up to 72 feet (21.95 m) long and 7 feet (2.13 m) wide and was primarily built for British Midland canals. In this case the limiting factor was the size of the locks. This is also the limiting factor on the Panama canal where Panamax ships were limited to a length of 289.56 m (950 ft) and a beam of 32.31 m (106 ft) until 26 June 2016 when the opening of larger locks allowed for the passage of larger New Panamax ships. For the lockless Suez Canal the limiting factor for Suezmax ships is generally draft, which is limited to 16 m (52.5 ft). At the other end of the scale, tub-boat canals such as the Bude Canal were limited to boats of under 10 tons for much of their length due to the capacity of their inclined planes or boat lifts. Most canals have a limit on height imposed either by bridges or by tunnels.

Canal Building Technique

Canals have to be designed and built to withstand enormous water pressure. In addition, they have to be watertight so that they will not leak and they have to be able offer protection from the risk of erosion caused by the flowing water.

The bottom of the canal is levelled off and compacted. A filter layer consisting of sand and gravel is then put in place and, on top of that, a sealing layer of fine soil, cement or asphalt. There is always a certain amount of leakage through the sealant and the filter ensures that the sealant does not get washed away. If fine soil is used as the sealing layer, it must be covered by an erosion resistant layer. It is highly important that the various layers are correctly compacted to avoid cracks in the sealing layer.

Even before the Revolutionary War gave New impetus to American expansionism, the colonial political and economic elites were deeply interested in the improvement of inland transportation. Vessels that plied offshore waters, small boats and rafts on the streams down to tidewater, and local roads and turnpikes served the immediate commercial needs of farmers and townspeople in the Atlantic coastal area. But the loftier dreams of planters, merchants, and political leaders—as well as of the common farmers who constituted by far most of the free population in British America—looked beyond the "fall line" that separated the rivers flowing to the coast from those that ran to the Ohio-Mississippi basin. A vast area for settlement and productivity—and riches—lay in the interior, and by the early 1790s demands for diffusion of New transport technologies and for investment in internal improvements were voiced frequently in both state and national political forums.

It was widely recognized that unless bulk agricultural commodities, which were the staples of a commercialized and expanding farm economy, could be carried cheaply. And over long distances, settlement and economic growth would be badly hampered in the region beyond the Appalachian Mountains. Then, too, there were opportunities for construction of short lines on the Atlantic seaboard to link already developed areas (coal mines, farming and lumber regions, and rising industrial sites), with the promise of immediate traffic and revenues. The latter were "exploitative" projects, tapping existing trade routes and resources; but the major east-west projects were "developmental," promoted with the goal of opening Newly or sparsely settled areas to economic opportunities. There was also a nationalistic or patriotic goal of canal promotion: to bind together far-flung sections of the young nation and to prove the efficacy of republican government.

And yet total canal construction in the United States up to 1816 totaled only 100 miles—the longest canal project being the Middlesex, which linked Boston's harbor with the farm region to the north. Other lines of some importance linked Norfolk, Virginia, to Albemarle Sound and connected the Santee River area to Charleston, South Carolina. Although many other canal projects were proposed up and down the Atlantic coast, progress was difficult because of shortages of capital and skepticism with regard to engineering feasibility projects. Moreover, regional or local jealousies notoriously worked against successful mobilization of governmental support in both the U.S. Congress and the state legislatures.

In the period from the mid-1820s to the Civil War, however, the United States underwent a vast expansion of canal construction, becoming the world's leading nation in both mileage of canals and the volume of tonnage carried on them. The canal lines were of crucial importance in the integration of a national economy, and they played a key role in the so-called "Transportation Revolution" that expedited both westward expansion and a robust industrialization process in the North and West.

Advantages, Disadvantages, and Construction Challenges

Canal technology proved especially attractive for several reasons. Since the 1760s, successful large-scale canal projects had been built in both Great Britain and France, and these canals had brought enormous economic advantages to the regions they served. The engineering advances pioneered in Europe gave American promoters confidence that they could build canals with equal success.

There was a downside to canal technology, too, though it was not always fully recognized in America. Difficult topography or uncertain water supply meant complex and highly expensive construction design. Canal building before the 1850s was mainly done with hand tools, augmented only by some primitive animal-powered machinery. A canal line had to be furnished with locks, permitting boats to pass through from one water level to another. The segments of line between the locks were of very gradual grade to permit controlled flow. At each lock, a gate at its higher level would be opened while the gate on the lower end was kept closed; once a boat entered the lock's chamber from the higher level, the upper gate was closed (holding the water flow back) while the lower gate was opened. As water ran out, the boat was carried down to the lower level, then passed through the open gate. For "upstream" movement, from the lower level to the higher, the process was reversed. The lock would be drained to the lower level, the boat would enter through the bottom gate, which was then closed, and water would then be admitted from the upper gate, lifting the boat up to the higher level. In steep areas, "flights" of locks, closely spaced, were necessary and often involved complex engineering; for transit of the boats, these series of locks meant a slow stretch and usually long waiting periods.

Locks varied in size. Lifts ranged from two to thirty feet, and there were great differences in the distances between gates as well as in the construction materials used. Masonry locks and metal or metal-trimmed gates were far more expensive—and more durable—than wooden gates and timber-supported rubble for the walls as found on some of the lines. The total rise and fall over an entire line was measured as "lockage," and served as an index of the difficulty of construction. For example, New York's Erie Canal route measured lockage of 655 feet, by contrast with Pennsylvania's lockage of 3,358 feet between Philadelphia and the Ohio River.

The size of the boats that could be accommodated, as well as the volume of water needed, were functions of the dimensions of the canal bed, or its "prism," as well as of the size of lock chambers. Prisms on American canals varied greatly, most of them ranging from forty to sixty feet in width at the top, with sloping sidewalls leading to a bottom of twenty-five to forty-five feet across. The Pennsylvania system was the most complex in engineering using inclined planes and steam-powered winches to drag boats out of the water and over some of the steepest hills.

To supply the line with flowing water, engineering plans had to include river connections, dams and reservoirs with feeder lines to the canals, and often massive culverts and aqueducts. Building the sidewalls to minimize loss of water through seepage was another challenging and expensive aspect of design. On many of the larger canals, such as the Ohio lines, engineers took advantage of fast-flowing feeder streams to design water-mill sites into the line.

Once a canal was in operation, moreover, maintaining navigation was a continuous challenge. Winter ice, droughts, floods, and breaches in the water-supply system would frequently cause navigational closings. Even in the best of circumstances, it was difficult to maintain regular schedules on the lines because of traffic bottlenecks at the locks and the continuous maintenance needed to keep water flowing.

Although steam-powered propeller craft were used on a few canal lines, this form of transport placed dangerous pressure on the canal walls. Hence, the use of horses or mules to haul canal boats was nearly universal, with the animals walking along the "towpath" alongside the line. Freight boats typically of 50-to 125-ton capacity operated at speeds of one to three miles per hour. On most lines they were owned by individuals or private companies, the line being a common carrier under the law.

In the short run, all the disadvantages of canal technology were more than offset by the cost savings for long-distance hauling, especially of bulk goods and produce. In the long run, however, innovations in steam technology and railroad engineering were destined to render many canals the losers in a New competitive age in transport that took shape in the late 1840s and the 1850s.

The Erie Canal

The great breakthrough came in 1817 with New York State's commitment to building the Erie Canal to connect the Hudson River at Albany with Buffalo on Lake Erie, a

Project far greater than any previously attempted in America. The Erie was important to subsequent canal development in several ways, most notably because it provided a model of public enterprise through its financing, administration, and implementation. The state raised capital through bond issues both in New York and in Europe, and supplemented these funds with tax revenues. Actual construction was overseen by a board of commissioners, some of whom were personally involved in the fieldwork, but the project became a celebrated "school for engineers," with most of the junior personnel learning their skills on the job under the tutelage of Benjamin Wright and James Geddes, two of less than twenty men who then constituted the profession in America. Many of the Erie engineers went on to direct canal surveys in other states.

The canal was divided into sections for purpose of construction, with private contractors taking on the work under the state engineers' supervision—a scheme that was emulated by nearly all the major canals subsequently built. It was an immediate commercial success once opened to its full length in 1825, leading the New York legislature to authorize a series of additional canals as well as the improvement and enlargement of the original line.

No nonmilitary enterprise in the United States had ever involved such expenditures as did the Erie, whose initial construction cost $6 million. The number of laborers employed was also unprecedented in any economic enterprise of the day. The state's construction expenditures energized local economies, giving part-time employment to farmers and creating sudden demand for stone, timber, mules, and oxen, and provisions for workers. Like canals and other public works throughout the country, moreover, the Erie attracted immigrant workers (mainly Irish and German) who were employed to do much of the most dangerous work.

The Erie's commercial impact on the rural countryside and on New York City's role as a center for trade with the interior and for exports to Europe—together with the rich stream of revenues from the tolls—heightened expectations everywhere in the country that other canals could produce equally spectacular fiscal and developmental results.

The Post-1825 Boom in Canal Building

Emulation of New York followed quickly. In 1825 Pennsylvania authorized a $10 million project, combining canal technology with the use of inclined planes. It was completed in 1834, tapping the Ohio Valley's farm country at Pittsburgh and giving Philadelphia trade advantages similar to those that its rival New York City had obtained from the Erie. The first of the western states to build a major line was Ohio, which authorized construction in 1825. Although still small in population and financial resources, Ohio, too, resorted to creation of a state enterprise and borrowed heavily both in the East and in Europe. Erie Canal engineers were brought in at first, but Alfred Kelley of Cleveland and Micajah Williams of Cincinnati,

Local entrepreneurs with no prior engineering experience, took principal charge of overseeing construction once the technical plans were adopted. Although administrative incompetence and corruption plagued the Pennsylvania project, Ohio's record was widely admired for its efficiency and strength of design. One line, the Ohio Canal, was completed in 1834 and extended from Cleveland on Lake Erie to Portsmouth on the Ohio River—the first water link between the Great Lakes and the great Mississippi-Ohio basin. A second line, completed in the mid-1840s, linked Cincinnati with Toledo to the north.

Other important lines begun or fully built prior to 1840 included the Delaware and Hudson Canal, a successful private line in the Pennsylvania coal country; the Delaware and Raritan Canal, also private, linking Philadelphia and New York; and the Chesapeake and Delaware Ohio Canal, which with substantial state support built a line, surveyed by the engineer William Strickland, through Maryland to link Baltimore with the Philadelphia port.

In the period 1815–1834, $60 million was invested in 2,088 miles of canals, with 70 percent of the funds coming from governmental sources, mainly the states. Most of the funds were borrowed at home and abroad. Also, Congress authorized the Army Engineers to conduct surveys for the states and federal companies; made some direct federal investments; and gave several million acres of public lands to Ohio, Indiana, and Illinois to subsidize their canal projects during this period.

In the decade following, 1834–1844, the "canal enthusiasm" continued to animate state governments and private promoters. Rivalries among states and competition among cities were intense, feeding the spirit of optimism. A new wave of canal construction followed, with the projects again heavily financed by loans from Europe and the eastern cities. Almost 1,300 miles of canal were built during this ten-year period. They cost $72 million, of which 79 percent represented public funds. In addition to major New state canal systems begun in Illinois and in Indiana (where the Wabash and Erie line would open another link for direct trade between the Ohio River and Lake Erie), and in Illinois, three of the pioneering state projects—the Erie, Ohio's two main canals, and the Pennsylvania system—were further expanded to satisfy sections of their states that had been left out of the original system designs. As the New canals were generally of larger dimensions than the first ones to be built, the carrying capacity for canal traffic doubled between 1834 and 1844. Until 1839, conditions of prosperity and expansion sustained the canal-building movement, and expenditures for the New canals stimulated overall economic growth.

Financial Problems and Railroad Competition

The 1837 financial panic and the 1839–1843 depression created enormous fiscal problems for many canal states, leading to defaults on state debts in Pennsylvania and Indiana. Because many of the expansion projects and new lines produced toll revenues far below expectations, moreover, there was widespread disillusionment with state enterprise; and this became a factor favoring railroads as an alternative to canals, especially given the much greater reliability of rail transport. In the Ohio-Indiana-Illinois area, by 1848 the proliferation of canal lines also produced intensified competition between the various Great Lakes and Mississippi River routes, now also served by steamboat lines on these connecting waters. The east-west and local railroads of the 1850s made matters worse. The result was heavy downward pressure on canal rates, consequently reduced revenues, and, soon, a scenario of operating deficits that placed an unwelcome burden on taxpayers.

Transport competition drove down rates so much that the period from the mid-1840s to the Civil War formed a distinctive "second phase" of the Transportation Revolution. By 1850–1852, for example, western canal tolls were less than a third the level of the 1830s, creating still further fiscal problems for the canal states and companies. Where private investment had been invited on a matching basis for "mixed" canal enterprises, the costs fell hard on the capitalists as well. But while revenues fell, ton-miles of canal transportation continued to expand on all the major lines throughout the 1850s.

During the period 1844–1860, a last major cycle of canal construction produced 894 miles of line at a cost of $57 million. Here again, governmental activism was crucial, with public funds accounting for two-thirds of the total expended. Much of this increase constituted the completion or improvement of lines built earlier; in addition, the still-successful Erie system in New York was further enlarged and upgraded. A large expenditure was made, too, on the Sault Ste. Marie Ship Canal, a short but massive deepwater project that connected Lake Huron with Lake Superior.

Although much of the canal system experienced operating deficits in the 1850s, the impetus these New facilities had given the economy had clearly warranted most of the capital invested. Commercialization of agriculture in the western states and other interior had been made possible, while eastern manufacturers and importers were afforded economical access to interior markets. Coal-mining and iron centers were linked, and consumer prices fell where the transport facilities had proliferated. In sum, the areas served by canals were enabled to build on comparative economic advantage; and, at least in the northern states, processing of primary products carried by the canals served as the origin of manufacturing growth that augmented urban commercial activity.

Railroad competition led to many closings of once-important canals; indeed, more than 300 miles of line were abandoned by 1860. A few of the canals did continue to carry heavy traffic after the Civil War. The most important to commerce in the twentieth century was the Atlantic intra-coastal waterway, which permitted vessels to transit offshore waters safely from New England to Florida. The Erie retained importance as a barge canal, as did some of the shorter coal-carrying lines. Some of the old canal lines became rights-of-way for railroads or modern roads; others were absorbed into the changing landscape as development went forward. In scattered locations, a few segments of the great canal lines are today preserved or restored for enjoyment of citizens seeking a glimpse of the once-glorious era of canal transport in America.

Prehistoric Engineering

Nobody in the New World moved water more efficiently than the Hohokam Indians. The canals they built to sustain their large and complex agricultural society are still considered an engineering marvel. The Hohokam people lived around the Mesa area of Central Arizona for almost 1,500 years. They appeared around 1 CE and disappeared about 1450. Their agricultural society flourished around the middle Gila River and Salt River. These sources allowed them to farm beans, squash, corn, and cotton. The Hohokam were skilled agrarians, managing the soil to replace lost nutrients. Additional food came from extensive gathering of wild plants, dry farming agave, and hunting deer and small animals.

Around 600 CE the Hohokam people began to construct irrigation canals. They excavated trenches up to 12 feet deep by hand, using a digging stick. The trenches drew off Salt River water and fanned into a network of smaller canals that brought a steady supply of water to fields. Between 1100 CE and 1450 CE, 500 miles of canals irrigated 110,000 acres. The food produced by this advanced irrigation system is believed to have supported up to 80,000 people—the highest population density in the prehistoric Southwest.

Hohokam canals possessed remarkable features that still stand as examples of engineering genius. First, they were wide at the mouth and carefully tapered, getting smaller as secondary branches drew water from the main channel. By shrinking the channel size as the flow decreases, the Hohokam were able to stabilize flow rate. A steady flow rate is key to creating a functional irrigation system. Water that moves too fast carries sand and silt that will eventually block the canal. Water that moves too slowly will not reach its destination. Second, canals covered miles of territory on a relatively steady grade, avoiding hills and valleys that could affect flow rate.

Though the Hohokam people are most celebrated for their canals, they made fire-hardened pottery and figures and jewelry from shell and turquoise. They built increasingly elaborate structures, such as the Big House in Casa Grande Ruins National Monument. The society also constructed large earthen platform mounds of unconfirmed purpose. Located in the center of villages, the mounds may have been used for housing elite members of the society or for ceremonial activities. Hohokam rock art, in the form of petroglyphs and pictographs, is found in mountain ranges and rocky areas throughout the region.

Mysterious Migration

The Hohokam disappeared from their villages near the Salt and Gila Rivers in 1450 CE. Their descendants survive as members of the Tohono O'odham Nation.

The three centuries leading up to the Hohokam's disappearance around 1450 CE had been a period of migration and social change, with old settlements and prestigious structures constructed in the Salt and Gila Basins being abandoned. The reasons for this mass abandonment remain unknown to this day. Perhaps disease or war threatened the population. Perhaps flood damaged the canal infrastructure, or drought brought hardship. There is also speculation that a great leader died, causing the people to drift away and migrate. We may never discover why the Hohokam decided to leave the area they had lived in for almost 1,500 years. But their descendants live on as members of the Tohono O'odham Nation.

Reportedly, canals were built in Egypt as much as 6,000 years ago. The first canal designed to accommodate seaworthy ships was built by the Pharaoh Sesosteris I4,000 years ago, linking the Red Sea to the Nile River and thus to the Mediterranean Sea. Remnants of this canal, which was alternately neglected and rebuilt half a dozen times, suggest that it was about 150 ft (46 m) wide, 16 ft (5 m) deep, and 60 mi (97 km) long.

The oldest canal still in use was begun in China in the fourth century B.C. for the purposes of collecting grain taxes and transporting troops. Rebuilt and enlarged about A.D. 600, the channel became known as the Grand Canal; its latest extension, completed about A.D. 1280, brought its length to 1,114 mi (1,795 km). It remains the longest navigable canal in the world.

For centuries, canals were dug using only hand tools. While building the Canal du Midi between 1665 and 1681 to link the Atlantic.

An aerial view of a canal and lock system.

Ocean and the Mediterranean Sea across southern France, gunpowder was first used to blast a tunnel. The invention of steam-powered machinery at the end of the seventeenth century helped fuel the "canal eras" in Europe and North America in the eighteenth and nineteenth centuries during which thousands of miles of canals were built. Horse-drawn plows continued to be used especially for final trimming of the channel to desired slopes and depths. Full mechanization of canal construction was achieved in 1946, when an American company built the first canal trimming and lining machines.

Locks

Flash locks, the first attempts to carry boats over difficult elevation changes on rivers or canals, date from the third century B.C. Greek historian Diodorus Siculus, writing 200 years later, described how Ptolemy II had improved the Nile-to-Red Sea canal by building a type of lock. Consisting of a single gate, flash locks carried boats downstream on a rush of water; boats headed upstream could be pulled forward while floating on the torrent released by opening the gate.

Flash locks were hazardous, and they used large amounts of water. A significant improvement came in A.D. 984, when the first double-gate lock was built on China's Grand Canal. Also called a pound lock (because it impounds water) this was the predecessor of modern conventional locks. Its gates were panels that lifted vertically. A similar lock with vertical-lift gates built in the Netherlands in

1373 also controlled its water level by partially opening either the upstream or the downstream gate. Pound lock operation was greatly improved in 1485 when an Italian lock was built with smaller, valve-controlled openings in the gates.

Leonardo da Vinci invented the miter gate in 1480. Two gates, each more than half as wide as the lock, swing on vertical hinges. In the open position, they are flush with the lock's walls. In the closed position, they meet in a V pointed upstream so that the higher water level presses against them to promote a tight seal.

The base of the upstream gate in a lock is higher than the base of the downstream gate. If the lock was very deep, water filling the chamber through an opening in the upper gate would create turbulence or even swamp a boat in the chamber. This problem was solved in France during the seventeenth century, when valve-controlled water channels for filling and emptying a lock were built

A lock is a rectangular chamber with gates at both ends; with both gates closed, the water level within the lock can be adjusted to match the canal water level on either side. Thus, a vessel entering the lock can be raised or lowered in order to enter the next level canal section.

Until the early nineteenth century, lock gates were made of wood, and lock chambers were lined with wood, stone, bricks, or turf. In 1827, cast iron was first used to build both a lock and its gates in Cheshire, England. Following the development of the Bessemer process for mass production of steel, that material was utilized for construction of locks and gates, both as primary elements and as reinforcing bars for concrete.

Raw Materials

Waterproof linings keep a canal's water from seeping into the ground. For many years, the best choice was puddle, a mixture of sand, clay, and water that dried to a waterproof state. Modern materials and additives that are more durable include concrete, fly ash, bentonite, bituminous materials, and plastic sheeting.

Locks are usually made of concrete, occasionally lined with steel. If construction of the lock exposes bedrock, the floor need not be lined. The gates are made by welding together steel plates and reinforcement beams. The vertical edges of the gates are fitted with effective sealing materials such as white oak. In 1999, a French company developed lock gates made of glass-fiber reinforced plastic laminate mounted in stainless steel frames.

Design

Early canals followed the most level surface route possible because large-scale earth-moving was so difficult and expensive. Better excavation equipment and lock construction capabilities permit the construction of shorter, more direct canal routes.

Because of geographic obstacles, sections of some canals are built in tunnels or on aqueducts (water-carrying bridges). It is important to fill and empty a lock chamber while producing minimal water turbulence. Modern designs place sluices in the gate sills or in the chamber's floor or walls. A lock may also be equipped with a submerged bubbler that releases air below the gate closure area; the resulting gentle turbulence keeps the area clear of debris that might prevent the gate from sealing properly.

Various gate designs are available for modern locks, and different types may be used on the upstream and downstream ends of a single lock. Miter gates are one of the most popular choices. Another common choice is the Tainter gate, a curved plate that rotates vertically. In this efficient design, which is used for valves in water-transfer culverts as well as main lock gates, water pressure actually assists in the gate's rotation. Flat gates that slide up, sideways, or down into the lock floor are other options, as are flat gates that are hinged at the bottom and curved gates that rotate horizontally into wall recesses.

William C. Gorgas was born October 3, 1854, near Mobile, Alabama. In 1875 Gorgas received a bachelor of arts degree from the University of the South. Desiring a military career he decided to enter the Army with a medical degree. After graduating from Bellevue Medical College in New York and interning at Bellevue Hospital, Gorgas was appointed to the Medical Corps of the U.S. Army in June 1880. Tours of duty followed in Texas and North Dakota, with nearly 10 years at Fort Barrancas in Florida—Gorgas was assigned this yellow fever area because he had previously had the disease and was immune.

After Havana, Cuba, was occupied by American troops in 1898, Gorgas took charge of a yellow fever camp at Siboney, soon becoming chief sanitary officer of Havana. Acting on information that a certain mosquito carried yellow fever, Gorgas quickly destroyed that mosquito's breeding ground, ridding the city of yellow fever.

In 1904, work commenced on the Panama Canal. Gorgas went to the Canal Zone to take charge of sanitation, succeeding in Panama and Colon. Gorgas came to be generally regarded as the world's foremost sanitary expert, and a number of foreign governments and international commissions sought his aid. His book Sanitation in Panama quickly became a classic in the public health field. In 1914 he was appointed surgeon general of the Army, serving in that capacity until his retirement four years later. Gorgas died on July 3, 1920, and is buried in Arlington National Cemetery.

The Manufacturing Process

Canal

- A thorough survey of the canal route must be made, not only for correct alignment but also for accurate depths of cuts and fills.

- Primary excavation is done with earth-moving equipment like bulldozers and excavators.
- A crawler-track mounted trimming machine is used to remove the final 12-18 in (30-46 cm) of soil to create the desired wall slopes and flat bed for the canal. Excavated soil is carried by conveyor and loaded onto trucks that will haul it away.
- A layer of highly permeable material may be spread over the walls and floor to promote drainage of groundwater under the canal.
- Grids of reinforcing steel are built and l1ifted into place on the canal walls and floor. Concrete blocks hold the grids above the surface so concrete can flow under them.
- Concrete is mixed, vibrated to remove air pockets, and applied with a slipform machine, encasing the reinforcing steel. Slipforming means pouring concrete between the canal surface and a form (mold) that slowly moves forward; the concrete hardens quickly enough to retain its shape by the time the form moves out of contact with it. Specialized machines place transverse and longitudinal expansion joints in advance of the slipform machine.
- Lock
- A temporary cofferdam is built around the proposed lock site. Steel sheet piles are driven into the ground to form a series of adjacent, vertical cells that extend above the waterline. The cells are filled with sand. Water is pumped out of the enclosed space to create a dry construction area. The lock site is excavated. If necessary, piles are driven into the ground to support the lock structure.
- Wooden forms are built to shape the floor and walls of the lock. Space for culverts and valve chambers is included within the forms, as are slots for the gate hinges and recesses to contain open gates. Additional forms are built to shape the approach walls that will guide vessels into the lock.
- After reinforcing steel cages are constructed in the forms, concrete is poured. When the concrete has cured (hardened), the forms are removed.
- Control valves for filling and emptying the lock are installed, along with hydraulic and mechanical equipment for operating the valves and gates.
- Gates are prefabricated and shipped to the site. Very large gates may be shipped in sections, which are welded together as they are installed in the lock.
- Accessories such as guardrails, mooring posts, and escape ladders are installed on the lock walls.

Byproducts/Waste

Disposing of the excavated material is one of the challenges of canal and lock construction. It may be used to construct embankments or be spread over the surrounding countryside where it is carefully landscaped for erosion control as well as appearance. Construction of the Divide Cut, a 29-mi (46-km) canal on the Tennessee-Tombigbee Waterway, in the early 1980s required the disposal of 150 million cubic yards (115 million cubic meters) of dirt.

During lock construction, excavated material can be used to fill cofferdam cells. When the cofferdam is removed, the material may be used to fill in behind the riverbank side of the lock wall.

Lock operation uses large amounts of water; for example, it takes 3.5 million gal (13 million L) of water to fill a 600-ft (180-m) long, 110-ft (34-m) wide lock with a 7-ft (2.1-nm) lift. On some canals and rivers, water supplies are limited and conservation is important. Rather than being released downstream as the lock is emptied, some of the water may be diverted to a side pond, where it can be stored and used to help fill the chamber for its next operation.

The Future

Because of the significant need to recondition or replace locks as well as build new facilities, the U.S. Army Corps of Engineers sponsors and conducts research on water-way construction. Scale models of proposed designs can be tested at its Waterways Experiment Station. Through its Innovations for Navigation Projects (INP) Research Program, it supports the development of new ways to build, repair, and operate canals and locks. Recent topics included improvements in underwater concrete and grout placement and development of low-density, high-strength concrete for building modular lock sections that can be installed without construction of a cofferdam. Other current research topics involve protection devices for lock walls and gates from barge impacts, improved designs for water intake and discharge systems, better equipment and control devices for operating gates, and techniques for enlarging (rather than replacing) existing locks that have inadequate capacity.

Canal - Construction and Operation

Engineers who are designing a canal must take several things into consideration. They must formulate the dimensions of the canal to accommodate the numbers and sizes of ships that are predicted to use the waterway. Natural obstacles in the path of the canal, such as rock formations, must also be modified, removed, or avoided. There must be adequate vertical clearance above the canal and the clearance afforded by pre-existing bridges must be able to accommodate the vessels that will use the canal. Finally, engineers must decide the scale and location of associated structures, such as bridges, tunnels, and locks.

The paths of most canals are affected by variations in the levels of terrain. Engineers compensate for these variations with either locks or inclined planes. A lock is a segment of the waterway that is closed off by gates at either end. When a vessel enters the lock, the front gate is already closed. The back gate is then closed behind the vessel, and the water level within the lock is raised or lowered to the level of the water on the outside of the front gate. Valves on the gates control the level of the water. Whereas locks are the most common means of compensating for elevation changes, the procedure is slow and uses a large amount of water. Inclined planes can be used to elevate and lower smaller vessels; they use no water and often allow more rapid passage of vessels. When vessels reach certain stations along the waterway, they are pulled out of the water and moved on trucks up or down the plane.

Canal operators must monitor the canal's supply of water. If the natural supply of water at the upper end of the canal is deficient, it must be supplemented by water pumped into the reservoirs. If nature supplies the reservoir with excess amounts of water, some of the water must be diverted from the canal. Otherwise, excess water may strengthen the current and disrupt canal operations.

Although canals are among the oldest civil works, they play a major role in commerce because they are by far the least expensive form of inland transportation yet devised.

Steps to complete Canal Construction

Identity cards had been issued to 2.02 lakh agriculture labourers under Tamil Nadu Agriculture Labourers Social Security scheme in the district so far, said the Collector, Har Sahay Meena.

Distributing identity cards to farm labourers at a function held at Theni-Allinagaram Cooperative Market here on Friday, he said agri-labourers and tiny and small farmers aged between 18 and 65 years had been included in the scheme.

Small farmers having 2.5 acres of wetland and five acres of dryland were eligible for this scheme.

The eligible beneficiaries need not pay any amount for joining this scheme. Each beneficiary, in case of death, would get Rs.1 lakh as compensation. A sum of Rs.50,000 would be given as compensation to the beneficiaries who had lost their limbs or eyes in accidents. The injured would get Rs.25,000 and family member of the beneficiary would receive Rs.10,000 in case of natural death of the beneficiary.

Marriage assistance of Rs.2,000 would be given to male member and Rs.5,000 to female member. Maternity assistance of Rs.2,500 would be given to women members and educational assistance ranging from Rs.1000 to Rs.6,500 to the wards of members.

All efforts would be made to complete the construction of 18th canal as early as possible, he said, adding that the work at Cumbum and Gudalur had been completed.

"We have been waiting for approval from the Forest Department for acquisition of land to construct a dam across the Vairavanar river."

Mr. Meena ordered the officials to expedite acquisition of land for constructing a canal at Melagudalur and Uthamapalayam areas.

On completion of the work, 44 old irrigation tanks and four new tanks would receive water and 4,704 acres would be irrigated.

This canal would pass through Gudalur, Cumbum, Uthamapuram Pudhupatti, Kombai and Thevaram, mostly dry areas, and connect with the Sutha Ganga stream.

Later, the Collector disbursed accident benefit scheme to four persons and identity cards to new members.

Design of Lined Canals

The lined canals are not designed making use of Lacey or Kennedy Theory because the section is rigid. Generally Manning's equation is used in design. To carry a certain discharge number of channel sections may be designed with different bed widths and side slopes. But it is clear that each section is not equally good for the purpose.

The section to be adopted should be economical and at the same time it should be functionally efficient. It has been found that the most suitable cross-section of a lined canal is a circular section with sloping sides. That is, the bed is not flat but it is an arc of a circle. This arc is tangential to the sloping sides.

Side Slopes

The side slope is selected in such a way that it nearly equals the angle of repose of the soil in the subgrade. Care is taken to ensure that no earth pressure is exerted on the back of the lining. From the knowledge of hydraulics it is clear that the section is economical when cross- sectional area is maximum for minimum wetted perimeter.

This condition is achieved when the centre of an arc lies at FSL of the canal. This section is also efficient in the sense that as the velocity of flow is higher silt carrying is also higher than a wide and shallow section. Thus the problem of silting is completely eliminated and functioning is efficient.

It may be mentioned here that such section with circular bed may be designed up to a discharge of 85 m/sec. When the discharge is more than 85 m/sec the section best suited is one with a flat bed and sloping sides with rounded corners. This section is certainly better than trapezoidal section, because it is more stable and economical to construct. Dimensions of a canal section with circular bed may be obtained from the equations given below. Figure 10.2 shows the canal sections of this type with two standard side slopes.

When r = 3.6 m or less, side slopes may be taken 1: 1

When r = 3.6 m side slopes may be taken 1.25: 1

Velocity of Flow:

Mean velocity of flow may be calculated using Manning's formula.

V = 1/N .R. S

where N is coefficient of rugosity and may be taken as 0.018

S is the slope of bed and expressed in fall of bed in m in 10,000 m length

For the given value of N formula may be reduced to

V = 0.556 R. S

where V is velocity of flow in m/sec;

R is hydraulic mean radius in m; and

S is bed slope expressed in metres/metre length,

since (10,000) is merged in the constant 0.556.

Following Table 10.1 gives the sectional data for the two side slopes:

Dimensions of a canal section with flat bottom may be obtained from the equations given below. Figure 10.3 shows the cross-sections of this type with two standard side slopes. It may be mentioned here that generally maximum velocity of flow is recommended to be 2 m/sec and Lacey's 'f is limited to 1.2. When r exceeds 55 m or when discharge is more than 85 m/sec, the canal should be designed with a flat bottom with fillets at corners. The Indian Standard 4745 in fact does not suggest lined canal with circular bottom.

Velocity of Flow

Mean velocity of flow may be calculated from Manning's formula for value of N = 0.018.

Then V = 0.556 R. S

Following Table 10.2 gives sectional data for the two side slopes:

It may be remembered that lined canal could be given steep slope to achieve recommended velocity of 2 m/sec and value of 'f' about 1.2. Critical velocity ratio is not applicable to lined canals. But to avoid possibility of sitting CVR should be aimed at more than unity.

Coefficient of Rugosity (N)

In general practice for lined canal average value of N may be taken as 0.018. For different types of linings the value of W varies. The values given for straight channels in Indian Standard 4745 are given in Table 10.3. When the alignment is not straight loss of head increases and a small increase in the value of W may be made to allow for additional loss of energy.

Freeboard

Freeboard is measured from the (FSL) full supply level to the top of lining. For lined canals having less than 10 cumec discharge 0.6 m free board is recommended. For bigger lined canals freeboard not less than 0.75 m is generally provided.

Bank Widths

The Indian Standard recommended following values for bank widths for main and branch canals:

Main canal in cutting and filling = 8.0 m

Branch canal in cutting = 6.5 m

Branch canal in filling, left bank = 6.5 m

Right bank = 5.0 m

Suitable provision of dowel, roadway, catch water drain, under drainage has to be made for every lined canal. Figure 10.4 shows three typical cross-sections of the lined canal in which provision of various components is illustrated.

The canal sections shown have trapezoidal shape. For circular bottomed canal arrangement will be similar. It may be noted that angle 0 shown in the section varies with the side slopes adopted. 9 for side slope 1: 1 has 45° value and for 1.25: 1 is 38° 402 as shown in Fig. 10.3. Maximum height of the spoil bank is limited to 6 m when the excavation is done by machinery. In case work is accomplished manually maximum height is to be restricted to 4 m only. Also when a canal section in filling involves more than 10.5 m filling each section shall be designed and tested for stability.

12

AGRICULTURAL DRAINAGE MANAGEMENT

Agricultural land drainage usually consists of surface or subsurface systems, or a combination of both. At the field scale, subsurface drain pipes and field ditches normally exit to an open main or collector drain. At the regional level, the latter then empties into a river or its tributaries. In some instances, depending on the character of the hydrological basin, main drains may dispose of drainage water to an evaporation pond, to a wetland, or to a saline agriculture/ agriculture-forestry system. Not all of these management options will necessarily be used in a single watershed.

Most agricultural producers improve the drainage on their land to help create a healthier environment for plant growth and to provide drier field conditions so farm equipment can access the farm field throughout the crop production season. Healthy, productive plants have the potential to produce greater yields and more food.

Agricultural drainage is the removal of excess water from the soil surface and/ or soil profile of cropland, by either gravity or artificial means. An agricultural drainage system is a system by which the water level on or in the soil is controlled to enhance agricultural crop production. It shows the field (or internal) and the main (or external) systems.

PURPOSE OF DRAINAGE SYSTEM

The function of the field drainage system is to control the water table, whereas the function of the main drainage system is to collect, transport, and dispose of the water through an outfall or outlet. In some instances, one makes an additional distinction between collector and main drainage systems. The two main reasons for improving the drainage on agricultural land are for soil conservation and enhancing crop production. Agricultural research has shown that agricultural drainage improvement can help reduce the year-to-year variability in crop yield, which helps reduce the risks associated with the production of abundant, high quality, affordable food. Improved access of farm equipment to the field provides more time for field activities, can help extend the crop production season, and helps reduce crop damage at harvest.

Field drainage systems are differentiated in surface and subsurface field drainage systems. Sometimes in irrigated, submerged rice fields, a form of temporary drainage is required whereby the drainage system is allowed to function on certain occasions only during the harvest period.

If allowed to function continuously, excessive quantities of water would be lost. Such a system is therefore called a checked, or controlled, drainage system. More usually, however, the drainage system is meant to function as regularly as possible to prevent undue waterlogging at any time and one employs a regular drainage system. In literature, this is sometimes also called a "relief drainage system."

Surface Drainage Systems

The regular surface drainage systems, which start functioning as soon as there is an excess of rainfall or irrigation, operate entirely by gravity. They consist of reshaped or reformed land surfaces and can be divided into:

1. Bedded systems, used in flat lands for crops other than rice.
2. Graded systems, used in sloping land for crops other than rice.

The bedded and graded systems may have ridges and furrows. The checked surface drainage systems consist of check gates placed in the embankments surrounding flat basins, such as those used for rice fields in flat lands. These fields are usually submerged and only need to be drained on certain occasions like at harvest time. Checked surface drainage systems are also found in terraced lands used for rice.

In literature, not much information can be found on the relations between the various regular surface field drainage systems, the reduction in the degree of waterlogging, and the agricultural or environmental effects. It is therefore difficult to develop sound agricultural criteria for the regular surface field drainage systems. Most of the known criteria for these systems concern the efficiency of the techniques of land leveling and earthmoving. Similarly, agricultural criteria for checked surface drainage systems are not very well known.

Sub-surface Drainage Systems

Like the surface field drainage systems, the subsurface field drainage systems can also be differentiated in regular systems and checked (controlled) systems.

When the drain discharge takes place entirely by gravity, both types of subsurface systems have much in common, except that the checked systems have control gates that can be opened and closed according to need. They can save much irrigation water. A checked drainage system also reduces the discharge through the main drainage system, thereby reducing construction costs.

When the discharge takes place by pumping, the drainage can be checked simply by not operating the pumps or by reducing the pumping time. In northwestern India, this practice has increased the irrigation efficiency and reduced the quantity of irrigation water needed, and has not led to any undue salinization.

The subsurface field drainage systems consist of horizontal or slightly sloping channels made in the soil; they can be open ditches, trenches, filled with brushwood and a soil cap, filled with stones and a soil cap, buried pipe drains, tile drains, or mole drains, but they can also consist of a series of wells.

Modern buried pipe drains often consist of corrugated, flexible, and perforated plastic (PE or PVC) pipe lines wrapped with an envelope or filter material to improve the permeability around the pipes and to prevent entry of soil particles, which is especially important in fine sandy and silty soils. The surround may consist of synthetic fibre (geotextile).

A subsoiler or mole plough is a tractor mounted implement used to loosen and break up soil at depths below the level of a traditional disk harrow or rototiller. Most tractor mounted cultivation tools will break up and turn over surface soil to a depth of 6" to 8" while a subsoiler will break up and loosen soil to twice those depths. Typically a subsoiler mounted to a Compact Utility Tractor will reach depths of about 12" and typically have only one thin blade with a sharpened tip.

The subsoiler is a tillage tool which will improve growth in all crops where soil compaction is a problem. The design provides deep tillage, loosening soil deeper than a tiller or plough is capable of reaching. Agricultural subsoilers, according to the Unverferth Company, can disrupt hardpan ground down to 24" depths.

Various manufacturer's brochures claim that crops perform well during hot and dry seasons because roots penetrate soil layers deeper to reach moisture and nutrients. Brochures further claim that in wet conditions, the water passes easier through the shattered areas, reducing the possibility of crops drowning.

Agricultural tractors will have multiple deeper reaching blades; each blade is called a scarifier or shank. Purdue University's Department of Agriculture indicates that common subsoilers for agricultural use are available with 3, 5 or 7 shanks. These units can be up to 15' wide, some models are towed behind tractors, others are tractor mounted to the 3pt hitch. A form of this implement with a single blade, a pipe-and-cable-laying plough, is used to lay buried cables or pipes, without the need to dig a deep trench and re-fill it.

Factors Contributing to Excess

Water Problems in Soils

The factors that contribute to excess soil water problems include: fine soil texture; massive soil structure; low soil permeability; topography; soil compaction; restrictive geologic layer; and excess precipitation.

Soil Texture : The sand, silt and clay composition of the solid mineral particles in a soil is called soil texture. For a loam soil texture, for instance, the mineral content might consist of 40 percent clay, 30 percent silt and 30 percent sand. Soil texture can have a dramatic effect on how well the soil holds water, and how easily water can move through the soil.

Fine-textured soils have a large percentage of clay and silt particles. These soils generally hold water well, but drain poorly. Coarse-textured soils have a large percentage of sand or gravel particles. These soils drain well, but have poor water-holding ability.

Soil Structure : The physical arrangement of the solid mineral particles of a soil is the soil structure. A granular structure helps promote the movement of water through a soil, but a structure that is massive lacking any distinct arrangement of soil particles usually decreases the movement of water.

Permeability : In general terms, the relative ease with which water can move through a block of soil is soil permeability. A soil's permeability can be affected by its texture, structure, human activities, and other factors.

Topography : The shape and slope of the land surface can cause wet soil conditions, especially around depressions where water tends to accumulate. Without an outlet, the water may drain away very slowly.

Geologic Formation : The geological formation underlying a soil can impact the drainage of water from that soil. For instance, a soil could have texture and structure properties that are beneficial to the movement of water. However, if the geologic formation underlying this soil consisted of dense clay or solid rock, it could restrict the downward movement of water, causing the soil above the formation to remain saturated during certain times of the year.

Compaction : Human activities may help create excess soil water problems. For example, operating equipment on a wet soil can compact the soil and destroy its structure. A soil layer that is compacted will generally have no structure, and most of the voids in this layer will have been eliminated.

Voids are open spaces between soil particles that can be filled with air, water, or a combination of both. Soil water will tend to accumulate above the compacted layer because movement of water through the compacted layer is severely restricted. If the compacted layer is located at the soil surface, very little water will enter the soil and much of the water will runoff, potentially creating a flooding and/or erosion hazard.

Precipitation : Ohio's average annual precipitation is 38 inches, based on a 50-year period of precipitation records. Even though the distribution of precipitation across the state varies, the state receives an abundant supply of precipitation in an average year. In an average month, most areas of the state receive two to four inches of precipitation.

This amount is adequate to sustain high crop yields. However, excessive rainfall, and/or winters with heavy snowfall, often produce excess soil water conditions. Furthermore, thunderstorms will frequently result in runoff because the rainfall rate is greater than the rate at which water can enter (infiltrate) into the soil.

The field drains (or laterals) discharge their water into the collector or main system either by gravity or by pumping. The wells (which may be open dug wells or tubewells) have normally to be pumped, but sometimes they are connected to drains for discharge by gravity.

Subsurface drainage by wells is often referred to as vertical drainage, and drainage by channels as horizontal drainage, but it is more clear to speak of "field drainage by wells" and "field drainage by ditches or pipes" respectively.

In some instances, subsurface drainage can be achieved simply by breaking up slowly permeable soil layers by deep plowing (sub-soiling), provided that the underground has sufficient natural drainage. In other instances, a combination of sub-soiling and subsurface drains may solve the problem.

A pipe-and-cable-laying plough is a trenchless method to bury cables or pipes. The machinery is a form of a subsoiler with a single blade. It is used to lay buried services of virtually any description, for drainage, water, electricity, tele-communications, gas supply etc.

A coil of the service pipe/cable is mounted on the tractor and is led down a guide behind the blade, and is left buried behind the plough, without the need to dig a deep trench and re-fill it. This process is normally used in rural areas where previously buried services will not be encountered and there are no hardened surfaces, e.g. tarmac concrete etc.

There are also specialised laying ploughs for cable laying behind traffic barriers, in stream or lake beds or even for the laying of submarine cables in deep sea, so-called sea ploughs. Sea ploughs are pulled behind cable ships and bury the cable in the sea bed. Burying submarine cables helps protect them from anchors, trawlers and other risks.

MAIN DRAINAGE SYSTEMS

The main drainage systems consist of deep or shallow collectors, and main drains or disposal drains. Deep collector drains are required for subsurface field drainage systems, whereas shallow collector drains are used for surface field drainage systems, but they can also be used for pumped subsurface systems. The deep collectors may consist of open ditches or buried pipe lines.

The terms deep collectors and shallow collectors refer rather to the depth of the water level in the collector below the soil surface than to the depth of the bottom of the collector. The bottom depth is determined both by the depth of the water level and by the required discharge capacity.

The deep collectors may either discharge their water into deep main drains (which are drains that do not receive water directly from field drains, but only from collectors), or their water may be pumped into a disposal drain.

Disposal drains are main drains in which the depth of the water level below the soil surface is not bound to a minimum, and the water level may even be above the soil surface, provided that embankments are made to prevent inundation. Disposal drains can serve both subsurface and surface field drainage systems.

Deep main drains can gradually become disposal drains if they are given a smaller gradient than the land slope along the drain. The technical criteria applicable to main drainage systems depend on the hydrological situation and on the type of system.

The final point of a main drainage system is the gravity outlet structure or the pumping station.

Applications

Surface drainage systems are usually applied in relatively flat lands that have soils with a low or medium infiltration capacity, or in lands with high-intensity rainfalls that exceed the normal infiltration capacity, so that frequent waterlogging occurs on the soil surface.

Subsurface drainage systems are used when the drainage problem is mainly that of shallow water tables. When both surface and subsurface waterlogging occur, a combined surface/subsurface drainage system is required.

Sometimes, a subsurface drainage system is installed in soils with a low infiltration capacity, where a surface drainage problem may improve the soil structure and the infiltration capacity so greatly that a surface drainage system is no longer required.

On the other hand, it can also happen that a surface drainage system diminishes the recharge of the groundwater to such an extent that the subsurface drainage problem is considerably reduced or even eliminated.

The choice between a subsurface drainage system by pipes and ditches or by tube wells is more a matter of technical criteria and costs than of agricultural criteria, because both types of systems can be designed to meet the same agricultural criteria and achieve the same benefits.

Usually, pipe drains or ditches are preferable to wells. However, when the soil consists of a poorly permeable top layer several meters thick, overlying a rapidly permeable and deep subsoil, wells may be a better option, because the drain spacing required for pipes or ditches would be considerably smaller than the spacing for wells.

When the land needs a subsurface drainage system, but saline groundwater is present at great depth, it is better to employ a shallow, closely spaced system of pipes or ditches instead of a deep, widely spaced system. The reason is that the deeper systems produce a more salty effluent than the shallow systems. Environmental criteria may then prohibit the use of the deeper systems.

Drainage Projects

In some drainage projects, one may find that only main drainage systems are envisaged. The agricultural land is then still likely to suffer from field drainage problems. In other cases, one may find that field drainage systems are ineffective because there is no adequate main drainage system. In either case, the installation of drainage systems is not recommended.

Reference: gives a general description of land drainage in the world and shows a paper on types of agricultural land drainage systems used in different parts of the world.

Drainage System Design

The analysis of positive and negative (side) effects of drainage and the optimization of drainage design in accordance to the drainage design procedures is discussed in the article on Drainage research.

There are several concerns about the sustainability of irrigation and drainage projects, and there are water quality problems related to the disposal of drainage water. There are also problems with land degradation due to irrigation induced salinity and waterlogging. There have been instances where saline or high nutrient drainage water has damaged aquatic ecosystems.

Drainage continues to be a vital and necessary component of agricultural production systems. In order to enhance the net benefits of drainage systems, more attention will need to be given to the water quality impacts of drainage water disposal. This document identifies potential problems and management options in the development, production, treatment and disposal of agricultural drainage water.

Effects of Excess Water

Excess water in the crop root zone soil is injurious to plant growth. Crop yields are drastically reduced on poorly drained soils, and, in cases of prolonged waterlogging, plants eventually die due to a lack of oxygen in the root zone. Sources of excess soil water that result in high water tables include: high precipitation in humid regions; surplus irrigation water and canal seepage in the irrigated lands; and artesian pressure. Waterlogging in irrigated regions may result in excess soil salinity, i.e., the accumulation of salts in the plant root zone.

Artificial drainage is essential on poorly drained agricultural fields to provide optimum air and salt environments in the root zone. Drainage is regarded as an important water management practice, and as a component of efficient crop production systems.

World food supply and the productivity of existing agricultural lands can only be maintained and enhanced if drainage improvements are undertaken on cropland currently affected by excess water and high water tables.

Drainage (both surface and subsurface) is not simply the conversion of wetlands, but the improvement of naturally inadequately drained cropland. It is complementary to irrigation and is viewed as an essential component of irrigated agriculture. The objective is to increase production efficiency, crop yields and profitability on naturally poorly drained agricultural lands.

It is difficult to quantify the effectiveness of most drainage methods. Drainage is best used as a short-term stabilizing method, due to the fact that, in the long-term, the drains need much maintenance and repair, which is often difficult to perform, and expensive. Drainage methods are split into various categories:

Shallow/Surface Drains

The aim of these is to reduce surface water and thereby reduce the porewater pressures at a much deeper level.

These are the easiest drains to repair but get damaged easily.

These drains will usually be either -

(a) lined or unlined ditches

(b) shallow, gravel-filled trenches

Commonly a herringbone or chevron shape is used for the drains on the surface as this is most effective in catching run-off water.

Deep Drains

These perform the function of modifying the shape of the seepage flow in the slope material.

They can be:

- Deep trenches (although they are more commonly shallow),
- Vertical bored drains - filled with sand or gravel,
- Horizontal bored drains - lined with perforated pipes.

Drains to Eliminate Construction Porewater Pressures:

These fall into two types:

- Drainage Blankets- Thses are installed either between layers of the fill material (which allows low permeability material to be used), or between the fill and foundation material.
- Vertical Drain Wells- Sand filled vertical drains have been discussed above. Thesedrains may be constructed using geofabric materials, cardboard, or merely sand alone. Their aim is to "catch" horizontal water flow.

Benefits of Drainage

The primary benefits of drainage go beyond the control of excess soil water and accumulation of excess salts in the crop root zone. The coincident environmental and socio- economic benefits associated with disease vector control and public health must be fully recognized. One of the major environmental benefits of drainage is its positive impact on improving the health of humans, plants and farm animals.

Drainage of wet, swampy areas has led to a reduction in mosquito breeding sites in all parts of the world. The effect has been a drop in the incidence and prevalence of important water related and mosquito transmitted diseases, e.g., malaria, yellow fever and filariasis.

Furthermore, drainage of stagnant water can eliminate foot-rot in large animals and, to a certain extent, the breeding environment of aquatic and semi-aquatic snails, which are the intermediate host of schistosomiasis.

Drainage also reduces or eliminates mildew infections and various root rots of plants. The overall impact of improved drainage has been an improvement in hygienic conditions, in the health sector and in the productivity of human beings. By growing high value food crops in well-drained soils, the health, nutrition and economic status of rural populations can be improved. There are also increased opportunities for employment, as new industries may develop in prosperous areas.

Where drainage is used to reclaim salinized and waterlogged lands, it is an environmentally beneficial practice, because the land is returned to its full productive potential. The adaptation of subsurface drainage systems to serve as sub-irrigation or controlled drainage systems leads to other benefits, i.e., the reduction of nitrate pollution.

The field-scale benefits of drainage can be summarized as follows:

i. Drainage promotes beneficial soil bacteria activity and improves soil tilth.
ii. There is less surface runoff and soil erosion on drained land.
iii. Improved field machine trafficability reduces soil structural damage. Soil compaction is reduced and less energy is required for field machine operations. Drainage also allows for more timely field operations. Consequently, the growing season can be lengthened and crops can achieve full maturity.
iv. Crop yields are increased because of improved water management and uptake of plant nutrients.
v. Higher value crops can be planted, and there is flexibility to introduce new and improved cropping systems.
vi. In general, land value and productivity are increased.
vii. Farm income is increased and income variability reduced.
viii. Drainage maintains favourable salt and air environments in the crop root zone.

Drainage Installation Methods

There are many different types of drainage solutions and drainage installation methods. Choosing the right one often can help alleviate a drainage problem. Ascertaining the exact drainage problem usually is a good first step. This may give an idea of what type of drainage solution should be used. Employing the draining installation methods properly can be a subsequent crucial step in solving the problem.

Generally, the most common type of drainage problem is the collection of excess water in a yard which can lead to flooding of the home, such as in the basement. The buildup of water can be a result of the shape of the landscape. Inclines and slopes in the land make some spots more prone to collect water. By correctly identifying these problem areas, one can then usually begin to examine appropriate drainage installation methods.

French Drain

A French drain is basically a ditch lined with rocks or gravel that helps drain water away from an area. It is generally used in gardening or to protect the foundation of a house or other structure from ground and surface water. A French drain may also be used as backup for retaining walls or to drain water away from a septic tank.

A French drain works on the principle of gravity, being slightly sloped down from the area to be drained to the area where one wishes to redirect the water. It is typically lined with perforated clay pipe and surrounded with a landscaping textile to prevent dirt or plant roots from clogging the system. Excess ground and surface water percolates into the French drain and is directed away.

The French drain is named after Henry French, a judge and farmer, and discussed in his 1859 book on the subject, Farm Drainage. However, the French drain predates Henry French by many centuries, as a version was used in ancient Rome. The system continues to be one of the most efficient and widely used methods of redirecting water, though it has been improved upon over the years.

Modern French drains are often installed well underground and hidden from view, for example by being covered with sod. A French drain can lead to a dry well, a structure that returns excess water to the supply of groundwater, or to a rain garden, an environmentally friendly invention that uses wetland plants to absorb excess water and return it to the atmosphere through transpiration.

French drains are moderately simple for the gardener or homeowner to build. Keep in mind that many countries, including the United States and the United Kingdom, regulate the use of French drain systems to keep certain contaminants out of the water supply. City codes and effects on your neighbors' property should also be considered before installing a French drain.

French drains are a prevalent drainage solution. The method for making a French drain generally involves digging a trench-like structure in the ground that then usually is filled with gravel or sand. The installation of the French drain is generally made to best facilitate the distribution of water away from problem areas. In this regard, determining the slope is critical before any actual installation.

Typically, French drains have a depth of 7 to 12 inches (about 17.75 to 30.5 cm) and a width of approximately 5 inches (about 12.75 cm). This usually can be done by hand. Gravel occasionally is added to the completed trench, filling it almost to the surface. Sand some times can be placed on top of the gravel. As the French drainage installation method is fairly basic, it often can be an elegant and simple way to eliminate draining problems.

Selecting Pipes for Drainage

Draining systems that use pipes is another method for solving drainage problems. A trench system needs to be dug so that the pipes can be placed inside. One can determine where to dig these trenches by examining the water flow and deciding a suitable place for relocation.

Selecting the different pipes to use can be an important part of this method. Flexible perforated pipe, usually 3 or 4 inches (about 7.62 or 10.16 cm) in width, commonly is used in home gardens. The installation of pipe drainage system may also decrease the buildup of sub-surface water.

Another method of draining installation is a rain barrel. The rain barrel generally is installed where there is an excess of rain water, such as under a roof trough. Rain barrels store the rain water, rather than entirely redistributing it. The stored water then can be used as drinking water or for watering plants.

Many times, drainage installation methods can be complicated. In these cases, there are many professional companies that offer drainage solutions and installation. Such firms generally should be able to correctly identify the exact drainage problem and then find the appropriate drainage solution.

Groundwater

Groundwater is water located beneath the ground surface in soil pore spaces and in the fractures of rock formations. A unit of rock or an unconsolidated deposit is called an aquifer when it can yield a usable quantity of water. The depth at which soil pore spaces or fractures and voids in rock become completely saturated with water is called the water table.

Groundwater is recharged from, and eventually flows to, the surface naturally; natural discharge often occurs at springs and seeps, and can form oases or wetlands. Groundwater is also often withdrawn for agricultural, municipal and industrial use by constructing and operating extraction wells. The study of the distribution and movement of groundwater is hydrogeology, also called groundwater hydrology.

Typically, groundwater is thought of as liquid water flowing through shallow aquifers, but technically it can also include soil moisture, permafrost (frozen soil), immobile water in very low permeability bedrock, and deep geothermal or oil formation water. Groundwater is hypothesized to provide lubrication that can possibly influence the movement of faults.

It is likely that much of the Earth's subsurface contains some water, which may be mixed with other fluids in some instances. Groundwater may not be confined only to the Earth. The formation of some of the landforms observed on Mars may have been influenced by groundwater. There is also evidence that liquid water may also exist in the subsurface of Jupiter's moon Europa.

The upper level of this saturated layer of an unconfined aquifer is called the water table or phreatic surface. Below the water table, where generally all pore spaces are saturated with water is the phreatic zone.

Substrate with low porosity that permits limited transmission of groundwater is known as an aquitard. An aquiclude is a substrate with porosity that is so low it is virtually impermeable to groundwater.

A confined aquifer is an aquifer that is overlain by a relatively impermeable layer of rock or substrate such as an aquiclude or aquitard. If a confined aquifer follows a downward grade from its recharge zone, groundwater can become pressurized as it flows. This can create artesian wells that flow freely without the need of a pump and rise to a higher elevation than the static water table at the above, unconfined, aquifer.

The characteristics of aquifers vary with the geology and structure of the substrate and topography in which they occur. Generally, the more productive aquifers occur in sedimentary geologic formations. By comparison, weathered and fractured crystalline rocks yield smaller quantities of groundwater in many environments. Unconsolidated to poorly cemented alluvial materials that have accumulated as valley-filling sediments in major river valleys and geologically subsiding structural basins are included among the most productive sources of groundwater.

The high specific heat capacity of water and the insulating effect of soil and rock can mitigate the effects of climate and maintain groundwater at a relatively steady temperature. In some places where groundwater temperatures are maintained by this effect at about 50°F/10°C, groundwater can be used for controlling the temperature inside structures at the surface.

For example, during hot weather relatively cool groundwater can be pumped through radiators in a home and then returned to the ground in another well. During cold seasons, as it is relatively warm, the water can be used in the same way as a source of heat for heat pumps that is much more efficient than using air.

Water Cycle

Groundwater makes up about twenty percent of the world's fresh water supply, which is about 0.61% of the entire world's water, including oceans and permanent ice. Global groundwater storage is roughly equal to the total amount of freshwater stored in the snow and ice pack, including the north and south poles. This makes it an important resource which can act as a natural storage that can buffer against shortages of surface water, as in during times of drought.

Groundwater is naturally replenished by surface water from precipitation, streams, and rivers when this recharge reaches the water table. Groundwater can be a long-term 'reservoir' of the natural water cycle (with residence times from days to millennia), as opposed to short-term water reservoirs like the atmosphere and fresh surface water (which have residence times from minutes to years). The figure shows how deep groundwater (which is quite distant from the surface recharge) can take a very long time to complete its natural cycle.

The Great Artesian Basin in central and eastern Australia is one of the largest confined aquifer systems in the world, extending for almost two million km2. By analysing the trace elements in water sourced from deep underground, hydrogeologists have been able to determine that water extracted from these aquifers can be more than one million years old. By comparing the age of groundwater obtained from different parts of the Great Artesian Basin, hydrogeologists have found it increases in age across the basin. Where water recharges the aquifers along the Eastern Divide, ages are young.

As groundwater flows westward across the continent, it increases in age, with the oldest groundwater occurring in the western parts. This means that in order to have travelled almost 1000 km from the source of recharge in 1 million years, the groundwater flowing through the Great Artesian Basin travels at an average rate of about one metre per year.

ISSUES

Overview

Certain problems have beset the use of groundwater around the world. Just as river waters have been over-used and polluted in many parts of the world, so too have aquifers. The big difference is that aquifers are out of sight. The other major problem is that water management agencies, when calculating the 'sustainable yield' of aquifer and river water, have often counted the same water twice, once in the aquifer, and once in its connected river.

This problem, although understood for centuries, has persisted, partly through inertia within government agencies. In Australia, for example, prior to the statutory reforms initiated by the Council of Australian Governments water reform framework in the 1990s, many Australian States managed groundwater and surface water through separate government agencies, an approach beset by rivalry and poor communication.

The time lags inherent in the dynamic response of groundwater to development have generally been ignored by water management agencies, decades after scientific understanding of the issue was consolidated. In brief, the effects of groundwater overdraft although undeniably real may take decades or centuries to manifest themselves.

In a classic study in 1982, Bredehoeft and colleagues modelled a situation where groundwater extraction in an intermontane basin withdrew the entire annual recharge, leaving 'nothing' for the natural groundwater-dependent vegetation community. Even when the borefield was situated close to the vegetation, 30% of the original vegetation demand could still be met by the lag inherent in the system after 100 years. By year 500 this had reduced to 0%, signalling complete death of the groundwater-dependent vegetation.

The science has been available to make these calculations for decades; however water management agencies have generally ignored effects which will appear outside the rough timeframe of political elections (three to five years).

Marios Sophocleous argued strongly that management agencies must define and use appropriate timeframes in groundwater planning. This will mean calculating groundwater withdrawal permits based on predicted effects decades, sometimes centuries in the future.

As water moves through the landscape it collects soluble salts, mainly sodium chloride. Where such water enters the atmosphere through evapotranspiration, these salts are left behind. In irrigation districts, poor drainage of soils and surface aquifers can result in water tables coming to the surface in low-lying areas. Major land degradation problems of soil salinity and waterlogging result, combined with increasing levels of salt in surface waters. As a consequence, major damage has occurred to local economies and environments.

Four important effects are worthy of brief mention. First, flood mitigation schemes, intended to protect infrastructure built on floodplains, have had the unintended consequence of reducing aquifer recharge associated with natural flooding. Second, prolonged depletion of groundwater in extensive aquifers can result in land subsidence, with associated infrastructure damage – as well as (thirdly) saline intrusion. Fourth, draining acid sulphate soils, often found in low-lying coastal plains, can result in acidification and pollution of formerly freshwater and estuarine streams.

Another cause for concern is that groundwater drawdown from over-allocated aquifers has the potential to cause severe damage to both terrestrial and aquatic ecosystems – in some cases very conspicuously but in others quite imperceptibly because of the extended period over which the damage occurs.

Overdraft

Groundwater is a highly useful and often abundant resource. However, over-use, or overdraft, can cause major problems to human users and to the environment. The most evident problem (as far as human groundwater use is concerned) is a lowering of the water table beyond the reach of existing wells. Wells must consequently be deepened to reach the groundwater; in some places (e.g., California, Texas and India) the water table has dropped hundreds of feet because of extensive well pumping.

In the Punjab region of India, for example, groundwater levels have dropped 10 meters since 1979, and the rate of depletion is accelerating. A lowered water table may, in turn, cause other problems such as groundwater-related subsidence and saltwater intrusion.

Groundwater is also ecologically important. The importance of groundwater to ecosystems is often overlooked, even by freshwater biologists and ecologists. Groundwaters sustain rivers, wetlands and lakes, as well as subterranean ecosystems within karst or alluvial aquifers. Not all ecosystems need groundwater, of course. Some terrestrial ecosystems – for example, those of the open deserts and similar arid environments – exist on irregular rainfall and the moisture it delivers to the soil, supplemented by moisture in the air. While there are other terrestrial ecosystems in more hospitable environments where groundwater plays no central role, groundwater is in fact fundamental to many of the world's major ecosystems. Water flows between groundwaters and surface waters. Most rivers, lakes and wetlands are fed by, and (at other places or times) feed groundwater, to varying degrees.

Groundwater feeds soil moisture through percolation, and many terrestrial vegetation communities depend directly on either groundwater or the percolated soil moisture above the aquifer for at least part of each year. Hyporheic zones (the mixing zone of streamwater and groundwater) and riparian zones are examples of ecotones largely or totally dependent on groundwater.

Aquifer drawdown or overdrafting and the pumping of fossil water increases the total amount of water within the hydrosphere subject to transpiration and evaporation processes, thereby causing accretion in water vapour and cloud cover, the primary absorbers of infrared radiation in the Earth's atmosphere. Adding water to the system has a forcing effect on the whole earth system, an accurate estimate of which hydrogeological fact is yet to be quantified.

Solution to Over-use of Groundwater

Irrigated agriculture is a big business in Asia. It accounts for 70 per cent of the world's irrigated land and 73 per cent of water used each year by the farming industry. As ageing large-scale surface irrigation schemes have become increasingly inefficient, and farmers have begun growing a wider range of crops requiring water on demand, the number of groundwater wells in India has exploded.

In 1960, there were fewer than 100,000 such wells; by 2006 the figure had risen to nearly 12 million. In India, a possible solution to overuse of groundwater is emerging, known as 'groundwater recharge'. It involves capturing rainwater that would otherwise run-off, and using it to refill aquifers.

Since 2000, the International Water Management Institute (IWMI) has been working with the Indian authorities to help improve the availability of water for agriculture in India. In 2006, India's finance minister invited IWMI to submit policy recommendations based on its research on groundwater depletion. One of the key recommendations was to instigate a programme of recharging groundwater across the 65 per cent of India that has hard-rock aquifers.

As a result, the Indian government allocated Rs 1800 crore to fund dug-well recharge projects (a dug-well is a wide, shallow well, often lined with concrete) in

100 districts within seven states where water stored in hard-rock aquifers has been over-exploited. These geological formations have a much lower capacity to store rainwater than alluvial areas with porous sand or clay rocks, hence being given priority. The money is sufficient to fund seven million structures to be installed on dug-wells to divert monsoon runoff.

The structures include a de-siltation chamber, plus pipes to collect surplus rainwater and divert de-silted water from the chamber to the well. As of the end of November 2009, funds amounting to Rs. 216.98 crore including Rs. 199.98 crore as subsidy to beneficiaries and 17 crore for IEC/Capacity Building activities had been released to the concerned states. Subsidies had been released to 566,637 beneficiaries.

Subsidence

Subsidence occurs when too much water is pumped out from underground, deflating the space below the above-surface, and thus causing the ground to actually collapse. The result can look like craters on plots of land. This occurs because in its natural equilibrium state, the hydraulic pressure of groundwater in the pore spaces of the aquifer and the aquitard supports some of the weight of the overlying sediments.

When groundwater is removed from aquifers by excessive pumping, pore pressures in the aquifer drop and compression of the aquifer may occur. This compression may be partially recoverable if pressures rebound, but much of it is not. When the aquifer gets compressed it may cause land subsidence, a drop in the ground surface.

The city of New Orleans, Louisiana, is actually below sea level today, and its subsidence is partly caused by removal of groundwater from the various aquifer/aquitard systems beneath it. In the first half of the last century, the city of San Jose, California, dropped 13 feet from land subsidence caused by overpumping; this subsidence has been halted with improved groundwater management.

Seawater Intrusion

Generally, in very humid or undeveloped regions, the shape of the water table mimics the slope of the surface. The recharge zone of an aquifer near the seacoast is likely to be inland, often at considerable distance. In these coastal areas, a lowered water table may induce sea water to reverse the flow toward the land. Sea water moving inland is called a saltwater intrusion. Alternatively, salt from mineral beds may leach into the groundwater of its own accord.

Mining

Sometimes the water movement from the recharge zone to the place where it is withdrawn may take centuries (see figure above). When the usage of water is greater than the recharge, it is referred to as mining water (the water is often called fossil water because of its geologic age). Under those circumstances it is not a renewable resource.

Pollution

Water pollution of groundwater, from pollutants released to the ground that can work their way down into groundwater, can create a contaminant plume within an aquifer. Movement of water and dispersion within the aquifer spreads the pollutant over a wider area, its advancing boundary often called a plume edge, which can then intersect with groundwater wells or daylight into surface water such as seeps and springs, making the water supplies unsafe for humans and wildlife. The interaction of groundwater contamination with surface waters is analyzed by use of hydrology transport models.

The stratigraphy of the area plays an important role in the transport of these pollutants. An area can have layers of sandy soil, fractured bedrock, clay, or hardpan. Areas of karst topography on limestone bedrock are sometimes vulnerable to surface pollution from groundwater. Earthquake faults can also be entry routes for downward contaminant entry. Water table conditions are of great importance for drinking water supplies, agricultural irrigation, waste disposal including nuclear waste, wildlife habitat, and other ecological issues.

In the US, upon commercial real estate property transactions both groundwater and soil are the subjects of scrutiny, with a Phase I Environmental Site Assessment normally being prepared to investigate and disclose potential pollution issues. In the San Fernando Valley of California, Real estate contracts for property transfer below the Santa Susana Field Laboratory (SSFL) and eastward have clauses releasing the seller from liability for groundwater contamination consequences from existing or future water pollution of the Valley Aquifer.

Love Canal was one of the most widely known examples of groundwater pollution. In 1978, residents of the Love Canal neighbourhood in upstate New York noticed high rates of cancer and an alarming number of birth defects. This was eventually traced to organic solvents and dioxins from an industrial landfill that the neighbourhood had been built over and around, which had then infiltrated into the water supply and evaporated in basements to further contaminate the air. Eight hundred families were reimbursed for their homes and moved, after extensive legal battles and media coverage.

Another example of widespread groundwater pollution is in the Ganges Plain of northern India and Bangladesh where severe contamination of groundwater by naturally occurring arsenic affects 25% of water wells in the shallower of two regional aquifers. The pollution occurs because aquifer sediments contain organic matter (dead plant material) that generates anaerobic (an environment without oxygen) conditions in the aquifer.

These conditions result in the microbial dissolution of iron oxides in the sediment and thus the release of the arsenic, normally strongly bound to iron oxides, into the water. As a consequence, arsenic-rich groundwater is often iron-rich, although secondary processes often obscure the association of dissolved arsenic and dissolved iron.

Ground Water Rule

In November 2006, the Environmental Protection Agency published the Ground Water Rule in the United States Federal Register. The EPA was worried that the ground water system would be vulnerable to contamination from fecal matter. The point of the rule was to keep microbial pathogens out of public water sources The 2006 Ground Water Rule was an amendment of the 1996 Safe Drinking Water Act.

Surface Drainage

In areas of heavy rainfall or low soil percolation, it is necessary to employ surface drainage systems to get rid of surplus surface water from the land. Commonly used surface drainage systems include shallow ditches, open channels, grassed waterways and sloped banks. These drainage systems drain away surface water quickly and efficiently, but do not make much of a difference to groundwater. If there are land drainage problems due to groundwater, they will need other solutions.

Surface drainage needs to be carefully planned by taking into account the land's topography, the amount of rainfall received, and the type of soil. In some cases, shallow ditches following the land contour may be enough, in others the ditches may need to be deeper or the land may need to be graded or sloped to facilitate soil drainage. The ditches may follow a random pattern or a parallel pattern.

Large fields with a good many depression areas do well with randomly patterned field ditches. These shallow ditches follow the land contours, do not impede the movement of farm machinery and drain away water from depressed areas. Parallel field ditches are excellent for agricultural drainage on flatter types of lands, and are also used as to prevent erosion on sloping farm lands. Water from the field ditches is drained away into side ditches. The side ditches then carry water to an outlet channel.

Surface drainage is often achieved by land forming and smoothing to remove isolated depressions, or by constructing parallel ditches. Ditches and furrow bottoms are gently graded and discharge into main drains at the field boundary. Although the ditches or furrows are intended primarily to convey excess surface runoff, there is some seepage through the soil to the ditches, depending on the water table position. This could be regarded as a form of shallow subsurface drainage. Surface drainage is especially important in humid regions on flat lands with limited hydraulic gradients to nearby rivers or other disposal points. There is also a need for good surface drainage in semi-arid regions which are affected by monsoons.

Subsurface Drainage

In this study, a two-dimensional numerical model for subsurface flow has been developed for the analysis of unsteady flow for subsurface drains. The governing equation is the two-dimensional Richard's equation in the mixed form. A strongly implicit finite-difference scheme has been used to solve the governing equation.

The model has been validated using the available analytical results of the one side drains. Validation of the model indicates good agreement between the results. It has been used to simulate a hypothetical case of subsurface drains with parallel drains. The present numerical model can be used to simulate the unsteady subsurface drainage problem having a side drain and parallel drains at both sides.

In areas with periodic irrigations or high intensity rainfall, the assumption of a steady recharge is no longer justified. Under these conditions, unsteady state solution of the flow problem must be applied. Therefore, knowledge of the water table and its fluctuations in the irrigated areas is essential in assessing the possibilities of waterlogging. In case a barrier is to be constructed to keep the polluted water on one side and fresh water on another, a lateral drainage can serve the purpose. The analysis of unsteady flow to subsurface drains has been carried out keeping this in view.

Surface drainage alone is seldom sufficient to remove excess water from the crop root zone. Deep ditches or subsurface pipe drainage systems enable a more rapid water table drawdown. The downstream ends of the laterals are normally connected to a collector drain. The required diameter of the pipe collectors increases with the area drained. Drain spacing is usually dependent on soil hydraulic conductivity and a design drainage rate coefficient. Depending on topography, land formation and proximity of a water receiving body, the collector may outlet by gravity to an open main drain or into a sump. In the latter case, the discharge is then pumped to another drain, or ultimately to a lake or stream.

In some flatter parts of eastern Canada, the eastern and mid-western United States, and parts of Europe, subsurface pipe drains are also used for sub-irrigation. In this case, in dry periods, surface water is introduced into the drain pipe system from an external source, and the water table is raised. Moisture then moves upward by capillary action to the root zone. Sub-irrigation is regarded as a highly energy and water efficient method of irrigation. In another process, known as controlled drainage, elevated water tables can be maintained with a control structure on the collector pipe.

Horizontal Subsurface Drainage Systems

These are used in irrigated arid and semi-arid regions to reclaim saline and waterlogged lands, and to maintain favourable long-term salt and water balances in the crop root zone. Salinity and waterlogging are caused by a build up of the water table due to deep percolation of normal excess water and canal seepage. Buried pipe drains are generally installed deeper in arid regions than in humid regions in order to control salinity.

Water in excess of plant evapotranspiration (ET) needs is always unavoidably applied during irrigation. This additional quantity of water applied is known as the leaching fraction. Naturally occurring as well as applied salts are then leached from the root zone by this water, and removed from the field via the pipe drains. Deeper drain installation ensures that salts do not rise too rapidly to the soil surface due to capillary action. Drainage also prevents waterlogging of the root zone.

The amount of irrigation water to be removed is generally less in arid than in humid regions. Vertical drainage by means of tube-wells is also used to control waterlogging and salinity in some parts of the world, e.g., India, Pakistan and central Asian republics. The primary purposes of tube-wells are the same as those of horizontal drains, and at the same time to extract groundwater for irrigation.

As a result of pumping, the water table is lowered, and salinization due to capillarity is minimized. This situation is ideal where the groundwater is not very brackish or saline, and is therefore suitable for irrigation. In areas where the groundwater is highly saline, the pumped water may be too saline for irrigation, unless mixed with fresher or less saline water.

Where the groundwater is too saline for crop production, it must be disposed of. Drainage does not have a direct impact on groundwater quality. It only serves to collect and transport excess water.

Improving Internal Drainage

Methods of improving the internal drainage of low permeability soils include: subsoiling, deep tillage, mole drainage, and biological practices, viz., cropping with deep rooted legumes (e.g., alfalfa) and crop rotations. In some parts of the world, deep rooted trees are used to lower the water table. There are usually no water quality hazards associated with these supplemental drainage practices.

Treating drainage water is normally one of the last drainage water management options to be considered. This is due to the high costs involved and to uncertainty about the treatment level achievable. The treatment of drainage water should be considered where all other drainage water management measures fail to guarantee safe disposal or where it is financially attractive. For subsurface drainage water containing very high levels of salinity, selenium and other trace elements, the treatment objectives are: (i) reduce salts and toxic constituents below hazardous levels; (ii) meet agricultural water management goals; (iii) meet water quality objectives in surface waters; and (iv) reduce constituent levels below risk levels for wildlife.

The treatment of agricultural drainage water presents a challenge due to the complex chemical characteristics of most drainage waters. The drainage waters are saline and of the NaCl-Na2SO4-type water. The waters conveyed by the San Luis Drain came from a single site in Westlands Water District in contrast to the evaporation pond waters that came from twenty seven sites.

The drainage water had to be treated with alum, and passed through a sedimentation pond and a chlorinated and filtration system. In spite of this level of pretreatment, the membranes tended to foul due to the precipitation of gypsum and calcite. The drainage waters are saturated with respect to calcite and gypsum. This same chemical fouling problem is being faced by the Yuma desalting plant off the Colorado River using drainage waters from the Wellton-Mohawk irrigation project.

The estimated cost of desalting is more than US$0.81/m3, too expensive for irrigated agriculture but possibly affordable for municipalities with freshwater shortages. This cost does not include the management and disposal of the brine water. However, a potential exists for partially treating the average 10-dS/m-drainage water to about 2-3 dS/m for use by agriculture and wildlife.

Trace Element Treatment

As the technology of reverse osmosis is experimental and expensive, cheaper methods of removing toxic trace elements are being pursued.

Chemical Processes

Chemical treatment processes refer to the use of chemicals to remove trace elements from polluted wastewater. Chemicals are frequently used for industrial wastewater treatment but are not effective in agricultural drainage water due to their often complex chemical characteristics (Lee, 1994). Chemical processes have been developed for the reduction of selenate to elemental selenium by means of ferrous hydroxide.

Under laboratory conditions, ferrous hydroxide was able to reduce and precipitate selenium by 99 percent in 30 min. In field studies, although 90 percent of the selenate was reduced, the reactor time required was up to 6 h. It appeared that dissolved bicarbonate, oxygen and nitrate influenced the reduction process.

Physical processes involve the adsorption of ions on natural and synthetic surfaces of active materials, including ion exchange resins. Box 13 provides an example of a mini-pilot plant for the removal of heavy metals.

Flow-through Artificial Wetlands

The figure given below shows the layout of a pilot project for removing selenium by flow-through wetland cells conducted in the Tulare Lake bed, a closed basin of the San Joaquin Valley. The goal was to remove selenium from drainage waters to a bird-safe level prior to disposal into evaporation ponds.

Tile drainage effluent containing about 20-ppb selenium from an adjacent farm was passed through a sand bed filter system and metered into the cells (15.2 × 76 m) with a variety of substrates (vegetation). The inflow water was measured twice a week by a totalizing meter. The water depth in Cells 1-7 was maintained at about 20 cm, and outflow was measured by v-notch weir. Cells 8-10 had variable water depths of about 20 cm, 60 cm where widgeon grass (Ruppia) was grown.

The target residence time for the flowing waters was 7 days for Cells 1-7, 21 days in Cells 8 and 9, and 14 days in Cell 10. These residence times were selected after preliminary runs for optimal removal. A residence time of three days was too short for selenium removal and a residence time of more than twenty one days did not increase selenium removal. Seepage rates in the cells were about 1 cm/d and evapotranspiration slightly greater than ETo (annual value about 1 600 mm).

The residence times achieved were reasonably close to target values considering the variability in monthly ETo. The selenium concentration in the outflow waters varied from 4.6 to 12.3 ppb. The ratio of outflow to inflow selenium concentration ranged from 0.25 to 0.68 (a small ratio indicates high selenium removal).

The ratio of outflow to inflow selenium on a mass basis ranged from 0.07 to 0.57 or 93 to 43 removal. The cell with open water had reduced selenium because algae and microbes naturally populated the cell and contributed to some selenium removal.

In terms of performance, the ratio based on mass of selenium is a good indicator. However, in terms of potential impact on birds, the outflow concentration and ratio based on concentration are better indicators.

The righthand-side terms of Equation 22 are mass fluxes, and mass (MSe) is defined as the product of selenium concentration and water volume, except for the volatilization term. Water inflow and outflow was monitored twice a week, water seepage estimated from the difference from inflow and outflow and ETcrop from ETo *Kc, where Kc is the crop coefficient. Volatilization of selenium by microbes and plants was monitored monthly. The DM/Dt is the mass flux of selenium accumulating in the control volume (cell) consisting of the sediments, organic detrital matter, fallen litter, standing water and standing crop.

The remainder of the selenium accumulated in the cell as selenium present in the sediments, organic detrital matter, fallen litter, standing water and standing plants. The values reported are the mass of selenium found in the cells in September 2000. About 11 percent of the total selenium could not be accounted for due to errors in sampling and monitoring over a four-year period, and the difficulties of analysing for reduced forms of selenium.

The sink mechanisms removing selenium from the floodwater were: adsorption of selenite (Se+4) to the mineral sediments mainly in the top 10 cm or so; selenium immobilized into elemental selenium (Se0) due to reduced conditions in the organic detrital layer; and organic forms of selenium (Se-2) tied up with the detritus and fallen litter. The principal removal mechanisms were adsorption and immobilization into elemental selenium and organic selenium.

The recommended selenium water standard to protect waterbirds is 2 ppb. None of these cells achieved that level of remediation but many cells certainly will reduce selenium toxicity. However, outflow waters from these cells contain organic selenium (17-33 percent of the total selenium), which is more toxic than inorganic forms to wildlife. These and other results are currently being reviewed to determine whether selenium removal flow-through wetland cells is a viable treatment option.

Evaluation and Selection of Treatment Options

The first steps in the selection of any drainage treatment process are: i) define the problem; ii) determine the reasons for the required treatment; and iii) determine what is to be achieved. The main reason for opting for drainage water treatment is normally the desire to reuse the drainage effluent or to conform to regulatory disposal requirements. For both purposes, specific water quality criteria apply.

In order to make a preliminary selection of suitable treatment processes, it is necessary that sufficient data be available. These data consist of historical data on the chemical constituents of the drainage water, seasonal flow variations and variations in the concentrations of the constituents of concern. Once combined with information on the targeted quality of the treated effluent, it is then possible to shortlist drainage treatment processes that are theoretically suitable.

The technical capability of the treatment process is an important factor in the selection of a treatment technology. However, it is important to consider economic, financial, social and institutional criteria in order to ensure the sustainability of the treatment facilities.

Environmental Impact Assessment

Many institutions require an environmental impact assessment (EIA) prior to construction activities associated with new projects or the rehabilitation of existing projects. The objectives are to identify potential adverse environmental effects, the magnitude of these effects, and to develop mitigative measures. Positive benefits are also identified. In cases where the adverse impacts far outweigh the expected benefits, the project may be completely redesigned or suspended. This is also the case where mitigative measures would be either too costly or technically not feasible. The EIA should be conducted in the earliest stages of decision making, when crucial decisions are still being deliberated.

According to Ochs and Bishay (1992), environmental effects can be classified as:

(i) direct and indirect, or first order and higher order. These are chain effects which are felt throughout, and possibly downstream of a catchment.

(ii) secondary. The primary activity of a drainage project may be extended to include secondary activities.

(iii) synergistic. These effects include an increased threat to the survival of certain species of wildlife that are under pressure in several ways as a result of the same project.

The ICID has developed an 'Environmental Checklist to Identify Environmental Impacts of Irrigation, Drainage and Flood Control Projects'. The World Bank has prepared an Environmental Assessment Sourcebook (World Bank, 1991). FAO has produced a paper on the steps in the EIA process and the major environmental impacts of irrigation and drainage projects. The purpose of these documents is to enable detailed environmental impact assessments of irrigation and drainage projects to be conducted. The ICID checklist provides a comprehensive list of environmental parameters which must be evaluated in an EIA. Table 1 shows the ICID checklist of possible environmental impacts of irrigation, drainage and flood control projects.

Ochs and Bishay (1992) listed the main steps in the EIA process as:

i. Scoping. This is a public process, involving the participation of all parties. It results in specific guidelines for inclusion in the environmental impact study (EIS). (EIS refers to the document, whereas EIA refers to the entire process).

ii. Drawing up the EIS. Potential adverse and beneficial impacts are identified. Proposed actions with alternatives are indicated and mitigative measures arc presented.

iii. Submitting the EIS for public review.

iv. Receiving advice from the reviewing agency.

v. Accepting the EIS.

vi. Choosing project components. The decision-maker chooses the project option to be implemented and the mitigative measures.

vii. Implementing the project.

viii. Monitoring. During and after project implementation, the actual environmental impacts are monitored and compared with the EIS predictions. This is useful for improving future predictions and project designs.

Matrices and checklists are some of the tools most commonly used in an EIA to assess the magnitude or relative weight of both positive and adverse environmental effects. It is hoped that the criteria on drainage water quality provided in the publication will further assist in the evaluation of EIA of irrigation and drainage projects.

Water Quality Impacts

The installation of drainage systems may result in changes to the associated ecosystem. These changes may be either beneficial or adverse.

However, there are potential adverse water quality impacts associated with drainage. The concentrations of salts, nutrients and other crop-related chemicals in drainage discharge vary with time and discharge rate. The use of fertilizers and pesticides in intensive agricultural production has sometimes led to damage to downstream aquatic ecosystems.

Drainage planners therefore need to analyse effluent for nutrients and pesticides. The nutrients of most concern are N and P. In addition, from time to time, natural trace elements from the soil itself may be harmful to the ecosystem. Effluent laden with N and P stimulates eutrophication in receiving water bodies. In addition to agricultural chemicals and trace elements, drainage water from irrigated areas frequently contains salts. The impact of salts on downstream users needs to be evaluated. Some soils are abundant in trace elements, and these could leach to the drainage system.

Small amounts of trace elements such as As, Cd, Hg, Pb, B, Cr and Se are harmful to aquatic species because of biological magnification. The environmental consequences of disposing of drainage water from California's irrigated San Joaquin Valley into the 470 ha Kesterson Reservoir (a closed basin) are well known. The reservoir was a waterfowl habitat, and concentrating Se in the drainage water caused fish species to disappear, and resulted in deformities in waterbird embryos.

The failure to construct an adequate outfall drain to the sea, the use of a waterfowl habitat for drainage water disposal, and the lack of proper water quality monitoring were the major reasons for these negative environmental impacts. Care must be taken to ensure that the disposal of drainage water does not interfere with the ecosystem's aquatic and terrestrial species. Concern has been expressed about damage by agricultural drainage water to estuarine fisheries in some countries.

Drainage water can be disposed directly to open surface water bodies, e.g., rivers, lakes, outfall drains, seas or oceans. Wetlands, evaporation ponds and solar evaporators are used as discharge points where there is no direct outlet to one of the open surface water bodies mentioned above.

This is generally the situation in hydrologically closed drainage basins. Some re-use of drainage water for irrigation is another management option, where sequential irrigation of increasingly salt-tolerant plants, ending with Salicornia, occurs. Drainage water can also be disposed of by injection to the deep groundwater.

If drainage water is disposed of to large, open surface water systems with significant dilution or assimilative capacity, then water quality problems are minimized. However, water quality problems may develop with repeated re-use, disposal in closed basins and injection to deep wells. In this last case, the concern is that if the drainage water contains sufficient amounts of salts, nitrates, bacteria and trace elements, the groundwater could become contaminated. The problem may be especially severe if the aquifer is also used for drinking water supplies downslope.

Minimizing Environmental Problems

The most efficient method of minimizing environmental problems is to implement source control practices at the farm level. In irrigated areas, this can be achieved by improved irrigation water management.

Higher irrigation efficiencies and lined irrigation conveyance structures will reduce the amount of drainage water which needs to be removed. Furthermore, timely and efficient applications of fertilizers and pesticides, used only when necessary, will reduce chemical leaching. Research by Madramootoo et al. (1995) has shown that intercropping corn with a legume or ryegrass reduces the amount of nitrates found in drainage water.

In many regions where irrigation water is scarce, drainage water is used to meet crop water requirements. Re-use is only sustainable if the drainage water is of sufficiently good quality. Some of the water quality concerns about drainage water re-use with plants of increasing salt tolerance are that: the effluent may be high in salt content (in irrigated lands); the drainage water can be contaminated with trace elements, toxic organic substances, industrial waste and municipal waste in open main drains.

Contaminated drainage water could lead to various problems including: impairment of soil physical and chemical properties, water-related health problems, and possible contamination of food products.

Ultimate disposal of drainage water to a river or sea is not always possible. Closed drainage basins present a unique environmental or water quality challenge. In such situations, evaporation ponds may be an appropriate means for disposing of drainage water. However, these ponds may eventually lead to other environmental problems. For example, toxic substances could accumulate in the ponds. Furthermore, in arid climates, as pure water evaporates from the pond, the concentration of the remaining water approaches that of brine.

The health of waterfowl, fish and other aquatic biota which use the pond could be negatively affected. Other environmental problems associated with evaporation ponds include: use of the pond to collect wastewater from homes; human health problems caused by consuming water from the ponds; and the need to ultimately dispose of the accumulated concentrated chemicals in the ponds.

In some cases, dry toxic materials may be spread by the wind. Furthermore, these ponds could become habitats for snails and mosquitoes, thereby causing malaria and schistosomiasis epidemics. In addition, if not properly managed, new waterlogged and saline areas will develop adjacent to the ponds.

Managing Drainage Water

There is now increasing interest in the utilization of natural and constructed wetlands to manage drainage water. Wetlands are particularly effective for removing sediment, N and P. Plant, soil and hydrologic parameters interact in a complex way to filter and trap pollutants, and to recycle nutrients. Certain tree and plant species have the potential to absorb pollutants.

Residence time, flow rate, hydraulic roughness and wetland size and shape are some of the factors which influence treatment efficiency. The water supply to the wetland must be sufficient to provide an excess to discharge and prevent salt accumulation.

In areas where soils, geologic and hydrologic conditions do not permit constructed wetlands, the saline agriculture/agriculture-forestry system may be appropriate for the disposal of drainage water. The operating principle is to successively re-use saline drainage water to irrigate crops and trees of increasing salt tolerance, and to discharge the final much reduced volume of water into a solar evaporator for salt crystallization.

The depth of water ponded in the evaporator is regulated to match the daily evaporation rate. The goal is to make the crystallization pond unattractive to flora and fauna. Another challenge of evaporation ponds is that evaporation is reduced as the ponds become more concentrated. This necessitates the use of additional land to maintain disposal (evaporation) capacity.

While there are instances where ecosystems have been damaged by poor quality drainage water, it is also possible to use subsurface drainage systems beneficially to improve water quality. By managing the water table, through controlled drainage or sub-irrigation, nitrate concentrations in drainage effluent can be significantly reduced. Downstream flood flows can also be reduced. Water table management could therefore be viewed as a best management practice. Skaggs and Gilliam (1981) and Madramootoo et al. (1993) have shown that sub-irrigation and controlled drainage not only increase crop yields, but also enhance denitrification. Water table management is best suited to flat lands, and can be achieved by a simple modification to the outlet of subsurface drainage systems.

It is important to recognize that while technologies may be available to minimize water quality impacts from drainage effluent, institutional mechanisms must also be put in place to ensure that the technologies can be implemented. Policies and programmes, including legal and monitoring aspects, require an institutional framework.

It is recognized that, in some instances, there may be insufficient data in the project design and planning stages to enable precise conclusions to be drawn about possible environmental impacts. For this reason, most chapters contain sections on environmental monitoring, so that project planners and designers can elicit the information required for proper decision making. For more detailed information on water quality monitoring, the reader is referred to the United States Environmental Protection Agency (EPA).

Purpose of Agricultural Drainage

The purpose of agricultural drainage is to remove excess water from the soil in order to enhance crop production. In some soils, the natural drainage processes are sufficient for growth and production of agricultural crops, but in many other soils, artificial drainage is needed for efficient agricultural production.

Surface drainage is the removal of water that collects on the land surface. Many fields have low spots or depressions where water ponds. Surface drainage techniques such as land levelling, constructing surface inlets to subsurface drains, and the construction of shallow ditches or waterways can allow the water to leave the field rather than causing prolonged wet areas.

Subsurface drainage removes excess water from the soil profile, usually through a network of perforated tubes installed two to four feet below the soil surface. These tubes are commonly called “tiles” because they were originally made from short lengths of clay pipes known as tiles. Water would seep into the small spaces between the tiles and drains away.

The most common type of “tile” is corrugated plastic tubing with small perforations to allow water entry. When the water table in the soil is higher than the tile, water flows into the tubing, either through holes in the plastic tube or through the small cracks between adjacent clay tiles. This lowers the water table to the depth of the tile over the course of several days. Drain tiles allow excess water to leave the field, but once the water table has been lowered to the elevation of the tiles, no more water flows through the tiles. In most years, drain tiles are not flowing between June and October.

A drain tile is a specialized pipe used in a drain tile drainage system. This type of drainage system is used to pull water away from the footings and foundation of a structure so that the water cannot penetrate the foundation. Water seepage can damage a structure over time, in addition to being inconvenient if the water ends up in a location such as a basement. Many hardware and home supply stores sell several styles of drain tile, and it can also be ordered directly from manufacturers.

The installation of a drain tile system starts with digging a trench around the footings of a home. A filter barrier is laid in the trench, and a drain tile is laid inside before being covered with gravel and covered with another filter. The filters prevent dirt and silt from entering the trench and clogging it, while the gravel provides a filter for water. Water will naturally drawn to the gravel because it is easier for water to flow through the gravel than it is for water to move through the surrounding soil. Perforations in the drain tile allow water to enter the pipe.

The system also has an outlet, allowing the water which collects in the drain tile to disperse. The outlet may be connected to a sump pump if a structure is on level ground or it is lower than the surrounding ground, or it may simply drain freely if a house is built on high ground, allowing the water to flow into storm drains or to trickle harmlessly downhill.

This type of drainage system is used to control the water table. As long as the water table lies below the drain tile, the drainage system is not needed, because the water is too low to enter the structure. If the water table starts to rise as a result of an event such as heavy rain, the drain tile quickly brings the water level back down, keeping water away from the foundations and footers of a home. Drain tile is usually installed as part of the process involved in building the foundation, although it may require periodic maintenance and replacement.

It is important to confirm that the trench and outlet are not clogged with dirt and silt. If the gravel becomes packed with dirt, it will be harder for the water to move through, which will result in less water reaching the pipe, and water may start to pool around a home. This can result in water seepage.

A silt trap is a designated area where water that is contaminated with suspended sediment as a result of construction activity or water runoff is contained. While the water is in a silt trap, the sediment can settle to the bottom of the trap until it can be removed. Silt traps can be made using silt curtains, silt fences or a series of shallow ponds to naturally filter the sediment from the water before it reaches a stream or clean body of water.

Environmental protection efforts, such as the silt trap, are often seen in conjunction with mining or construction. Activity from these industries can result in the production of grain-size particles, stone dust and other components that create suspended sediment when caught up in water runoff during rainfall. As the rainwater carries these particles and other pollutants to streams, rivers and lakes, the suspended sediment can cause a serious issue for the fish and other wildlife that inhabit these waters.

The silt trap is designed to prevent the ecological disaster that could follow if these waterways were suddenly to become filled with suspended sediment as a result of mining or construction activity. Using the laws of gravity and the physics associated with liquid suspensions, the silt trap allows the denser, heavy particles of the suspended sediment to drop out of the suspension by creating an artificial break in the current. These particles are easily picked up by water runoff moving with a certain amount of speed, but when the water's current is slowed or completely eliminated by the silt trap, the heavier particles of sediment naturally drop to the bottom of the trap.

If left unchecked, silty soil and suspended sediment could travel long distances in fast-moving water. The potential for damage to aquatic wildlife is very great. By incorporating the silt trap into a mining or construction site, these industries can prevent unnecessary pollution or damage to the local watershed, thus helping the environment.

In some cases, the silt trap might also serve an additional purpose. Particularly in the mining industry, the silt trap becomes the last opportunity to reclaim precious metal ores that otherwise might have been lost. Settling tables, sluices and gold pans are time-tested methods of reclaiming gold, and these methods all work on the same principles as the silt trap. Many mining companies that produce ore in areas where precious metals are found might also regularly remove the accumulated sediment in their silt traps and process it to reclaim the precious metals that can be found.

Drainage Ditch : Pathway for Water to Flow

Excess water can erode landscaping, damage home foundation and create a very muddy and messy situation for a homeowner. To prevent water buildup and collection in a yard, create a drainage ditch that will naturally allow excess water to flow away from the yard and home.

A drainage ditch creates an avenue or pathway for water to flow, which corrects the issue of standing or collecting water. To create the best flow of water, the best angle for a drainage ditch is downhill or sloping to create a gravitational pull to flow the water away from the yard. A drainage ditch can be left with a gravel or stone bottom or can be planted with grass to disguise the ditch if desired.

Digging a Drainage Ditch

Remove grass, plants and vegetation from the path of the drainage ditch. Use the shovel to clear the top of the ground or a hoe to help remove tough rooted grass or plants.

Dig the drainage ditch using the shovel. When digging, place the removed soil into the wheelbarrow for use later, do not discard the soil. The ditch needs to be 1 foot wide and starting at 1 foot deep at the highest point of the ditch. As one progress in digging the ditch, it must slightly slope downward. Remember to keep the ditch sloping downward for proper drainage of water.

Line the bottom of the ditch with the large stones or large-sized gravel. If using gravel, make certain the gravel is big enough that the water can't move it and create a blockage. The stones or gravel will prevent the water from eroding the bottom of the ditch and enlarging or deepening the ditch.

Planting over Ditch

Lay a flat, thin layer of landscaping cloth over the stone or gravel bottom of the drainage ditch. This step is only required if you plan to plant grass over the ditch to disguise its appearance.

Create a 1-inch layer of soil over the landscaping cloth. This soil can be new potting soil or can simply be soil that was removed when digging the ditch.

Seed the soil with grass seeds and water. The grass will grow over the drainage ditch. Because the ditch still has pitch and slope, water will flow away from the yard and home, but the drainage ditch will be less obvious in the yard.

Operations and Timing

On average, about two-thirds of annual precipitation is used by crops in the eastern Corn Belt. The rest falls at a time when it does not meet crop needs. Monthly precipitation remains fairly constant throughout the year, while evapotranspiration (a combination of evaporation from soil and transpiration from the crop), is much higher from June to September.

From January to May, and from October to December, precipitation is greater than evapotranspiration, creating a water surplus. The surplus results in excess water in the crop root zone and the need for drainage. Drainage is primarily a concern in the periods prior to the growing season (January to April) so that crops can be planted at the optimum time.

Even in humid areas, crop water needs often exceed precipitation in July, August and September. Although some of the moisture deficiency is met by stored soil moisture, irrigation can often increase yields.

Equipment Used

Designing and installing a drainage system is a complex process. Every field is unique and usually requires an individual design. Drainage depends on topography, crops that will be grown on the field, and soil type. Every soil type has different properties that affect its drainage. Agronomists and engineers have developed recommendations for drainage depth and spacing in specific soil types based on years of experience and knowledge of soil properties. Drainage contractors use these recommendations to design drainage systems that economically and effectively drain a particular field.

Drainage plows that can be pulled by farm tractors are becoming more popular. But most farmers hire contractors to design and install their tile drainage systems because of the knowledge, skills and experience needed to install a successful system.

Potential Environmental Concerns

The major concerns related to drainage are:

- Loss of wetlands, and
- Increased loss of nitrate through tile drains.

Wetlands

Much of the Midwest landscape consisted of wetlands before large-scale drainage began in the 19th century. Although enormous public health and economic benefits have resulted from the draining of these wetlands over the last 150 years, there have also been negative impacts on the environment. Wetlands have an important hydrologic function in regulating water flow and maintaining water quality, as well as providing habitat for water-based wildlife. Recognition of their value has changed the way our society thinks about and protects wetlands.

Drainage improvements today are rarely for the purpose of converting existing wetlands to agricultural production. Improved drainage is usually aimed at making existing agricultural land more productive. Some fields have drain tiles that were installed 100 or more years ago, and are broken or plugged.

In many fields, only a few of the wettest spots were originally drained, while the entire field would benefit from improved drainage. More tiles are often added to improve drainage efficiency, with the goal of increasing production.

Water Quality

Poor Drainage

Drainage has both positive and negative effects on water quality. In general, less surface runoff, erosion, and phosphorus is lost from land that has good subsurface drainage than from land without drainage improvements or with only surface drainage.

Nitrate loss can be quite high from drained land. Because nitrate is very soluble, it flows easily through the soil and into tile lines. Nitrate flow from subsurface drains is one of the main sources of nitrate in streams and rivers in the Midwest. Concern about hypoxia, or low oxygen levels, in the Gulf of Mexico has increased concern about this nitrate source. Concentrations of nitrate in tile drains are usually quite high (10-40 mg/l).

Pesticides can also flow into subsurface drains, but usually only in very low concentrations. Pesticides move more easily in flow over the soil surface than through the soil, so the highest concentrations of pesticides in tiles are often in fields that have surface inlets into the drains. In fact, subsurface drainage may actually reduce pesticide loss to rivers and streams because it reduces surface runoff.

Best Management Practices

Traditionally, the goals of drainage were to:

1. Maximise crop yield and
2. Minimise costs of drainage installation.

Reducing water quality effects of drainage is becoming a third objective in drainage design.

Nitrate loss is the biggest water quality concern related to tile drainage. Several new technologies can reduce nitrate loss. Controlled drainage keeps the water table high during the off-season when crops are not growing. The high water table increases the rate of denitrification (a process that converts nitrate to harmless nitrogen gas (N2) as soon as the saturated soil warms up in the spring) and reduces nitrate loss to the environment.

Controlled drainage can be combined with subirrigation to improve yields while protecting water quality. Subirrigation is irrigation back through the subsurface drain tiles. Subirrigation may be economical when fields are relatively level and need to be drained anyway, since additional infrastructure consists mainly of increased numbers of tiles the pumping system.

One system being developed in Ohio combines a wetland for water treatment and a pond serving as a reservoir for subirrigation with a drainage system. This system has been shown to increase yields and reduce water quality impacts of drainage, although it is costly.

Tile Drainage

Tile drainage is an agriculture practice that removes excess water from soil subsurface. Whereas irrigation is the practice of adding additional water when the soil is naturally too dry, drainage brings soil moisture levels down for optimal crop growth. While surface water can be drained via pumping and/or open ditches, tile drainage is often the best recourse for subsurface water. Too much subsurface water can be counterproductive to agriculture by preventing root development, and inhibiting the growth of crops. Too much water also can limit access to the land, particularly by farm machinery.

In terms of access, most modern agriculture depends on the usage of large machinery—tractors and implements—to prepare the seedbed, plant the crop, carry out any cultivation and applications during the growing season, and ultimately, to harvest the crop. Operating most machinery in excessively wet conditions may result in soil degradation because of excessive soil compaction, and inhibit the operation of the machinery (ie, "getting stuck").

Increased Crop Yields

Most crops require specific soil moisture conditions and do not grow well in wet, "mucky" soil. Even in soil that isn't "mucky" the roots of most plants will not grow much deeper than the water table. Early in the growing season when water is in ample supply, plants are small and don't require much water. During this time, the plants do not need to develop their roots to "reach" for the water.

As the plants grow and use more water during the growing season water becomes more scarce. During this time, the water table starts to fall. Plants suddenly need to start developing roots to reach to the water. During dry times, the water table can fall faster than the plants can develop roots to "reach" for the water. This can seriously stress the plants.

By adding drain tile, the water table is effectively lowered, and plants can properly develop their roots. The lack of water saturation allows oxygen to exist in the soil around the roots. Drain tile prevents the roots from being under the water table during wet periods that could cause excessive plant stress. By removing excessive water, crops use water they have more effectively.

An increase in crop yield can be simply summarized by the following: Simply by forcing the plants to have more developed roots, the plants can more effectively absorb more nutrients and water.

The same principal is employed by containers that hold house plants: they have drain holes in the bottom to allow oxygen to the roots, and prevent soil saturation. By placing drain tile under a field in a grid style layout, the same effect can be employed on a several hundred acre field.

Plumbing of Drain Tile

In a tile drainage system, a sort of "plumbing" is installed below the surface of agricultural fields, effectively consisting of a network of below-ground pipes that allow subsurface water to move out from between soil particles and into the tile line. Water flowing through tile lines is often ultimately deposited into surface water points—lakes, streams, and rivers—located at a lower elevation than the source. Water enters the tile line either via the gaps between tile sections, in the case of older tile designs, or through small perforations in modern plastic tile.

Soil type greatly affects the efficacy of tile systems, and dictates the extent to which the area must be tiled to ensure sufficient drainage. Sandier soils will need little, if any, additional drainage, whereas soils with high clay contents will hold their water tighter, requiring tile lines to be placed closer together.

Both Cato and Pliny have described tile drainage systems, in 200 BC and the first century AD, respectively. According to the Johnston Farm website, tile drainage was first introduced to the United States in 1838, when John Johnston brought the practice from his native Scotland to his farm in Seneca County, New York. Johnston labored to lay 72 miles (116 km) worth of clay tile on 320 acres (1.3 km2). The effort paid off by increasing his wheat yield from 12 bushels per acre to 60 bu/acre. Johnston, "Father of American Tile Drainage", continued to advocate tile drainage throughout his life, attributing his success as a farmer to the formula "D,C, and D" (dung, credit and drainage).

The expansion of drainage networks was an important technical aspect of Westward Expansion in the 19th century. Although land in the United States was parceled out in accordance with the Public Land Survey System as established by the Land Ordinance of 1785, development, especially of agricultural land, was often limited by the rate at which it was made capable for cultivation. For example, although Iowa was made a state in 1846, maps depicting land ownership show below-average population densities in the northwestern region as late as the 1870s, a corner of the state that today is still noted for its high water table and numerous lakes, marshes, and wetlands.

States thoughout the region faced similar limits to agricultural intensification. Many states offered government incentives to improve land for farming. For example, legislation in Indiana prompted an Act of Congress in 1850 that provided for swamplands to be sold at a discount to farmers on the condition that they drain the land and bring it into agricultural productivity. To facilitate this process, most states set up government agencies to oversee and regulate the installation of tile drainage systems. Even today, ballots for elections in rural America often include candidates for local drainage supervisory boards.

The decades following the American Civil War saw rapid expansion of drainage systems. For example, historical literature from Ohio notes that in the year of 1882, the number of acres drained was about equal to the area of land drained in all previous years. In the 1930s, the Civilian Conservation Corps contributed to the tile network throughout the Midwest, much of which is still in use.

Advances in Drainage Technology

Throughout the twentieth century, the technology of tile installation remained similar to the methods first used in 1838. Although, cement sections later replaced the original clay tiles, and machines were used to dig the trenches for the tile lines, the process remained quite labor-intensive and limited to specialized contractors.

The introduction of plastic tile served to reduce both the cost of tile installation, as well as the amount of labor involved. Rather than set individual sections of cement tile end-to-end in the trench, tile installers had only to unroll a continuous section of lightweight, flexible tile line. Towards the end of the twentieth century, when large, four-wheel-drive tractors became more common on American farms, do-it-yourself tile implements appeared on the market.

By making tile installation cheaper and allowing it to be done on the landowner's schedule, farmers are capable of draining localized wet spots that may not create enough of a problem to merit more costly operations. In this way, farmers may enjoy increases in crop yield while saving on the capital costs of tile installation.

Perhaps the most useful implement in drainage history was James B. Hill's Buckeye Traction Ditcher, which laid drainage tiles at a record pace. Hill's ditching machine drained the Great Black Swamp in Ohio, vast stretches of Louisiana and Florida's swampland.

Effects of Tile Drainage

Unfortunately, the ability for farmers to install their own tile can be problematic. First of all, private installations may reduce the ability of local drainage supervisory boards to regulate tile installation, which in some areas of the country requires proper documentation before a contractor can continue. This leads to the second potential conflict, the unintentional interruption of existing tile networks.

Most "do-it-yourself" tile plows do not dig trenches but rather split the soil enough to squeeze the tile line in; thus, a farmer would not be aware if he breaks a line of tile that might serve his neighbors, as well Mutual tile lines are often dictated by topography rather than land ownership, and the location of many old, but still effective, tile lines are unknown. The potential for across-the-fence disputes are obvious.

Ecologically, the expansion of drainage systems has had tremendous negative effects. Hundreds of thousands of wetland species experienced significant population declines as their habitat was increasingly fragmented and destroyed. Although market hunting within the Central Flyway was a contributing factor in the decline of many waterfowl species' numbers in the early decades of the twentieth century, loss of breeding habitat to agricultural expansion is certainly the most significant. Early maps of midwestern states depict many lakes and marshes that are either nonexistent or significantly reduced in area today. Channelization, a related process of concentrating and facilitating the flow of water from agricultural areas, also contributed to this degradation.

In bypassing the natural flow of water from the surface to the water table, drainage systems often prevent the natural filtration of water provided by soils and wetlands. Thus, drainage systems pose a threat to surface water sources by directly depositing water laden with fertilizers, eroded soil, agrochemicals, and other types of agricultural runoff pollutants. In very flat areas, where the natural topography does not provide the gradient necessary for water flow, "agricultural wells" can be dug to provide tile lines sufficient outlet. In these cases, it is the groundwater that stands to be polluted by unfiltered tile output.

Intensive Livestock Operations (ILO) have led to challenges of livestock effluent disposal. Livestock effluent contains valuable nutrients, but the misapplication of these materials can lead to serious ecological problems, such as nutrient loading. Injecting effluent directly into the ground is one method employed by manure applicators to improve nutrient uptake. Drainage tiles may increase injected manure seepage into surface waterways from manure injection because liquid manure seeps through soils and then drains out of the field and into waterways via drainage tiles.

Today, a number of state and federal initiatives serve to reverse habitat loss. Many programs encourage and even reimburse farmers for interrupting the drainage of localized wetholes on their property, often by breaking tile intakes or removing the tile completely. Land-owners are often partially or fully compensated for forfeiting the ability to grow crops on this land. Such programs and the cooperation of land-owners across the country have had significant positive effects on the populations of a wide variety of waterfowl.

BIBLIOGRAPHY

- Froriep, Siegfried (1986): "Ein Wasserweg in Bithynien. Bemühungen der Römer, Byzantiner und Osmanen", Antike Welt, 2nd Special Edition, pp. 39-50
- Grewe, Klaus (2008): "Tunnels and Canals", in: Oleson, John Peter (ed.): The Oxford Handbook of Engineering and Technology in the Classical World, Oxford University Press, pp. 319-336, ISBN 978-0-19-518731-1
- Moore, Frank Gardner (1950): "Three Canal Projects, Roman and Byzantine", American Journal of Archaeology, Vol. 54, No. 2, pp. 97-111
- Schörner, Hadwiga (2000): "Künstliche Schiffahrtskanäle in der Antike. Der sogenannte antike Suez-Kanal", Skyllis, Vol. 3, No. 1, pp. 28-43
- Serban, Marko (2009): "Trajan's Bridge over the Danube", The International Journal of Nautical Archaeology, Vol. 38, No. 2, pp. 331-342
- Tudor, D. (1974): Les ponts romains du Bas-Danube, Bibliotheca Historica Romaniae Études, Vol. 51, Bucharest: Editura Academiei Republicii Socialiste România, pp. 47-134
- White, K. D. (1984): Greek and Roman Technology, London: Thames and Hudson, pp. 110-112; 227-229, table 6
- Wikander, Charlotte (2000): "Canals", in Wikander, Örjan (ed.): Handbook of Ancient Water Technology, Technology and Change in History, Vol. 2, Leiden: Brill, pp. 321-330, ISBN 90-04-11123-9
- Burton, Anthony (1995) [1989], The Great Days of the Canals, Twickenham: Tiger Books, ISBN 1-85501-695-8

- Calvert, Roger (1963), Inland Waterways of Europe, George Allen and Unwin
- Edwards-May, David (2008), European Waterways - map and concise directory, 3rd edition, Euromapping
- Hadfield, Charles (1986), World Canals: Inland Navigation Past and Present, David and Charles, ISBN 0-7153-8555-0
- Needham, J (1971), Science and Civilisation in China, C.U.P. Cambridge
- Rodda, J. C. (2004), The Basis of Civilization - Water Science?, International Association of Hydrological Sciences
- Blair, John (ed.) (2007). Waterways and Canal-building in Medieval England. Oxford: Oxford University Press. ISBN 978-0-19-921715-1.
- Broadbridge, S.R. (1974). The Birmingham Canal Navigations. Volume 1: 1768-1846. Newton Abbot: David & Charles. ISBN 0-7153-6381-6.
- Volume 2 was never published.
- Burton, Anthony (1995). The Great Days of the Canals. London: Tiger Books International. ISBN 1-85501-695-8.
- Burton, Anthony (1983). The Waterways of Britain: A Guide to the Canals and Rivers of England, Scotland and Wales. London: Willow Books, William Collins and Sons & Co Ltd. ISBN 0-00-218047-2.
- Hadfield, Charles (1966). The Canals of the West Midlands. Newton Abbot: David & Charles. ISBN 0-7153-4660-1.
- Hadfield, Charles (1981). The Canal Age (Second ed.). David & Charles. ISBN 0-7153-8079-6.
- Lindsay, Jean (1968). The Canals of Scotland. Newton Abbot: David & Charles. ISBN 0-7153-4240-1.
- Malet, Hugh (1961/1990). Bridgewater: The Canal Duke 1736-1803, 3rd rev ed, paperback. Nelson, UK: Henton Publishing Co. ISBN 0-86067-136-4.
- Paget-Tomlinson, E. (2006) The Illustrated History of Canal & River Navigations: Landmark Publishing Ltd ISBN 1-84306-207-0
- Palmer (chairman), Robert (1955). "Canals and Inland Waterways, Report of the Board of Survey". British Transport Commission.
- "Reader's Digest Library of Modern Knowledge". London: Reader's Digest. 1978..
- Rolt, L.T.C. (1944). Narrow Boat. London: Eyre Methuen. ISBN 0-413-22000-1.

INDEX

A

Abandoned 109, 188, 202, 267

Abundant 63

Acidic 109

Adequate 92

Aldabra 53

Aleutian 28

Alkaline 92

Amphibians 53

Ancient 87, 117-118, 260

Antarctic 44, 54-56, 58-66

Aquatic 5, 106, 109, 168, 310

Aquifer 96-102, 104, 152, 296-298, 300-302

Aquitard 96, 101, 296, 301

Arsenic 102, 303

Artesian 202, 297-298

Atlantic 32-46, 55-56, 164, 178, 264, 269

Atmosphere 3, 41, 108, 134

B

Bacteria 107, 218, 224

Baseline 198

Basins 10, 23, 29, 44, 164-167, 178-181, 311

Beyond 99, 228, 269

Bioreactor 149

Birmingham 264

Bothnian 35

Brindley 263

Britain 262, 267

Bucket 117, 121, 242

Budhil 81

Bukephala 87

Buried 208, 211-212, 287, 289

C

Canyon 74, 77

Carbon 220, 222, 225

Cartridge 218

Chemical 18, 104, 111, 113, 152, 306, 311

Chitral 88

Chlorine 218, 222

Cholistan 84

Cisterns 207, 211

Clearance 280

Climate 21, 24, 29, 38-39, 41, 47-48, 60, 153, 189

Concrete 210-211, 215, 277, 279-280

Converges 178

Conveyance 215-216, 250

Crabeater 63

D

Deionized 186, 192

Detection 187, 193, 200-201

Disposal 6, 290-291, 310-312

Ditches 291, 303

Domestic 18, 239

Drainage 46, 72, 77, 83, 156, 160, 163-168, 178-182, 255, 285-287, 289-296, 303-322

E

Eastern 29, 37-39, 44-45, 49, 88, 164-165, 179, 304

Ecology 167, 181

Ecosystem99, 299-300

Effluent 91, 149-150, 152, 308, 310, 322

Elevation 74, 81-82, 164, 178, 196

Endorheic 165, 179-180

Eocene 44

Equator 25, 28, 37, 48-49

Equinox 56

Eroded 80

Estuaries 106

Expedition 32, 36, 55

Extensive 14, 267, 299

F

Faucet 123, 137, 183, 196, 229, 238, 240

Filtration 150, 216, 218, 220-221, 224, 242, 253

Fixtures 183, 188, 194, 204-205, 236

Floristic 53

Flowing 13, 79, 81, 87, 90, 132

G

Gallon 208-209

Galvanized 210

Gandaki 74-75

Gauges 166, 180

Germany 262, 267

Giardia 221

Gilgit 86-87

Gravel 251, 295, 314, 316

Gudalur 282

H

Habitat 51, 139, 176, 310, 322

Hemisphere 25, 28-29, 45, 48

Horizontal 293, 305

Hudson 166, 180, 264

Hydrologic 3, 7, 10, 21, 164, 178, 312

I

Imposed 204

Indo-Burma 53

Indoor 194, 205, 219

Instances 111, 285, 289

Instead 56, 123, 125, 154, 242

Irrigation 6, 15-17, 21, 67, 119, 122-125, 133, 136, 141-142, 144, 147, 149-151, 185, 192-193, 197-200, 202, 213, 219, 226-228, 242-243, 245, 260, 275, 287, 291-292, 305, 309, 311, 318

K

Karakoram 87

L

Latitude 37, 41, 49, 60

M

Mammals 53, 63

Marginal 33, 45-46

Marine 31-32, 50-51, 61, 64-65, 91, 106

Microbes 113, 307

Microns 221, 224

Mitigative 309

N

Nonpotable 151

Nutrients 106, 110, 126, 130, 140, 310

O

Offshore 44

Originates 39, 72, 86

Osmosis 15, 148, 150, 224

P

Pathogens 102, 104, 107, 148, 150, 219, 221

Peninsula 46, 165, 179

Permeable 132, 134, 291

Phreatic 96, 296

Polygon 166, 180

Retrofit 204

Riparian 103

S

Saline 14, 92, 291, 305, 312

Sankey 262-263

Sargasso 40

Scarcity 16, 23, 119

Scrutiny 104

Seafloor 61-62

Sediment 49, 80, 105-106, 109-110, 113, 140, 167, 181, 220, 314-315

Seepage 14, 228, 292, 307, 314

Selenium 306-308

Shafts 248, 253

Subduction 30, 44

Submerged 30, 286

Subsurface 96, 134, 285-287, 289-291, 296, 304-305, 313, 318-319

Suburban 108-109

Sumatran 53

Surcharge 203-204

Surplus 92, 316

T

Tenants 200

Thompsoni 64

Tillage 287

Tonnes 63, 127, 129

Topography 131, 167-168, 182, 288

Topsoil 139

Toxins 106

Trapway 184

Trench 25, 36, 54, 253-254, 295, 314

Triassic 42

Tributary 71-74, 82-83, 86, 88, 178

Tropical 29, 41, 50

Tsangpo 76-77

Turbidity 106

Turbines 136

Turbulence 278

V

Valley 71, 73, 78, 82, 86-89, 104, 255, 259, 302

Valves 196, 279

Velocity 131-132, 283

Volcanic 30

W

Waterfowl 310, 322

Watershed 12-13, 88, 130, 156-163, 167-177, 181

Weddell 57, 59-61, 63

Westerlies 29

Wetland 268, 312

X

Xeriscape 197-198